# Challenging Perspectives on Mathematics Classroom Communication

*A volume in*
International Perspectives on Mathematics Education
*Series Editor:* Leone Burton

## International Perspectives on Mathematics Education

Leone Burton, *Series Editor*

---

2004 *Mathematics Education within the Postmodern*
Edited by Margaret Walshaw

2003 *Which Way Social Justice in Mathematics Education*
Edited by Leone Burton

2002 *Research Mathematics Classrooms: A Critical Examination of Methodology*
Edited by Simon GoodChild and Lyn English

# Challenging Perspectives on Mathematics Classroom Communication

*Edited by*
**Anna Chronaki**
*University of Thessaly, Greece*

*and*
**Iben Maj Christiansen**
*University of KwaZulu-Natal, South Africa*

80 Mason Street • Greenwich, Connecticut 06830 • www.infoagepub.com

**Library of Congress Cataloging-in-Publication Data**

Challenging perspectives on mathematics classroom communication / edited by Anna Chronaki and Iben Maj Christiansen.
p. cm. – (International perspectives on mathematics education)
Includes bibliographical references.
ISBN 1-59311-152-5 – ISBN 1-59311-151-7 (pbk.)
1. Mathematics–Study and teaching–United States. 2. Mathematical notation–United States. 3. Communication in education–United States.
I. Chronaki, Anna. II. Christiansen, Iben Maj. III. Series.
QA13.C435 2004
510'.71'073–dc22

2004022159



Printed in the United States of America

*This book is dedicated to Angelos, Dylan, Keaton, and Maria Rafaela, our four children, who gave our lives more depth and perspective, while we were writing this book and bringing them into the world. It is dedicated to Lionel and Vassilis, our husbands, who supported us in our struggle to make the actual book fit as much as possible with our ideas and ideals. It is also dedicated to all the students and teachers who feel lost in learning and teaching mathematics, despite or because of all academic writings about it.*

# CONTENTS

# SERIES EDITOR'S PREFACE

A book on communication in the mathematics classroom is an important addition to this Series. As theoretical perspectives in mathematics education have broadened to include the social and the political, communication has moved more and more towards the centre of our understanding of what happens in classrooms. That is not to say that communication was not always seen as important. Teachers' communicating clearly and concisely to learners was, and is, regarded highly both by the teachers themselves and by their pupils. However, this one-way form of communication, validated by some questioning of certain pupils, is a mockery of what we mean when we talk of communicating. As Anna Chronaki and Iben Christiansen point out in the first chapter of this book, communication not only encompasses "*what is happening and said between participants in a classroom setting but also what are the underlying values, ideologies and politics that influence the practice.*" They point to how such communicative practices affect "*the formation of identities, the ways in which participants make sense of their experiences, what counts as valid activity, and the extent to which the participants can claim to belong to specific communities of practice.*"

But they have an additional intention in this book. They want to open up the topic of communication to challenge and debate, to investigate the areas of communicative practices that are problematic and help to identify challenges to research. Because of this, their authors were asked to offer challenges to perspectives, theories and practices in the teaching and learning of mathematics.

This is not a small agenda but they are helped in meeting it by the comprehensive range of the authors of chapters in the book. Authors are

*Challenging Perspectives on Mathematics Classroom Communication*, pages ix–xi located across many nations and disciplinary backgrounds, with foci on different levels and purposes of education. They include those concerned with the role of technology and the manner in which it has been viewed separately from its cultural context, as well as those who focus upon the inadvisability of separating language use from the situation in which it is used, whether this be spoken or written. They do not shy away from subjecting to critical analyses the political and cultural contexts within which mathematics learning is placed, nor from the social justice perspectives induced by such discussions.

Additionally, the Editors offer readers the perspectives of four well-established researchers whose work has taken them into the field of communication, although from different theoretical and methodological directions. These colleagues read the manuscript and wrote reflective commentaries that are included as the third part of the book. It is unusual for a book not only to offer readers chosen content, but also to offer critical comment on that content. The books in this Series have, in different ways, tried to engage readers by drawing them into the debates current in the field. These Editors chose the route described and readers should find interesting and challenging the issues that are taken up and explored. For example, Anna Sfard critiques the book from a meta-discursive perspective asking how well the chapters are, themselves, communicative. Helle Alrø and Ole Skovsmose use ideas they have developed elsewhere to link dialogue, intention, reflection and critique. They draw attention, in particular, to the mathematics of any communicative engagement. Stephen Lerman takes an overview of the field and the research implications of following the agendas set out in these pages. Given that the same manuscript sparked all three commentaries, the book itself becomes an exemplar of the facets of communication to which commentators and authors draw attention, demonstrated by the differences, and the similarities, in what the commentators have to say.

As a field of research, communication is recognized in this book as difficult, not well understood, inter-dependent with context, experience, theoretical and methodological positions and, of course, language. The authors of these chapters are the first to acknowledge that there are no "answers," especially not "right" answers, and for some in mathematics education, this is itself a great challenge. On the other hand, they offer the reader nuanced and layered visions of how their take on communication helps to expose the dangers and difficulties of previous perspectives. At the same time, they accept that it is in the very nature of research that further dangers and difficulties are likely to be provoked in the future.

A number of comparatively recent ideas are discussed in this volume. I have already mentioned identity and how classroom experiences influence and contribute to the learner's identity as mathematician. Inevitably too, the agency of the learner and the learner's voice are foci of interest, as is the role of power in controlling and directing who achieves and who fails at mathematics. Most of all, in these pages we have an examination of the discourse of mathematics classrooms, how it is constituted, whose access to it is problematic and why, what are its own practices and what are the socio-cultural practices that it supports. These are important issues central to the learning of mathematics of all students, successful or not. They have not been featured widely in the mathematics education field until recently and it is my hope that this volume will offer a new and useful resource to those who are struggling to clarify the picture. As many authors in the book point out, we are not engaged in a search for any holy grail. Rather, we are all on journeys, seeking together, and sometimes finding, new ways of describing what we have seen, as well as ways we might continue. Welcome aboard!

—Leone Burton

# part I

## INTRODUCTION

The first part of this book aims to provide a coherent introductory discussion concerning the varied perspectives used to research communication in mathematics classrooms. It consists of one chapter that provides an overview of main studies in the field, and places the chapters of this volume within the context of that discussion. The outline of this research study follows three themes that, in our view, present three streams of perspectives. The first theme considers communication as registers, representations, and contexts, the second as social interaction, social setting, and classroom activity, and the third as practice, community, identity and politics. These three streams of perspectives, although interconnected, reflect distinct interests, ideologies, and methodologies in studying communication in mathematics classrooms.

The reason for putting together this edited book is to acknowledge studies that either consider communication as part of their analyses or study communication per se. It aims to initiate a dialogue about the varied perspectives utilized to theorize communication in the field of mathematics education, to reflect on the multiplicity of views and to highlight potential challenging entry points.

CHAPTER 1

# CHALLENGING PERSPECTIVES ON MATHEMATICS CLASSROOM COMMUNICATION

## From Representations to Contexts, Interactions, and Politics

**Anna Chronaki**
***University of Thessaly, Greece***

**Iben Maj Christiansen**
***University of KwaZulu-Natal, South Africa***

We have chosen to put together this book for two reasons. First, we acknowledge that an increased focus on communication has influenced and shaped work in a range of perspectives and areas in mathematics education research and practice. Over the past decade and a half, we have witnessed more and more studies that—although not directly concerned with analyses of communication—are considering communication as an integral part of pedagogy and didactics in mathematics classrooms, mathematics education curricula, and broader educational structures. The notion of communication opens up to embrace not only what is happening and what is being said by the participants in a classroom setting, but also conveys the

*Challenging Perspectives on Mathematics Classroom Communication*, pages 3–45

underlying values, ideologies and politics that influence the practice, and thereby the formation of identities, the ways in which participants make sense of their experiences, what counts as valid activity, and the extent to which the participants can claim to belong to specific communities of practice. Second, we would like to initiate a dialogue concerning these areas of study, and to reflect on how they describe, define and conceive of communication. Despite recent developments, studies of mathematics classroom communication still largely assume particular perspectives on learning, on persons, on mathematics, and on research. Our main focus, as the title of the book implies, is to challenge predominant perspectives and to highlight through the chapters, and through reflection and discussion over the chapters, what continue to be unanswered questions, the silenced issues, and the research challenges.

The chapters of this book are organized in three parts. Part I, containing this chapter, provides an introduction to the book, as well as a review of multiple perspectives of research about communication in mathematics classrooms. In this chapter, we place the chapters of the book in the context of an overview of major studies concerning mathematics classroom communication over the past decades. This serves as a backdrop for a thematization of research in the area. Thus, we have formulated three themes. The first theme approaches communication mainly as register, representations, and contexts. The second theme places more emphasis on communication as social interactions, social setting, and classroom activity. Finally, the third theme introduces a new perspective of communication as practice, community, identity, and politics. These three themes provide a typology of major research perspectives relevant to mathematics classroom communication. They can also be seen as a tool to discuss how communication in mathematics classrooms has been researched over the last decades, and to highlight new directions. Having said that, we must emphasize that we do not approach these thematic areas as discrete; they interconnect. Part II contains nine contributing chapters, which are further organized into groups that represent the three themes outlined above. Part III contains three reflective commentaries. These commentaries are written by experienced academics in the field of mathematics education, Anna Sfard, Helle Alrø, Ole Skovsmose, and Stephen Lerman, who all have an interest in language and communication as contexts to research and to conceive mathematical learning. Each of these commentaries provides a useful reflection across individual chapters and to the work of the book as a whole.

The three themes in Part II are not viewed as discrete although they are presented separately. The first theme (Theme I: Register, Representations, and Contexts), deals with how representational media and language come into use when humans try to communicate their mathematical ideas. The issues discussed include: how knowledge production and communication are transformed by how media are used by humans and not media alone (Chapter 2); the influence of the integration of different representational media in geometry classrooms (Chapter 3); and the language use in developing and interpreting mathematical texts in different practices or situational contexts (Chapter 4).

Throughout this introduction, the word "context" is employed with a number of meanings, from referring to the recontextualization of everyday situations in themes or tasks, over referring to the context of particular discourses and practices of the mathematics classroom, to referring to the broader social, political and cultural context in which the school situation is embedded. We will preface our use of "context" throughout the text, using "task context," "situational context," and "political context" or "cultural context," respectively. To some extent, "situational context" and "socio-political/cultural context," will overlap with each other, as well as with the notions of "social setting" and "practice." Since the terms are used so extensively, we have chosen to include all of them, refraining from a lengthy discussion of possible differences in interpretation.

The second theme (Theme II: Social Interactions, Social Setting and Classroom Activity), focuses on how interactions between teacher and pupils or amongst pupils influence the patterns and values of communication in mathematical activity. The focus is on micro-analyses of classroom activity and the topics discussed include: the development of mathematical arguments (argumentation) as part of productive communication in mathematics classrooms, facilitating concept development, reasoning and problem solving (Chapter 5); the understanding of mathematical induction in co-operative settings (Chapter 6); and a range of conflicts and harmonies amongst emotional, cognitive, and social factors, that can influence mathematical activity (Chapter 7).

Finally, the third theme (Theme III: Practice, Community, Identity, and Politics), consists of chapters which discuss communication and pupils' access to mathematics learning in a broader sociological, cultural, and political context. Topics include: an exploration of how mathematics colonizes pupils' lives in Togo, thereby also giving insights into the colonization of pupils' lives elsewhere in the world (Chapter 8); an examination of the discourse and politics of race and task context in school mathematics as experienced in a boys school in South Africa (Chapter 9); and a model for a pedagogy based on critical communication and addressing the complexities of UK mathematics classrooms (Chapter 10).

We now move into a brief discussion of the concept of communication and then into an overview of current trends and major research studies within each of the three themes mentioned above. A section follows where the focus of the book and the possible "readings" of the collection of chapters are discussed. We end with a few words about the editorial process.

## COMMUNICATION

In a very broad sense, communication is about sharing. Etymologically, the word has its root in the latin verb *communicare*, which literally means "to make common" or "to share." However, most dictionary definitions are based on the metaphor of information "exchange," "transport" or "transmission" (e.g., Encyclopedia Britannica, which defines communication as "...the exchange of meanings between individuals through a common system of symbols" quoted in Sfard and Kieran, 2001, pp. 46–47). We also witness a current emphasis on the improvement of "communicative skills" as "skills in information exchange" as part of the discourse of an "Information Society" through mainly, but not only, the effective use of information communication technologies in varied sectors of employment, public services, and education (Castells, 1996; Castells et al., 1999).

Anna Sfard makes a strong critique of an oversimplified use of such "folk" definitions of communication for research in mathematics education. She explains that the metaphor of "exchange" implies that information in the form of ideas and feelings can be objectified and moved from one person to another almost un-problematically. In contrast, she argues that constructing meaning is not a private, but a public affair, and thus communication of mathematical meaning is not about exchanging mathematical information, but about co-constructing information and knowledge (see Sfard, 2000, 2001; Sfard & Kieran, 2001). In her view, the need to communicate drives the construction of new mathematical objects and, as such, the process of communication is inseparable from the process of cognition. Further, communication taken broadly is only partly about sharing information with the intention of sharing knowledge or competencies; it primarily implies sharing beliefs, attitudes, values, rituals and behaviors, thereby building and shaping feelings, relationships, identities, communities, practices, ideas, and knowledge. Furthermore, communication presupposes social interaction in one form or another, and thus assumes the resolution or acceptance of any conflicts or differences among participants.

Communication is an integral part of classroom and schooling processes (Mercer, 1995; Edwards & Mercer, 1987). It is now being made more explicit, as it entails a distinctive aspect of educational programs, reflected in curriculum reforms in some countries. A number of the mathematics curricula in elementary schools encourage active use of teacher-pupil and pupil-pupil communication (expressed in talking and writing) as pedagogic forms, and emphasize pupils' development of communicative skills (Cockroft, 1982; NCTM: National Council for Mathematics Teachers, 1989; Undervisningsministeriet: Danish Ministry of Education, 2001). Underlying these guidelines for curricular change is the hypothesis that the quality of communication influences the quality of teaching and learning mathematics (Cobb & Bauersfeld, 1995; Lampert & Cobb, 1998; Pirie, 1998).

Still, how is communication in the mathematics classroom being conceived? Nerida Ellerton and Philip Clarkson (1996) trace the interest in aspects of language and communication in the field of mathematics education back to the writings of Bryne fifty years ago. Bryne claimed that, "words are links in the chain of communication" and "symbols" of mathematical meanings. They also draw attention to Aiken's review in 1972, which focused on classroom discourse, and to Austin and Howson, who in 1979 presented a framework for discussing classroom communication between teachers and learners considering the participants' backgrounds of language, culture and reasoning models (Ellerton & Clarkson, 1996, p. 987).

Recently, Candia Morgan (2000) and Anna Sfard (2001) both reflect on current and earlier discussions concerning language and communication in mathematics classrooms, and observe that research has moved away from a conception of mathematics communication seen only in terms of its specialist vocabulary and symbolism, often subordinated to natural language as exemplified by Austin and Howson in 1979. They agree that researchers now place more emphasis on language "in use" rather than on "correct use" of mathematical language. However, Anna Sfard points out that "...the old infrastructure has already been shaken, but the new foundations are not yet fully shaped" (2001:3). She indicates specific weaknesses in terms of conceptual tools, methodological frameworks, and pedagogical considerations that still need attention, but at the same time she admits that addressing these types of needs requires time and collective efforts. We are still at the beginning of what may prove to be a very challenging but worthwhile route.

Candia Morgan further observes that the majority of studies still concentrate on micro analysis of teacher–pupil conversations in lessons "...as dependent solely on their current social setting ... without considering their histories and futures" (2000, p. 97). The chapters in this book reflect the striving to engage with the social complexities of mathematics classroom communication at the levels of classroom, school and community and take research on mathematical communication a step further. They include studies on how ideologies, values, beliefs, existing practices, and expectations influence communicative activities.

## COMMUNICATION: REGISTER, REPRESENTATIONS, CONTEXT(S)

Communication in mathematics classrooms is ultimately based on understanding and utilizing the symbols, language and rituals of the school mathematics register. In this section, we discuss work which focuses on communication as a question of how a mathematics register develops, as well as work which considers what might be the role of representations and contexts in teaching and learning mathematics.

### Teaching and Learning a Mathematics Register

Mathematics education programs have as an integral goal to develop learners' competencies in communicating in a school mathematical language, and using mathematical symbols in what is considered an appropriate manner. Whatever the underlying intention, this often boils down to becoming acculturated into a culture of doing school mathematics (reading, speaking, writing, drawing, as well as behaving and valuing), which should not be confused with doing mathematics or using mathematics in relevant out-of-school situations. The task of developing these competences (cf. Niss, 1999) presents one of the main challenges and problems of everyday teaching.

David Pimm studied the mathematical register and its use in mathematics classrooms in the United Kingdom (1987). He showed that students can be misled by mathematical vocabulary to attach meaning to signs which are not in accordance with the mathematical concepts with which they are intended to be dealing. Sometimes, students create impressions which later turn out to be incomplete or misleading, such as "all numbers are either even or odd" or "multiplication makes bigger."

In *The Nature of Mathematical Knowledge* (1984), Philip Kitcher presented his hypothesis that mathematical knowledge develops through the modification of mathematical practices (which to Kitcher are the practices of mathematicians), consisting of five components: a language, a set of accepted statements, a set of accepted explanations, a set of questions selected as important, and a set of meta-mathematical views (p. 163). He considers as especially important one particular type of inter-practice transition, that is related to the development of mathematical language. As an example, here only presented in a very simplified form, think of the imaginary unit $i$ or $\sqrt{-1}$. If the referent of "number" is fixed by using the available paradigms, it restricts the referent to the reals. In that case, $i$ fails to refer to anything. If "numbers" are instead thought of as those entities which can be added, subtracted, multiplied and divided, then developing analogies of the ordinary arithmetic operations implies that *i* is a number (pp. 174–176).

As we see it, a parallel can be drawn to the learning of mathematical concepts in school. Heinz Steinbring argues that for mathematical symbols, vocabulary, and other representations to function as signs, they cannot stand by themselves—just like symbols and vocabulary of "natural language," they must refer to a *reference domain* from which they receive meaning, i.e., they have the referent of the sign fixed (2000, p. 82). In other words, they must be associated with some type of experiences. A concept that has been exemplified in various domains and furthermore has the potential to be exemplified in new domains, can develop—as seen with the historical expansion of the concepts of number and function, for instance. On the other hand, the domains in which a concept has been exemplified can restrict the concepts that students construct.

Rudolf vom Hofe has clarified these points in his discussion of "basic ideas" (*Grundvorstellungen* in German) versus "individual images" (1995, 1998). Generally, not one but a number of "basic ideas" underlie a mathematical concept. As an example, the "basic ideas" of multiplication are either (1) repeated addition—putting the same number of things together several times, (2) enlarging something (possibly by a non-integer measure), or (3) determining the number of possible combinations of a given number of items. vom Hofe uses the term "basic ideas" prescriptively, seeing them as the teacher's tools for the purposes of communication in the classroom. The teacher will specify the "basic ideas" underlying a given mathematical concept, procedure or result and transpose these basic ideas into a learning context. If, for instance, the mathematical idea with which the teacher wants the students to engage is multiplication of fractions, repeated addition is not a suitable basic idea, whilst enlarging something may be. The teacher looks for a context where "enlarging by a non-integer factor" plays a natural part (e.g., enlarging photos). This context activates students' individual images, their assumptions, and their models of explanation. In mathematics teaching, *individual images* then become a descriptive tool; a notion with which to capture what sense students make, in contrast to what sense the teacher intended them to make, as was prescribed through the "basic ideas." The students use their individual images to grasp the learning context, which again may lead them to develop "basic ideas" and then eventually (through comparing, reflecting, abstracting, generalizing, synthesizing, etc.) engage with the mathematical concepts.

Obviously, a number of stumbling blocks exist. For one, "inappropriate" individual images can be activated by a learning context, leading to incomplete or inadequate understanding of the basic mathematical ideas that the teacher intended to teach. Second, the teacher may use inappropriate basic ideas, which restrict the development of concepts or lead to a particular understanding which may later prove limiting –cf. the examples from Pimm (1987) mentioned above. These stumbling blocks do not only apply to the development of an understanding of mathematical concepts, but also to broader thinking processes such as mathematical reasoning. For instance, Brian Rotman (1988) has pointed out that a proof is not only a chain of arguments and conclusions expressed in symbolic language; it is based on an underlying principle or narrative which must be grasped in order for the proof to make sense.

## Representations, Representational Media and Contextualizations

*Representations* (concrete, verbal, symbolic, graphic or real-life narratives) are considered important to conceptual development and an essential part of communication in mathematics education classrooms. Their role can be seen as twofold. On the one hand, the teacher can use a number of graphic, iconic, mental or physical representations to communicate ideas about mathematical concepts with students. For example, Claude Janvier has grouped representations of mathematical concepts that could be used both to describe and prescribe classroom activities, such as: situations, verbal descriptions, tables, graphs, and formulae (Janvier, 1987). On the other hand, students themselves can construct their own representations (verbal, iconic or symbolic) or "translations" across representations that become an expression of their understanding and learning progress. For instance, Martin Hughes (1986), Marit Johnsen-Høines (1987), and Maulfry Worthington and Elizabeth Carruthers (2003) describe how even very young children express in writing their understanding of number and simple arithmetic operations (addition, abstraction, equation) using representations of varied iconic and symbolic types.

Information and communication technologies can provide dynamic media, enabling students and teachers to manipulate and program visual representations of mathematical ideas. Despite the fact that computers and adequate software are not widely accessible to the majority of learners and teachers, we cannot ignore that the development of particular educational software (e.g., Logo, Cabri, Geometry Sketchpad, etc.) or tools (e.g., the graphic calculator) offer new possibilities for teaching and learning mathematics. These possibilities include not only representing mathematical concepts visually (Dubinsky & Tall, 1994), but also linking multiple representations (Kissane, 2002), dynamic manipulation of mathematical objects (Laborde, 1993), active construction of mathematical knowledge and programming (Hoyles & Sutherland, 1989), testing mathematical hypotheses (Andersen, 2003; Mudaly, 2004), informing proof (de Villiers, forthcoming), critical and reflective use of mathematical information in themed activities or modeling (Christiansen, 1996; Chronaki, 2003), and establishing communities of practice (Andersen, 2003). However, technology is not a factor that by itself enhances (or for that matter hinders) learning. The outcome of using technology depends not only on the type of hardware or software, but ultimately on the role technology develops in close interaction with the roles of teachers and learners within the pedagogic context of classroom communication (Chronaki, 2000b).

This idea is strongly argued by Marcelo Borba, in Chapter 2. He introduces the notions of *technologies of intelligence* and *humans-with-media* as ways of overcoming the dichotomy between humans and technology. He considers how technology together with relevant pedagogies reorganize communication in the classroom. His case study refers to the experiences of a group of biology majors in higher education in Brazil, who use graphic calculators to study the notion of derivative. Borba claims that it is neither the technology, nor the teacher, nor the pedagogy that makes new forms of communication evolve. Instead, the students-with-technology transformed the students' ways of learning and communicating (e.g., having a visual representation of function, making conjectures, checking possible replies, etc). In this sense, more conventional technologies in the classroom, such as "orality" and "writing," are not replaced by the new technology, but play a collaborative role in the communication process.

The increasing presence of specialized educational media such as hands-on materials, manipulatives and computer programs aims to facilitate the construction and communication of mathematical knowledge. David Pimm (1995) discusses various representational media used for the purposes of introducing or experimenting with concepts in geometry, arithmetic and algebra such as: paper folding, posters, diagrams, abaci, calculators, tiles, geoboards, cuisenaire rods, and the Dienes multi-base arithmetic blocks. Paul Cobb, Erna Yackel and Terry Wood (1992) explain how such external representations are part of the communication emerging amongst teachers and students and rely heavily on the participants' subjective interpretations. As a result, the students do not necessarily see the concepts we intend them to see. For example, Labinowicz (1985), found that many third graders see 600 when they look at the 10x10x10 Dienes' block intended to represent 1000. Cobb, Yackel and Wood make a strong case that students will not discover mathematics simply by dealing with these representations (1992). Lucio Meira (cited in Ainley, 2000), uses the concept of "transparency," as introduced by Jean Lave and Etienne Wegner (1991), to address pupils' difficulties in accessing mathematical knowledge when representational media are in use. Janet Ainley (2000), based on Meira, studies the transparency of graphs and indicates that their meaning in many cases is not easily apprehended by students.

In Chapter 3, Triandafillos Triandafillidis and Despina Potari, study the integration of written and verbal forms of representations in situations that involve the exploration and construction of models of three-dimensional real objects (i.e., vinegar and soda bottles) by primary school pupils in Greece. They explore the extent to which the integration of different representations can change existing discourses in the mathematics classroom. Triandafillidis and Potari find that, although the new media can challenge the ways pupils communicate their understandings concerning the properties of geometrical objects towards more expressive modes, the acceptance of their choices depends on what teachers value as proper mathematical knowledge. Older students have internalized this, and for them, developing mathematical meaning is a complex process of negotiating meaning and task (Christiansen, 1997).

"Reading" mathematics through external representations is not a straightforward process. Sometimes, instead of facilitating the development of concept understanding, they create confusion or even block learning. Gerard Vergnaud (1987, p. 229), conceptualizes the problem of interpreting representations as an interactive "translational" play between three entities: the referent, the signified and the signifier (the *referent* is the real world to be experienced by the learner who acts on it; the *signified* is the cognitive level consisting of internal representations where the learner recognizes invariants, draws inferences, generates actions and makes predictions; the *signifier* is the symbolic system, such as natural and mathematical language, that the learner uses for communication purposes). Richard Lesh, Tom Post and Merlyn Behr (1987), focusing on this interpretation problem, suggest that the competence to interpret mathematics across representations is linked with broader competencies in learning mathematics, such as simplifying, generalizing, describing, acting out, writing, reading, symbolizing, concretizing, and formalizing.

Though not stressed in his chapter, Borba's study also illustrates the effect on the communication and cognitive processes of making modeling activities central in the classroom (cf. Christiansen, 1996). It is one way to link representations that can be expanded into formalizing, symbolizing, abstracting, etc. On the other hand, it implies connecting mathematics with other contexts or practices, which can both facilitate learning and give rise to new problems.

## From Representations to Contexts

Recent discussions, from situated cognition theorists (e.g., Lave, 1988), sociologists of mathematics education (e.g., Dowling, 1998) and poststructuralists (e.g., Walkerdine, 1988), have conceived the problem of interpreting mathematical information in varied situational contexts as based on the wrongful assumption that "transfer" between contexts (or practices) is unproblematic. *Transfer* refers to the use of mathematical knowledge and competencies from one situational context in another (e.g., the application of mathematical knowledge in everyday activities, the use of cross-curricular work, the recontextualizing of out of school themes in pedagogic practices). For example, in her 1988 book, Jean Lave took a strong position when she claimed that "transfer" from "in" to "out" of school practices generally does not occur. Paul Dowling (1998) analyses the employment of "realistic" task contexts in UK mathematics textbooks, and explains how this can result in a mythologized image of the subject of mathematics as a powerful universal language. Jeff Evans (1999) finds that transfer of learning across situational contexts is not impossible, but it is problematic and undependable. However, based on Valerie Walkerdine (1988), he claims that it is possible to build bridges between such contexts (or practices) through looking for inter-relations between the discourses involved, constructing "chains of meaning" (Evans, 1999, p. 30).

The "reading" of mathematical representations and registers also lies in the cultural contexts through which the readers understand (see Saxe, 1991)—a point captured by Tamsin Meaney in Chapter 4. Resolving the ambiguity of mathematical meanings carried through natural language or symbolic systems is not always a matter of the accurate use of verbal, visual, static or dynamic representations (through media or software). In fact, all representational media seem both to limit and to open up possibilities for the construction of meaning. Representations are rooted in and take meaning from the cultures where natural language (or the formal language of teaching) lies. For instance, Alan Bishop, in his review of cross-cultural studies, describes languages where generalized numbers do not exist, and where notions about, for example, vertical directions are absent from the vocabulary (1988, pp. 20–59). Vicky Zack also refers to a number of languages not using particular discourse patterns such as "why–because" and "if–then" (1999, p. 134). In contrast, some languages have more meaningful names for geometrical concepts. For example, Khuzwayo (1999), refers to a discussion with a South African multi-lingual teacher who finds that concepts such as angle and quadrilateral make more sense to students in Afrikaans than in English due to the more descriptive names.

In Chapter 4, Tamsin Meaney, based on the work of Michael Halliday, adopts a social semiotic perspective, in order to explore the relations between linguistic choices made by participants in a mathematical activity, and the mathematical texts produced. She analyses two contrasting mathematical texts: an e-mail dialogue between two mathematicians, and a classroom discussion amongst a group of students who try to solve a task posed by the teacher. These two situational contexts and the associated conversations differ significantly in terms of the purposes and intentions of the participants in relation to the task, their role and mathematical content, but also in their competencies to use a mathematical register eloquently. Meaney claims that the participants' perceptions are determined by their culture, including their beliefs about mathematics. Her analysis raises important questions for educators working in multicultural and multilingual classrooms. For example: How can existing knowledge be taught without devaluing pupils' self-expression? How can teachers handle teaching the mathematical register and at the same time making explicit how appropriate linguistic choices are made?

In conclusion, we find that the research increasingly has recognized that the cultural context, the natural language and the classroom discourse influence students' reading of symbols, their appropriation of the mathematics register(s), as well as their understanding and construction of concepts and their representations. From focusing on the relations amongst referent, signified, and signifier, the focus has increasingly become what influences these relations. This is a move from seeing mathematical language as separate from its use and the context in which it is being used. A mathematical register does not develop in a social vacuum. It is used alongside the everyday registers (in either mother-tongue or an official language) of teachers and learners. In this sense, a mathematical register develops along with metaphoric and metonymic expressions, reference material, and representations that reflect (and are rooted within) the culture and discourse of a particular group of people.

Although neither the "translation" between concrete and symbolic experiences, nor the transfer across situational contexts are easy processes, the use of multiple representations and contexts in mathematics classrooms not only may further desired learning; they are unavoidable. However, though the use of "realistic" themes in textbooks, assessment items, or lesson planning signifies a desire to present mathematics as friendly and close to real life, it may also cause a gap between the context of the "theme" and the mathematical activity itself (Christiansen, 1996; Chronaki, 2000a). Discussing how to address this problem, Jeff Evans (1999) suggests the use of context-sensing (or contexting) questions as a tool for unraveling different aspects of the contexts involved and bridging the related discourses.

Viewing mathematical learning through register, representations, and contextualizations has received quite a lot of attention in mathematics education, providing us with insights in how signs further meaning making, how signs limit the intended meaning making, and how the development of meaning and of symbolizing is intertwined with classroom communication. It has provided us with insights into the role of signs in prevalent forms of classroom communication, as well as in classrooms where the communication has been altered with the specific purpose of exploring possible changes. And over the past decade, it has provided us with valuable insights in the problems of using realistic task context or expecting transfer across situational contexts.

In our view, the progression towards embracing knowledge as situated, yet acknowledging the possibility of exploiting overlaps in practices or linking situations through constructing chains of meaning between discourses, has taken this area forward by a giant leap. It embraces learners as whole people, with unbreakable links amongst emotions, body, intellect, social belonging and spiritual feeling. Within this school of thought, with its focus on the unitary self of the individual learners, we would like to see this further utilized in discussions of learning and communicating the mathematical register, in linking signifier, signified and referent, in determining what are appropriate communicative contexts through which "basic ideas" can be shared, and in understanding learners' own mathematical images.

## COMMUNICATION: SOCIAL INTERACTIONS, SOCIAL SETTING, CLASSROOM ACTIVITY

The process of developing mathematical communication takes place in a social setting that involves activity, including interaction amongst human beings and human interaction with tools (i.e., conceptual, material, historical). Although the "social" is regarded important within many schools of research in mathematics education, the connection between social interaction, communication and learning has been approached in rather different ways (see de Abreu, 2000; Sierpinska, 1998; Cobb & Bauersfeld, 1995 for more details concerning conceptual and methodological differences amongst various frameworks). Along with exploring the role of language, the organization of classroom communication has been analyzed through the lenses of the organization of interactions. Among the issues being researched are: how people work and interact in groups, what is the influence of their interaction on learning mathematical concepts, what is the influence of an expert, what is the role of language and register in the interaction, what patterns of talk and norms of behavior emerge, and even what alternative patterns can be encouraged. We will briefly outline some dominant schools of thought, such as groups that adhere to constructivist, interactionist and socio-cultural perspectives.

## Social Interactions and Individual Learning (Constructivist Perspectives)

Researchers of the Geneva group, or the *neo-Piagetians* (Perret-Clermont et al., 1991) base their work on Piaget's view that "... social interaction is necessary for the development of logic, reflexivity, and self-awareness" (quoted in Cobb & Bauersfeld, 1995, p. 6). Unlike Vygotsky, who claimed the primacy of the social over individual learning, they argue that interpersonal interaction triggers socio-cognitive conflict between learners, which, in turn, gives rise to individual cognitive conflict. As learners try to resolve these conflicts, they become aware of their activity and construct increasingly sophisticated systems of thought.

*Radical constructivism* is also grounded in Piaget's view of learning as a "self-organization" process based on assimilation, accommodation and search for viability through interaction with others. It has been advanced by the work of Ernst von Glasersfeld, who adheres to the premise that truth (and language use) is always subjective and related to the effective or viable organization of one's own activity (von Glasersfeld, 1995). Communication is thus seen as a complex process of mutual adaptation wherein individuals negotiate meanings by continually modifying their interpretations.

The *Purdue Problem-Centered Mathematics Project* by Paul Cobb, Grayson Wheatley, Terry Wood and Erna Yackel, follows the above ideas, exploring mathematics teaching and learning in classrooms organized in so-called *inquiry mathematics teaching* (Cobb, 1991, p. 13). Inquiry mathematics teaching is characterized by specific communicative patterns that encourage students' active construction and negotiation of mathematical meaning (see Cobb, 1995 for an extensive discussion). In their research, adhering to what is known as a *socio-constructivist* approach, they take as units of analysis pupils' work in small groups and teacher intervention. A basic premise is that interaction and communication among pupils or pupils and teachers are seen as creating opportunities for negotiating meanings and developing taken-to-be-shared understandings (Cobb, 1995). They are interested in how socio-mathematical norms are constituted and stabilized in the mathematics classroom and how they promote learning. The term *socio-mathematical norms* describes a set of values with regard to mathematical activities in classrooms, for example what counts as insightful or elegant mathematical solutions, what might be seen as acceptable ways of talking, writing, interacting etc. (see also Kitcher's description of a mathematical practice).

Their work has led to a focus on the development of a *reflective discourse* in mathematics classrooms. Paul Cobb, Kay McClain and Joy Whitenack write that through reflective discourse, "...mathematical activity is objectified and becomes an explicit topic of conversation" (1997, p. 258). Although "reflective abstraction" is an individual process, reflective discourse is present in the social processes in the classroom, and, as they suggest, appears to support students' reification of their mathematical activity (Christiansen, Nielsen and Skovsmose, 1997 discuss reflections and reflective discourse in another sense, namely reflections on applications of mathematics and on students' perceptions of learning mathematics.) Paul Cobb, summarizing much of the work done in the team, also links this to the development of appropriate forms of notation: "In general, we have found that the development of classroom discourse and the development of ways of symbolizing and notating go hand in hand and are almost inseparable" (cited in Sfard et al., 1999, p. 47). On this basis, they argue for inquiry mathematics teaching, as an alternative to currently dominant practices.

The above approaches, rooted in Piagetian theory though with some influence from other schools of thought, use mainly a psychological analysis for investigating the individual's subjective interpretations. As such, their potential to conceptualize and explore the role of the "social" in classroom mathematical activity has been questioned. For example, Stephen Lerman (2000) argues that the study of the "social," including language use and communication, takes a secondary place in studies such as these. And although Solomon (1989) agrees that social interaction can raise cognitive conflict, she claims that pupils' solutions are not always compatible with "the taken-as-shared" mathematical symbols and practices of the wider community. In other words, she challenges the view that "negotiation of meaning" and "symmetrical relations" can ever become norms of classroom communication and learning. Instead, along with Walkerdine (1988), Solomon claims that what counts as a mathematical problem and as a solution is deeply social and that the roles taken by teachers, pupils and curricula always entail "power" and "control" over what is being learned and communicated. However, recent advances in the Geneva group, known for its neo-Piagetian or psycho-social approach (de Abreu, 2000, p. 2), involve the coordination of traditional psychological methods with anthropological ones and exhibit an interest in investigating the role of the social more deeply. The aim is to explore how the macro and micro-social context(s) interact and influence pupils' interaction and learning through "...the evocation of social experience by symbolic means" (de Abreu, 2000, p. 13). Also, a collaboration between the Purdue Group and the Bielefeld Group (Heinrich Bauersfeld, Götz Krummheuer & Jörg Voigt) tries to co-ordinate a psychological and an interactionist analysis of the classroom micro-culture, arguing that the individuals' positioning and the classroom culture are reflexively constituted (Cobb & Bauersfeld, 1995).

## Patterns of Interaction in Classroom Communication (an Interactionist Analysis)

From the *interactionist* perspective ascribed to by the Bielefeld group, communication is perceived as a process of often implicit negotiations in which subtle shifts and slides of meaning frequently occur outside the participants' awareness (see Krummheuer and Voigt, 1991; Cobb & Bauersfeld, 1995). For the study of social interactions and communication, the local classroom micro-culture is taken as the reference point. Götz Krummheuer and Jörg Voigt (1991) explain that the work of their team at Bielefeld constitutes an evolution from earlier work in mathematics education, which focused either rather narrowly on content or on the psychology of learning, to a focus on interactions in the classroom, manifested in the extensive use of *interaction analysis* of everyday school situations. They summarize their theoretical constructs this way: "In the social interaction in mathematics instruction, pupils and teacher interpret educational objects and processes on the basis of different subjective realms of experience and within different frames" (p. 18, our translation). In their work, the notion of frames refers to the habitual patterns of interpretation:

> These frames assimilate through modulation, without necessarily overlapping. A provisional work "space" (or a working interim) thus is produced in the process of developing understanding. Its conflict potential is disarmed through routines. Through a "train of compulsion" in the social interaction, the routines are linked together into patterns of interaction. (p. 18, our translation)

The aim is mainly to describe patterns of communication and interactions *per se* as they occur in the classroom culture and not to prescribe how these should be organized. Sierpinska explains that for the interactionist, communication is not meant in a transitive form; "Meanings are not in people's heads to be transmitted from one person to another. People do not have to mean what they say, but what they say definitely means something, not just to 'others' but something in the given culture" (1998, p. 52). The role of communication is seen as fundamental and as a prerequisite for language/register acquisition. Heinrich Bauersfeld (1995) conceives communication as a series of "language games" specific to mathematics classrooms, generally characterized by the competence in what he calls *technical "languaging."* He cites Ludwig Wittgenstein, who writes that "[t]he term 'language game' is meant to bring into prominence the fact that the speaking of language is part of an activity, or of a form of life" (quoted in Bauersfeld, 1995, p. 279). Further, he prefers using the term "languaging" instead of "language use," because it makes explicit the relationship between language games, classroom micro-culture, and consideration of individual differences. The processes of talking, writing, drawing or constructing can be seen as an accomplishment of "language games" and "languaging," performed by teachers and students in their attempts to negotiate "taken-as-shared" meanings and signs of mathematical activity. As examples of encouraging a flexible and multidimensional "languaging," he mentions that teachers can offer "open" tasks, whole class discussions, ample chances for students to demonstrate their own ways of thinking, and an early introduction to the process of argumentation.

Other contributions of this group involve an analytic ethnography of argumentation (Krummheuer, 1995) and a detailed exploration of thematic patterns of interaction (Voigt, 1995). Götz Krummheuer sees argumentation, not only in the Aristotelian sense of rhetoric (i.e., a process accomplished by a single person confronted with an audience that is to be convinced), but as a basic communicative aspect of everyday activities, offered also in classrooms, such as arguing, explaining, justifying, illustrating, exemplifying and analogizing. This view is also elaborated by Nadia Douek in Chapter 5. Jörg Voigt has offered a detailed analysis of patterns of interaction in mathematics classrooms. He explains that communication and especially "negotiation of meanings" is fragile, and there is always a risk of disorganization. He argues, thus, that patterns of interaction function to minimize this risk. Voigt (1985) compares two different patterns, the *elicitation pattern* and the *discussion pattern*. In the elicitation pattern, the students offer an answer and solution to a problem posed by the teacher, and the teacher evaluates or guides students towards a definite argument, solution or answer. In the discussion pattern, on the other hand, the teacher asks students to report on how they arrived at their answers and solutions, and then the teacher contributes with further questions, hints, reformulations or judgments.

Along similar lines, Terry Wood (1995) discusses the *funnel pattern* and *focus pattern*. In the funnel pattern, the teacher uses a series of guiding questions that narrow students' queries until they arrive at the correct answer. In contrast, the essential aspects for solving a problem are brought to the fore in the focus pattern, where the teacher's role is to indicate what are the critical features of the problem, to avoid providing the solution, and to summarize important parts that lead to the solution. Voigt and Wood seem to suggest that patterns such as "discussion" and "focus," tend to promote specific communicative norms in the mathematics classroom which characterize an "inquiry model of mathematics teaching," mentioned earlier (see Cobb & Bauersfeld, 1995; cf. Christiansen with Jørgensen & Geldmann, 2000 for an example from mathematics teacher education).

Further, Heinz Steinbring (2000) supports an interactionist analysis with a strong epistemological basis, arguing for the reciprocity of social interactions and the epistemological constraints of mathematics (its signs, symbols, the language means etc.). He explains that:

> ...Only when mathematical symbols and signs are interpreted as intentionally expressing relationships and structural connections, mathematical communication could become a vivid social process in which all partners have to construct their mathematical understanding by actively interpreting the signifiers conveyed by other communication partners. (Steinbring, 2000, pp. 87–88)

Consistent with this view, the work by the Purdue group has developed the notion of socio-mathematical norms, the Bielefeld group has coined the term "thematic patterns," by which interactions that involve mathematical activity are analyzed, and the socio-culturalists base their analysis on Vygotsky's conception of how scientific concepts are developed. These works all attempt to co-ordinate the social and the individual in their analyses without ignoring the development of disciplinary knowledge.

## The Primacy of the Social Setting in Learning (a Socio-Cultural Approach)

Whereas theorists from the interactionist perspective propose that individual students' mathematical activity, the curriculum, and the classroom microculture are reflexively constituted, those who work within the socioculturalist perspective, argue that the "social setting" has a primary importance over what is learned and how it is learned (see de Abreu, 2000; Lerman, 2000). This premise is based on Vygotsky's claim that all intellectual development evolves from the interpersonal (i.e., interacting with others in a historical, cultural and political context) to the intrapersonal (i.e., engaging in subjective actions and thinking); in other words, from the social domain to the individual (Vygotsky, 1978, 1986; Wertsch, 1985). Vygotsky says:

> Any function in the child's cultural development appears twice, or on two planes. First, it appears in the social plane, and then on the psychological plane. First it appears between people as an interpsychological category, and then within the child as an intrapsychological category. (quoted in Wertsch & Stone, 1985, p. 164)

Communication is realized via a complex net of interactions in the social setting and is regarded as an important route for intellectual development and for the historical constitution of human consciousness. Alexei Leont'ev argued that thought develops through purposeful activity, and in particular activity that is oriented around some practical and object-specific goal (1981). A number of theoretical constructs such as the notions of *cultural tool, mediation,* or *zone of proximal development* consist of key analytic tools for explaining aspects of development and instruction.

A series of "cultural tools" can have either a material, conceptual or historical nature and can contribute to the social construction of knowledge through becoming thinking or psychological tools in the activity of problem solving (Mellin-Olsen, 1987). Maria Bartolini-Bussi (1996) and Anna Chronaki (1998) have analyzed what might be the nature of such "tools" in situations involving the teaching and learning of perspective drawing or geometric transformations in primary or lower secondary classrooms. They identify "tools" such as the use of photographs, acquaintance with methods of real life drawing used by artists or architects, construction of classifications for what changes and what not in perspective drawings or in geometric movement, application of specific methods for construction, etc. As a result, they argue that the semiotic "mediation" of tools is realized in the classroom in manifold ways (e.g., a tool developed elsewhere is drawn into new problems, a new tool is created by pupils, old tools can be introduced through the introduction of historical examples of methods, or new tools can be suggested as means for focusing observation and thinking). Christiansen (1996) discusses realistic modeling as a "tool" in problem solving in high school mathematics, and what it means to students' construction of knowledge and interactions in peer groups and with the teacher.

The *zone of proximal development* (ZPD) was developed by Vygotsky in the 1930s as a means to examine the potential of instruction in purposefully organized educational practices. He defines the zone as "...the distance between the actual developmental level as determined by independent problem solving and the level of potential development as determined through problem solving under adult guidance or in collaboration with more capable peers" (Vygotsky, 1978, p. 86). The role of an "expert" is thus becoming the focus of attention to discuss issues such as: how semiotic mediation forms internalized self-regulative speech (Bruner, 1985), how an adult and a child who take different perspectives in communication can interact in specific situations (Hundeide, 1985), and how the ZPD becomes a general mechanism through which culture and cognition interact (Cole, 1985). Language use, tool use, motives, actions and operations within the boundaries of the activity are thus seen as central units for the analysis of communication and interactions, exploring in depth the values of motives implicit in any activity. According to Benjamin Lee, Vygotsky has argued that the development of higher forms of thinking, such as scientific thought, demands that the individual "...differentiat[e] the various levels of generality and communication in his language" and that "...communication and generalization are inextricably linked" in the formation of scientific concepts and thinking (Lee, 1985, p. 7).

According to Nadia Douek in Chapter 5, the teacher plays a crucial mediating role in organizing and orchestrating pupils' activity in order to facilitate argumentation that will support conceptualization. She proposes a number of classroom communication practices that encourage this interaction, and can be organized by the teacher, such as: individual oral argumentation with the teacher, collective classroom discussion managed by the teacher, and individual written argumentation as reaction to other students' texts. Argumentation is considered as a particular development of communication that presupposes a "contradictory listener," the development of consciousness, and the organization of concepts into systems. Although she does not develop the argument in this chapter, Douek also claims that argumentative communication in classrooms facilitates competencies in mathematical proving.

This topic is taken up by Inger Wistedt and Gudrun Brattström in Chapter 6. They problematize classroom communication, exploring the merits and limitations of peer discussion and cooperative work in developing the competency to prove by induction through investigating the particular case of non-algebraic tasks. They argue that active participation alone does not safeguard that all students will grasp the cultural and cognitive practice of their activity. Wistedt and Brattström argue that such knowledge rests on a meta-theoretical level and this type of knowledge allows students access to the practice that enables them to understand and change their views of what constitutes mathematical knowledge within a particular domain. Wistedt and Brattström observe that students as new-comers in a practice cannot access this level of communication, and, as a result, they cannot contextualize their discursive activity, resolve conflicts or understand the meaning of particular mathematical activities. Thus, they stress the importance of a more experienced member who interacts with students to enable a challenge to pre-existing assumptions and to encourage inclusion of what, to students, are alternative ways of proving—in this case by means of geometric reasoning.

When acknowledging the primacy of social interactions in the learning process, we cannot ignore the influence of interpersonal relationships, particularly the emotional aspects. Chapter 7, by Dave Hewitt, highlights the communication and learning that takes place in the interplay of social, cognitive and affective factors rooted in the relations between individuals. Through various classroom examples, he illustrates how these issues are sometimes in harmony and other times in conflict, and he reflects on when cognitive, social and emotional aspects work for or against mathematical learning. Hewitt talks from a teacher's pedagogic perspective about the existence of all three aspects, and he stresses the conflicts and/or harmonies that can occur. His analysis does not adhere strictly to any of the above perspectives, yet demonstrates the value of pedagogic descriptions and acknowledges the importance of affective factors.

Valerie Walkerdine (1988, 1998) in her seminal work, has argued about the important relation between subject positioning in the school practice and his or her cognitive development. Through examples, she discusses how classroom activity is being co-constructed amongst learners and educators in a complex interplay of varied discourses that bring to the foreground cultural and linguistic conflicts. As a result, the process of learning is "emotionally charged" as Jeff Evans (1999) argues and involves not only ideas and strategies but also values and feelings. The cognitive and the affective are linked, and as such the potential for "learning transfer" depends on relationships of signification amongst contexts, discourses and practices. Anna Sfard and Carolyn Kieran (2001) also argue against the split between cognition and affect and explain the essential roles of a meta-discursive dimension (e.g., emotions, values, feelings, intentions, motives, or interests) and a discursive focus in mathematical communication.

The work discussed above implied that this recognition of students as individuals with dreams, hopes, plans, etc., is largely ignored. This goes beyond saying that students have sentiments and that their learning is both cognitive and affective. It fundamentally is about the recognition of students as subjects. In that sense, it relates directly to considerations of students' control over various aspects of the educational situation, and its potential to lead to double binds for the students, as also discussed by Stieg Mellin-Olsen (1991).

## The Individual versus the Social Debate

In this section so far, we have discussed three perspectives that have provided useful studies concerning classroom communication and learning. Anna Sierpinska (1998) has provided a similar categorization in her discussion of dominant theoretical frameworks, but we differ in order. She starts discussing constructivism, then socio-culturalism and ends with interactionism, suggesting that this last perspective provides a more promising direction for research on language and communication. Although we also start with constructivism, agreeing that studies in this theory put less emphasis on communication, we keep interactionism in the middle and end with socioculturalism. However, our aim, with this reverse order, is not to suggest the primacy of one perspective over another. Instead, we want to emphasize the value of each approach and the need for more collective work.

We have also outlined how "social interactions" play a different role in communication in these three theoretical perspectives. On the one hand, the neo-Piagetians, the socio-constructivists and the interactionists are concerned with the analysis of classroom communication and view "social interaction" as the plane of verbal and written discourse. For them, social interaction plays either a secondary or a reflexive role in relation to the individual's actions. Communication is seen more as a process of negotiations of meanings amongst individuals. The socio-culturalists, on the other hand, argue for the primacy of social interaction (both in the historical and collective sense) in the development of learning and its crucial role in organizing productive communicative situations. The "social" is conceived in its human, material or historical dimension and can be realized through the semiotic mediation of more knowledgeable others, peers or tools. Communication and language is seen as a major channel for the constitution of meaning which does not depend solely on individuals' ability to negotiate meanings but mainly on the availability and access to historical, cultural and cognitive tools, as well as the potential for interaction with more knowledgeable others.

The emphasis on the "social" versus the "individual" debate as a key issue characterizing differences amongst these perspectives has dominated current discussions. While this has provided useful thinking and constructive work within frameworks, it in our view also presents a trap for mathematics education research for two reasons. First, even if one was truly primary over the other, it seems to be impossible to determine which that is. The chicken or the egg ... Thus, pragmatism warns us to act with caution in assuming the one framework before the other. Second, the discussions seem to insist on the correctness of one framework at the expense of the other. This excludes the possibility of seeing the frameworks as complementary. In our view, the blind spot of the one framework is where the other framework resides. Let us make a parallel to physics: the electron behaves as a wave in certain circumstances and as a particle in others. It cannot be seen as both at the same time; the descriptions "wave" and "particle" exclude each other. This forces us to see the electron as something in its own right which cannot be comprehended fully by either description but which has characteristics that in our present framework are mutually exclusive. This makes it clear that "wave" and "particle" are only metaphors for aspects of the electron. In a similar way, we can see the social as primary or see the individual as primary; these are complementary descriptors, and it is often worthwhile to consider what assuming the one or the other will imply, rather than taking one for granted.

According to Valerie Walkerdine (1988, 1998), classroom communication situations bring to the foreground varied discourses that reflect the participants' cultural and linguistic backgrounds, personal histories, positioning, and emotions. Communication is situated in a much wider context than the classroom walls—it is not only what teachers, pupils and textbooks represent, but also the discourses, ideologies, and values that work unconsciously and influence their actions, utterances, motives, and evaluations. Thus, organizing the process of communication in order to facilitate desired learning requires much more than the careful organization of social interactions. It requires embracing learning as involving aspects of identity formation, belonging to communities, and acting with a historical, social and political context. As pointed out in the previous section, the work within this perspective could benefit from considering the goals and motives of students, in line with *Activity Theory*. While the group of research contributions described in this section has expanded the perspectives of the first group, they still largely assume an a-historical perspective of classroom interactions, ignoring aspects of subject positioning in terms of cultural and social capital. As the silences of the educational reality are increasingly being voiced, this dimension has come to play a bigger role in mathematics education research. Methodologically, this also calls for an approach that draws in ethical values as a potential source of evidence (cf. William, 1998). These aspects are considered as equally important elements in a practice perspective on learning, and this will be discussed in detail in the next section.

## COMMUNICATION: PRACTICE, COMMUNITY, IDENTITY, POLITICS

In this section, we consider communication from the perspective of social or sociological theories of learning, taking into account the nature of the practices in which communication evolves, and which at the same time give structure and meaning to all the activities involved, including communication. Whereas Shirato and Yell (2000) see communication as a practice in its own right, "the practice of producing meaning," the way we use "practice" in this section is to refer to activity imbued with structure and meaning through its historical and social context. Etienne Wenger argues that "practice" denotes much more than being engaged in activity; it is activity in a community that shares a repertoire reflecting the history of the particular community of practice:

> The concept of practice connotes doing, but not just doing in and of itself. It is doing in a historical and social context that gives structure and meaning to what we do. In this sense, practice is always social practice. Such a concept of practice ... includes the language, tools, documents, images, symbols, well-defined roles, specified criteria, codified procedures, regulations, and contracts that various practices make explicit for a variety of purposes. But it also includes all the implicit relations, tacit conventions, subtle cues, untold rules of thumb, recognisable intuitions, specific perceptions, well-tuned sensitivi-

> ties, embodied understandings, underlying assumptions, and shared world views. (Wenger, 1998, p. 47)

In the following sub-sections, communication is discussed as an integral aspect of practices in which the communities of learners, their developing identities, and the politics involved are all considered. Exploring communication in mathematics classrooms through these notions involves considering the aspects of developing mathematical registers and organizing social interactions but these are seen as inseparable from a wider context of situated activities.

## Practices of Mathematical Activity

It is not a new idea in the philosophy of mathematics to focus on the practice of mathematics rather than on mathematical knowledge alone. For example, Ludwig Wittgenstein (1953) pointed out:

> Of course, in one sense mathematics is a branch of knowledge—but still it is also an activity [...] It is what human beings say that is true and false: and they agree in the language they use. That is not agreement in opinions but in form of life. (quoted in Ernest, 1991, p. 31)

The discipline of mathematics is full of tacit conventions and rules of thumb that are often not formulated in clear logical forms but expressed and accepted through social procedures within the professional communities of mathematicians. Alan Bishop (1988) has explored the practice of mathematicians and users of mathematics in different cultural sites, described through the prevalent values of these practices. His work illustrates just to what an extent mathematical practices of all sorts are imbued with cultural values. Leone Burton and Candia Morgan (2000) have identified different styles in the ways mathematicians express and justify mathematical arguments in their texts. They explain that the mathematical practices are reciprocally related to their ways of using mathematical language and their conceptions about mathematical knowledge and learning.

The practices of professional mathematicians, are not the same as the practices of school mathematics, and both differ from the practices of applying mathematical knowledge in everyday practices such as shopping and cooking or in technological practices within science, engineering, social science or the arts. All these practices produce distinct discourses. Tamsin Meaney, in Chapter 4, compares the communication amongst mathematicians and amongst school pupils working on particular mathematical problems, and reviewing her examples from the perspective outlined above, we find many of Wenger's points exemplified. It is clear how the repertoires of the two practices are distinctly separate; we see the duality of participation and reification in making meaning; we recognize how the repertoires shape the negotiation of meaning taking place, and yet they are still ambiguous enough to require such negotiation to take place in order for the participants to experience meaning. The practices differ in the type of reifications with which they engage. The mathematicians engage with mathematical concepts and symbols as if they were existing objects, while the students engage with the written task text, attempting to decode it to give direction to their activity. Thus, the social enterprises have different purposes; in *Activity Theory* terms, they are different activities entirely.

The students cannot apprehend the practice of their mathematics classroom and its social interactions without taking into account the mathematical meaning they derive from their experiences in this context (Steinbring, 2000), but neither can they develop mathematical meaning from experiences in the classroom without apprehending the patterns of participation in the mathematics class. In line with this, Paul Cobb finds that: "...in helping students explain their task interpretations, the teacher is simultaneously initiating and guiding the renegotiation of the socio-mathematical norm of what counts as an acceptable explanation" (quoted in Sfard et al., 1999, p. 47). Through this negotiation, the belonging to a classroom community is also cemented, but yet the question of what this "negotiation" really means to students and how it takes place remains open.

The connections between school mathematical knowledge and various forms of everyday practices (e.g., shopping, street vendors, etc.) as well as their re-contextualization in pedagogic discourses (e.g., textbooks, tests, etc.) have been analyzed by a number of researchers (Cooper & Dunne, 2000; Dowling, 1998; Lave, 1988; Nunes, Schliemann & Carraher, 1993). They identify that although the tasks appear to require similar mathematical activity, there were major differences in the approach taken in school and everyday settings, respectively. For instance, Jean Lave studied how adults use mathematics in supermarket shopping activities such as deciding on the best buy, and found that the shoppers rarely used mathematical operations learned in school. She highlights that the practices of school mathematics frame a very particular social learning context (Lave, 1988). Paul Dowling (1998) and Cooper and Dunne (2000) point out how the recontextualizing of mathematical applications from everyday contexts into school contexts (e.g., via textbooks or test items) serves to differentiate among esoteric mathematical knowledge and knowledge required for making sense of the application of mathematics in the realistic context. In particular, pupils from lower working class backgrounds have difficulties identifying the required mathematical aspect of such tasks and cannot link the produced discourses of mathematics to the realistic contexts.

Generally, the practices of school subjects play a significant role in shaping participants' perceptions of the actions required in a given activity. For example, Jan Wyndhamn (1993) found that when pupils were asked to solve the same task (putting stamps on a letter), they did unnecessary calculations in the mathematics classroom but not in the social sciences classroom. However, we need to stress that there is not only one practice of school mathematics. Different practices can be identified when comparisons are made across countries, across school communities, and across (as well as within) classrooms. Often, there exist a number of parallel practices within the classroom itself, among the students in particular—practices where what counts may be the color of one's new pencil or the ways in which the official curriculum is resisted (Alrø & Skovsmose, 2003). So while students are involved in becoming part of a community of mathematics learners, they are also forming their identity as mathematics learners, which can vary across a spectrum from the good student to the rebel who is part of a community resisting the submission to schooling. These choices of participation and identity often reflect the conflicts between the school discourse and broader socio-political discourses to which the students belong such as class, race and gender.

In Chapter 8, Wenda Bauchspies narrates the life of students who resist the mathematical practices offered in mathematics classrooms in Togo, West Africa. She discusses the meaning of mathematics and communication in mathematics classrooms for those students and teachers who live within a complex set of conflicting experiences from a French colonial educational system and African contexts. Taking a sociological perspective, her analysis brings to the foreground implicit repertoires and negotiation of meanings constituting communication in the Togolese classroom. The teacher as the primary actor in the classroom teaches "the practice of numbers," while students play a secondary, silent role, and respond mainly when asked. But, at the same time they develop their own communicative and "resistance" codes (e.g., being silent towards the teacher but talkative and cooperative with classmates). More than giving insights into the power, cultural and social relations in Togolese mathematics classrooms, Bauchspies' explorations allows us to look at classrooms in our home countries and see the colonization of our students through the imposition of mathematical ideas upon them. For example: the lack of "higher level" mathematical thinking, the emphasis on rote skills, the disciplining of the body so that to engage in the practice (e.g., sharing shoes in order to create an image of a proper student, or manipulating certain situations in order to create personal advantages).

As we outline in the next section, the re-shaping of identity has been addressed in the socio-political theory of Basil Bernstein. For immigrant students, or students who study in a language different from their own, this is reflected not only in the dominant language but also through different practices in their new mathematics classroom that can provide stumbling blocks to participation and negotiation of meaning (de Abreu, Bishop & Presmeg, 2001; Adler, 2001; Gorgorió & Planas, 2000).

## Identity and Political Dimensions in Pedagogic Practices

Practice, and in particular "pedagogic practice" is used by Basil Bernstein as a way to talk about learning not only within the school, but also learning in other arenas such as doctors and patients, architects and planners, etc. In this sense, his notion of practice also involves learning as doing, learning as becoming part of a particular community, learning as negotiating meaning, and learning as producing and reproducing identities. He explains: "...the notion of pedagogic practice which I shall be using will regard pedagogic practice as a fundamental social context through which cultural reproduction-production takes place" (Bernstein, 2000, p. 3).

Bernstein approaches the cultural reproduction-production (including gender, race and social class), through considering the structure and logic of the pedagogic discourses, and the forms of communication amongst agents involved in the practices. Bernstein (2000, p. 4) views this process as highly political and asks: "...how does power and control translate into principles of communication and how do these principles of communication differentially regulate forms of consciousness with respect to their reproduction and the possibilities of change"? Bernstein analyzes "pedagogic practice" in the school context through the components of curriculum, pedagogy and evaluation. Students' access into these pedagogical practices is argued to be influenced not by an inert cognitive ability, but primarily by their social and cultural capital that determines their subject positioning (cf. also Walkerdine, 1988).

Bernstein (1990, 2000) has developed a detailed theory linking a series of concepts. These include: classification and framing (applied in exploring power and control relations amongst knowledge, pupils and teacher), contextualization and recontextualization (applied in exploring how disciplinary knowledge becomes part of pedagogic discourses in the form of tasks, textbooks, test items), and the rules of discursive order such as instructional and regulative (or socially based) discourse. Together, these concepts constitute a *language of description* that enables a detailed analysis and interpretation of how relations of power and control through interactions and communication unfold in pedagogic practices. Bernstein's perspective has recently influenced the work of a number of researchers in the field of mathematics education who are interested in highlighting the socio-political context of mathematics learning in school communities (see

Lerman and Zevenbergen [in press] for an outline of how the theory is currently used in mathematics education research). The notion of "practice," in the ways exemplified by Wenger or Bernstein, helps us to appreciate the deeply social, historical and political nature of classroom communication and learning. Wenger (1998) continues this line of thinking by intricately linking identity formation, development of meaning, practice participation, and the social aspects of belonging to a community—all seen as aspects of learning.

As much as the community component of learning has to do with belonging, it has to do with becoming. In the process of coming to belong to a particular community, the identity of the agent is also shaped. Wenger (1998, p. 5) sees identity as "a way of talking about how learning changes who we are and creates personal histories of becoming in the context of our communities," and he states that:

> Building an identity consists of negotiating the meanings of our experience of membership in social communities. [...] Talking about identity in social terms is not denying individuality but viewing the very definition of individuality as something that is part of the practices of specific communities. (pp. 145–146)

In the mathematics classroom, students may develop identities such as "the troublemaker" or the "disciplined," "the genius" or "the slow learner," and so forth. These identities may hold one value attached by the teacher and another by fellow students. Chronaki (in press), based on the study of two ethnic minority girls through their interaction with an adult student teacher on school arithmetic tasks, discusses how the educator's expectations and the students' positioning influence how their identities can potentially be shaped and developed into "learning identities."

Dalene Swanson, in Chapter 9, explores how a group of black students in South Africa experience mathematics instruction and reconstruct their identities as learners in a new school context. These students, through scholarships, are "given" access to participate in the mathematics classrooms of an affluent historically white school. Their move from the home school to the new environment means a degrading in the evaluation of their mathematics competencies. They are positioned in a lower stream of mathematics classes due to their lack of fluency in English and their supposed lack of experience, which could inform their participation in the higher stream classes. As a result, their opportunities for full participation in core mathematical activities are limited, and their identities in some respect are cemented as "lacking." This strongly reflects the predominant view on black students in the South African Apartheid education system (Khuzwayo, 1999). The agency of school mathematics and the role of the wider social and political context (regulative discourse) in establishing subjectivity within the classroom discourse are pivotal to this discussion. Swanson argues that the main reason that these children remain disadvantaged is their lack of awareness about the communication gap created between teachers, students and pedagogic practice. Students often do not understand what they are required to do in particular test items or tasks, and as a result they cannot produce a successful reply.

According to Swanson, the main reasons do not reside in students' ability to handle language, but in the pedagogic discourse as expressed through the use of mathematical content, materials, test items as well as the more subtle regulations of participation in classroom practice. All these may or may not allow students access to the regulating discourse of the school mathematics practice, and thus influence whether they succeed or fail. In that sense, teachers' expectations, informed by their perceptions of students' backgrounds, strongly influence their interaction with students. In the process, the identity element of a student concerning his or her mathematical ability is constructed, reflected in performances and participation. This is similar to R.P. McDermott (1993), who claims that the reified category "learning disability" acquires a certain proportion of children, as long as this categorization is given life in the organization of tasks, skills, and evaluations of schools.

## Belonging and Becoming as Integral Elements of Communication

Communication is an essential part of the practices of mathematics classrooms. The "practice" perspective enables us to relate communication with the root ideas of belonging (participating in a community) and becoming (developing an identity). Bernstein's theory promotes the analysis and clarification of the social and political character of such processes. He argues that having access to the rules of a practice (e.g., the mathematics classroom activity) or being able to develop a corresponding identity (e.g., the successful learner of mathematics) are highly influenced not only by what happens in the teaching context (instructional discourse) but also by the discourse(s) played at the school and community level (regulative discourse). As Swanson, in Chapter 9, claims:

> Classroom mathematics communication is, therefore, not merely about 'transmission' of mathematical ideas within a neutral context, or a conduit of 'mathematical language' outside of school mathematics. Rather, it is a set of activities, interactions or practices which are socio-culturally and politically

> situated and serve to produce and reproduce, or contest, certain relations of power and control from within the broader social domain.

As a result, communication in mathematics classrooms is part of the history and future trajectories of the pedagogic practice in all dimensions of content, pedagogy and assessment. Both Wenda Bauchspies and Dalene Swanson, in this volume, explain how communication in mathematics classrooms is influenced by discourses of the wider social context, and in what ways such influences are realized by participants (teachers and pupils) in the classroom practice. Their accounts urge us to challenge predominant assumptions about communication in mathematics classrooms. A deeper understanding of the politics of context is required, including: how dominant views about mathematical knowledge and its recontextualization in tasks and activities for use in the classroom are promoted and encouraged; how certain identities of mathematics learners are constructed or resisted; what the visions for the mathematics classroom pedagogic practice are; whose interests such a practice serves; and what the means for realizing such a vision are. All these dimensions play a significant role in how the nature of classroom communication evolves and develops.

In this volume, a number of chapters explore the changing (or not) of the mathematical activity in classroom practices based on attempts on the level of curriculum to change the practice of communication itself. For example, Tony Cotton argues that adopting a *critical communication* pedagogic model can change drastically the classroom discourse to serve the interests of under-represented groups in terms of gender, race, and social class. In Chapter 10, he outlines a pedagogic model of critical communication, based on ideas offered in the perspectives of Bernstein, critical mathematics education, and social justice that can be used in mathematics classrooms. He claims that the model can serve a vision of pedagogy in mathematics education that takes into account issues of gender, race and class and views mathematics teaching as the basis for personal and social development. Tony Cotton defines critical communication as the competency

> ...to critique reports that base their arguments on the use of mathematical data in the media. Similarly they can draw on mathematical arguments to counter claims from those that hold power over them. They can negotiate the processes through which they will come to increase their mathematical understandings, and they can analyze the administrative structures that construct a common sense view of what it is to learn mathematics.

The model includes aspects such as: seeing learning as a shared endeavor; exhibiting openness and honesty; planning for democratic, collaborative and inclusive decision making within the classroom; viewing pupil control of their learning as positive; showing trust of learners; giving care and responsibility to learners; expecting pupils to learn through challenge and variety.

Cotton makes a concerned effort to make his research democratic and useful to the students by trying to ensure "that every action I took to move my research forward should in itself be a worthwhile activity for the children involved." However, it remains a problem for research which wants to embrace issues around identity formation, subject positioning, class, race, gender, and so forth, that the researcher often writes from a position very different from those about whom he or she writes. We wish to see more research within this school of thought that engages with the methodological issues connected with voicing the silences in education and addressing issues of positioning in a way which embraces students' subjective realities.

## FOCUS AND READING

This book is not meant to be an historic account or a complete review of related studies in the field. Neither is it a book that strives to promote a specific model for classroom communication. It is not a book about "effective communication" or "how to do things." Furthermore, it does not bring together a series of chapters reporting research on classroom communication per se. Instead, it offers a collection of studies (the contributed chapters in Part II) that analyze and discuss communication in mathematics classrooms from varied perspectives that approach communication within the complexity of the pedagogic context. It also provides reflections on the value and importance of these perspectives for understanding mathematics classroom communication (the reflective commentaries in Part III).

We have grouped the chapters around the three themes through which we have discussed the significance of the contributions in the field of mathematics education and reviewed research studies in the area. This was to align them with what we consider an important strand of development in the field. Obviously, we could have chosen to group the contributions in other and radically different ways.

For example, an alternative grouping may be based on the theoretical roots and assumptions of each study. In this respect, some chapters reflect studies related to the trends of constructivism and socio-cultural theory, such as the chapter by Nadia Douek. Other chapters are studies that borrow concepts from the broader field of sociology. Specifically, Dalene Swanson addresses the sociology of mathematics education through the theories of Bernstein, Dowling and Walkerdine. Wenda Bauchspies draws on social anthropology, Tamsin Meaney on social semiotics and Tony Cotton on social justice. The rest of the chapters favour a more eclectic use of theories in their analyses. Marcelo Borba uses ideas from Levy and

Tikhomirov to discuss the use of technology in communication in mathematics classrooms, and Dave Hewitt provides observations analyzed through the lens of an experienced teacher, seeing learning as related to the emotional, cognitive and social nature of the opportunities offered in the classroom interactions. Grouping the chapters this way provides us with a useful overview of the various theoretical frameworks that authors use to analyze and interpret their data, and it can provide new directions within each theoretical framework. However, it to some extent disguises possible interrelations and complementarities across theories. In contrast, our grouping within the three themes provides a better view of what contributions different theories have to offer on particular issues.

The question remains: how can the collection of research studies in this volume help us to see communication differently? What are the new answers or the new questions regarding the issues of communicating in, with and about mathematics? How do the contributions challenge the underlying assumptions and perceptions about communication in mathematics classrooms? What unanswered questions remain?

It is difficult to recognize shifts of paradigms in a field which is evolving and expanding in so many ways, and where so many theories exist and interact. What at first appear to be minor shifts in perspective, can later be recognized as major developments in the field. For instance, Anna Sfard (1996) has argued that while learning in older mathematics education texts is mainly seen as *acquisition* (in a particular non-participatory sense), newer texts reflect a *participation* view on learning. The participation perspective, seeing learning as increased participation in relevant practices rather than as acquisition of knowledge, has evolved gradually. This evolution is also reflected in the three themes discussed above. The chapters in this volume do not claim to represent any grand new theories or paradigm shifts, but they do reflect developments in the field over the past decades and as such produce a partial snap-shot of a field in movement.

Reading through the contributions in this volume, we witness an emphasis towards seeing communication in mathematics classrooms as part of pedagogic practice, as closely related to human interactions, and as part of striving to establish a particular discursive culture (e.g., promoting certain values such as collaboration, negotiation of mathematical meanings, dialogue, critical thinking, etc.). There is a noticeable shift from approaching communication as a toolkit of techniques and skills. Communication involves learners and teachers as whole human beings, encompassing not only talking and symbolizing but also feeling, valuing and imagining. The contributors argue that communication is closely connected with the contexts, the cultures and the practices of the mathematics classrooms. Their work challenges the underlying assumption of most mathematics classrooms that is based on a view of communication as unproblematic *transfer* of information (see Evans, 1999; Sfard, 1996). Further, they challenge predominant perspectives that approach mathematics teaching and learning through a "representational frame of mind," and take a direction where the social and the individual are seen as connected. Finally, they reject assumptions that communication is either value-free or that interlocutors share similar values, ethics, and purposes.

In all, the contributions in this volume can be seen as raising a voice concerning research in mathematics classroom communication that includes at least the following dimensions:

a) *The role of media and innovative representational modes in changing patterns of communication in the mathematics classroom is not value-free:* Marcelo Borba argues that media such as the graphic calculator always need to be seen in relation to the humans who use them. He sees their role as dynamic in changing classroom discourse (even when not using the technology). This contribution makes it clear how much we still have to learn about the influence of media on our discourses and practices. In contrast, Triadafilidis and Potari notice that although the use of innovative representational media in the classroom encourage pupils to express themselves, the classroom communication patterns do not change. It seems that the role of the teacher (and the instructional interventions) remains an essential parameter.

b) *Communication needs deliberative organizing when the purpose is to develop mathematical competences or to challenge predominant views on mathematical knowledge:* In particular, Nadia Douek argues that productive communication as a foundation for conceptual development must be carefully organized, with the teacher taking a deliberative role. Inger Wistedt and Gudrun Brattström explain that a positive tone of communication in peer group settings is not enough to make university students embrace alternative perspectives concerning mathematical competencies. Interaction with a more knowledgeable other is seen as an essential parameter in the process of communication. It is certainly important to explore further those practices that promote productive communication and seriously take into account the development of mathematical knowledge.

c) *Culture, power, practice, identity and politics influence communication in the mathematics classrooms:* Tamsin Meaney presents a model to illustrate how any practice, including the work of mathematicians or pupils in a mathematics classroom, filters through the language/register used, with its cultural connotations necessarily forming the discourse. Wenda Bauchspies explains that power relations are reflected in the communication in mathematics classrooms, but at the same time the structures intended to maintain the status quo are cleverly manipulated by the students so as to gain some control over their positioning. Interestingly, this shapes these students' identities in new ways, perhaps counter to what they desired, namely as the more difficult learners who have managed to manipulate themselves into more desirable positions yet do not seem capable of exploiting this opportunity in furthering their education. Dalene Swanson documents the extent to which the students are positioned and their identity forced into place through the communication with and about them. The remnants of Apartheid's "Fundamental Pedagogics" continues to acquire black students into its category of less able, though it is now disguised through being seen as a result of limited exposure to what are deemed relevant contextual experiences as well as language difficulties. Therefore, it appears a challenge to explore more relevant methodologies that enable us to research how communication at the macro-social level of school and community influence communication at the micro level of classroom.

*d) Mathematics classroom communication needs to be understood through the participants' perspective:* This is embraced by Dave Hewitt, who acknowledges the interplay between students' emotional, cognitive and social reasons for how they choose to engage in mathematical activities. In a few cases, this has been linked to the inner somatic experience of engaging with mathematics (Slammert, 1993), or to the grounding of mathematical ideas in bodily experiences (Núñes, 2000). Tony Cotton's interest in doing research *for* the students is reflected in methodological choices where the students' perspective informs the research format. Whereas most work in the humanities wrestles with issues of objectivity and validity, as also touched on by Bauchspies, Cotton insists that the methodology must reflect the social change aim of the research. This is an area which deserves more attention in mathematics education research. So far, Renuka Vithal (2000) provides a reflexive account of her engagement with a research methodology that respects the participants' perspective as well as reflects a social-cultural-political change perspective.

## A NOTE ABOUT THE PROCESS

The idea for this monograph came about during one of Anna's visits to Aalborg University. There, she presented her perspective on constructivism to the doctoral students at the *Centre for Educational Development in University Science.* Anna's point was to look at constructivism as a discursive practice in education, rather than as a learning theory per se. Various theories or perspectives on education have, at various points in time, come to dominate educational discourse. While such theories/perspectives would not gain such prominent positions if they did not have a lot to offer, they at the same time frame and limit both the research and educational practices (through dialogue with the research as well as through managerial initiatives).

Through our discussions after Anna's presentation to the students, we came to acknowledge how research with a different starting point or applying a different perspective than what is considered mainstream, had been important to us. We could only agree that our task as researchers is not as much to provide guidelines for practice as it is to challenge the habitual. That implies bringing forward "alternative" discourses such as less prevalent perspectives. We have each worked on doing so in a number of areas. Mathematics classroom communication is one we both have worked within. It was Anna's suggestion to comprise a monograph with the aim of bringing forward alternative perspectives on mathematics classroom communication.

In the invitation for contributions we sent out in March 1999, we made the point that a prevailing research position in the field is to focus on a socio-cognitive approach where classrooms are analyzed as independent micro-cultures, communication is considered a rational process, mathematical content is not challenged, and students' intentions and values are not given due focus. We asked for papers which would directly challenge existing perspectives, theories or practices, or which offer novel perspectives. In particular, we requested papers with a socio-political approach to classroom communication, with an embodied approach to classroom communication or with novel methodological approaches. It was also our wish to have papers from countries less represented in the research literature. In other words, it was an ambitious project into which we ventured.

It has been a challenge to find less known contributors from Africa, South America, Asia and Eastern Europe. As Iben resides in South Africa, she had a number of potential authors in mind from the region, but they all declined the offer on grounds of being too engaged with teaching, administration, political work or completing thesis work. Even invitations to write joint papers were declined. Perhaps this reflects a general concern—potential contributors from countries in political and economical transition have too much on their plate to contribute to the international research community, which then continues to be dominated by white, male, European and North American voices. The community should be aware of this, as it means a loss to all. Our contributions do cover the areas we requested, though it becomes apparent that the socio-political approach still forms the most dominant alternative discourse. Thus we can see that our attempt to engage the community on challenging perspectives has only been partially fruitful.

The papers have been through a very extensive review process. Each paper was sent to two other authors, as well as to two or three external reviewers. In most cases, the reviewers managed to complete the task, so that each paper was reviewed by three to five people, besides ourselves and Leone Burton. The many rewrites took time, and often required more involvement from our side than we had originally expected. In the process, which now has lasted $3\frac{1}{2}$ years, we have had to undertake the immense extent of the editing process, as well as the difficulties that accompany bringing together our own very different positions in writing this introduction.

We also had to go through the most challenging, rewarding, time consuming and demanding times in our lives. Anna was the first of us to enter parenthood, when little Angelos joined her and Vasilis in January 2001. This was about the time when Iben fell pregnant with twins. Dylan and Keaton were born into her and Lionel's care in September 2001. Which was about the time that Anna became pregnant again. The final addition (so far) was Maria Rafaela, in July 2002. So while the production time of the monograph has been longer than anybody involved ever imagined (and certainly more than we would all have liked), it has indeed been a time of production and contributing to the world in the most wonderful ways.

It has been a rewarding but time consuming task to put together this introduction. We have gone back and forth on the ways in which to organize the many approaches and perspectives prevalent in the field. Accordingly, we more than anybody know that the themes are our constructs, based on our reading of the research accumulated during the past decades, and our personal experiences and positions. Because of this, we have asked four prominent yet critical researchers in the field to comment on the chapters, and on the book as a whole. Stephen Lerman, Anna Sfard, Helle Alrø, and Ole Skovsmose have offered their commentaries in the final Part of the book.

We want to take this opportunity to thank first and foremost Leone Burton for offering us the editorial responsibility, and especially for her attempts to be patient with us and the frequent delays. Of course, we are most grateful to all the contributors, who have always been receptive to our comments and suggestions. Finally, we want to thank all the reviewers: Barbara Allen, Charlotte Krog Andersen, Bill Barton, Liz Bills, Morten Blomhøj, Laurinda Brown, Tony Brown, Guida de Abreau, Simon Goodchild, Patricio Herbst, Yusuf Johnson, Henri Laurie, Romulo Lins, Candia Morgan, Stig Andur Pedersen, Sal Restivo, Anna Sfard, Anna Tsatsaroni, Renuka Vithal, Rudolf vom Hofe, and Vicky Zack. Without them, this monograph certainly would be less complete in more ways than one.

## REFERENCES


Adler, J. (2001). *Teaching mathematics in multilingual classrooms.* Dordrecht, The Netherlands: Kluwer Academic Publishers.

Ainley, J. (2000). Exploring the transparency of graphs and graphing. In T. Nakahara & M. Koyama (Eds.), *Proceedings of the 24th conference of the international group for the psychology of mathematics education,* Vol. 2 (pp. 9–16). Hiroshima, Japan: Hiroshima University

Alrø, H., & Skovsmose, O. (2003). *Dialogue and learning in mathematics education: Intention, reflection, critique.* Dordrecht, The Netherlands, Kluwer Academic Publishers.

Andersen, C.K. (2003). *Faglige potentielle medlæringer i universiteternes matematikundervisning.* Doctoral dissertation, Dansk Center for Naturvidenskabsdidaktik and Institut for Matematiske Fag, Aalborg University, Denmark.

Bartolini-Bussi, M. (1996). Mathematical discussion and perspective drawing in primary school, *Educational Studies in Mathematics 31,* 11–41.

Bauersfeld, H. (1995). "Language Games" in the mathematics classroom: Their function and their effects. In P. Cobb & H. Bauersfeld (Eds.), *The emergence of mathematical meaning: Interaction in classroom cultures.* Hillsdale, NJ: Erlbaum

Bernstein, B. (1990). Class codes and control. (Vol. IV). The structuring of pedagogic discourse. London, UK: Routledge.

Bernstein, B. (2000). *Pedagogy, symbolic control & identity,* UK: Rowman & Littlefield.

Bishop, A. J. (1988). *Mathematical enculturation: A cultural perspective on mathematics education.* Dordrecht, The Netherlands: Kluwer Academic Publishers.

Bruner, J. (1985). Vygotsky: A historical and conceptual perspective. In J.V. Wertsch (Ed.), *Culture, communication, and cognition: Vygotskian perspectives.* (pp. 21–34). Cambridge, UK: Cambridge University Press.

Burton, L. & Morgan, C. (2000) Mathematicians' writing, *Journal for Research in Mathematics Education 31*(4), 429–453.

Castells, M. (1996) The information age: Economy, society and culture: (Vol. I) the Rise of the Network Society. Oxford, UK: Blackwell.

Castells, M., Flecha R., Freire, P., Giroux, H.A., Macedo, D. & Willis, P. (1999) *Critical education in the new information age.* Lanham, UK: Rowman & Littlefield.

Christiansen, I. M. (1996). *Mathematical modelling in high school: From idea to practice.* (R–96–2030) Aalborg, Denmark: Aalborg University Department of Mathematics and Computer Science.

Christiansen, I. M. (1997). When negotiation of meaning is also negotiation of task: Analysis of the communication in an applied mathematics high school course, *Educational Studies in Mathematics, 34*(1), 1–25.

Christiansen, I. M., Nielsen, L. & Skovsmose, O. (1997). Ny mening til begrebet refleksion i matematikundervisningen? In J.C. Jacobsen (Ed.): *Refleksive Læreprocesser,* (pp. 173–190). Copenhagen, Denmark: Politisk Revy.

Christiansen, I. M. with Jørgensen, A. and Geldmann, M. (2000). 'I begyndelsen var Optagetheden', *Skrift fra Center for forskning i matematiklæring,* no. 15, Roskilde University Centre, Denmark.

Chronaki, A. (1998) Exploring the socio–cultural aspects of maths teaching: Using "tools" in creating a maths learning culture. In O. Björkqvist (Ed.), *Mathematics teaching from a constructivist point of view.* (pp. 61–85). Finland: Åbo Academi University Publications.

Chronaki, A. (2000a). Teaching maths through theme-based resources: Pedagogic style, “theme” and “maths” in lessons. *Educational Studies in Mathematics, 42,* 141163.

Chronaki, A. (2000b) Computers in classrooms: Learners and teachers in new roles. In B. Moon, M. Ben-Peretz & S. Brown (Eds.), *Routledge Companion to Education.* (pp. 558–572). London, UK: Routledge.

Chronaki, A. (2003) Mathematical learning as social practice: Basic dimensions for the design of an online context. In A. Gagatsis & S. Papastavridis (Eds.), *Proceedings of the 3rd Mediterranean congress on mathematics education.* (pp. 705–716). Greece: Hellenic Mathematical Society and Cyprus Mathematical Society.

Chronaki, A. (in press). Learning about learning identities in the school arithmetic practice: The experience of two young minority gypsy girls. In E. Elbers & G. de Abreu (Eds.), The social mediation of learning in multiethnic schools, *(special volume) European Journal of Educational Psychology.*

Cobb, P. (1991). Reconstructing elementary school mathematics. *Focus on Learning Problems in Mathematics 13*(2), 3–32.

Cobb, P. (1995). Mathematical learning and small-group interaction: Four case studies, in P. Cobb & H. Bauersfeld (Eds.), *The emergence of mathematical meaning: Interaction in classroom cultures.* (pp. 25–130). Hillsdale. NJ: Erlbaum.

Cobb, P. & Bauersfeld, H. (Eds.). (1995). *The Emergence of Mathematical Meaning: Interaction in Classroom Cultures,* Hillsdale, NJ: Erlbaum.

Cobb, P., McClain, K. & Whitenack, J. (1997). Reflective discourse and collective reflection. *Journal for Research in Mathematics Education 8*(3), 258–177.

Cobb, P., Yackel, E., & Wood, T. (1992). A constructivist alternative to the representational view of mind in mathematics education. *Journal for Research in Mathematics Education 23*(1), 2–33.

Cockcroft, W.H. (1982). Mathematics counts: Report of the committee of inquiry into the teaching of mathematics in schools. London, UK: Her Majesty’s Stationary Office (HMSO).

Cole, M. (1985). The zone of proximal development: When culture and cognition create each other. In J.V. Wertsch (Ed.), *Culture, Communication, and Cognition: Vygotskian Perspectives.* (pp. 146–161). Cambridge, UK: Cambridge University Press.

Cooper, B. and Dunne, M. (2000). *Assessing children’s mathematical knowledge: Social class, sex and problem solving.* London, UK: Open University Press.

de Abreu, G. (2000). Relationships between macro and micro sociocultural contexts: Implications for the study of interactions in the mathematics classroom. *Educational Studies in Mathematics, 41*(1), 1–29.

de Abreu, G., Bishop, A. J. & Presmeg, N.C. (Eds.). (2001). *Transitions between contexts of mathematical practices.* Dordrecht, The Netherlands: Kluwer Academic Publishers.

de Villiers, M. (in press). The role and function of quasi-empirical methods in mathematics. *Canadian Journal of Science, Mathematics and Technology Education 4*(3).

Dowling, P. (1998). *The sociology of mathematics education: Mathematical myths/pedagogic texts.* London, UK: Falmer.

Dubinsky, E. & Tall, D. (1994). Advanced mathematical thinking and the computer. In D. Tall (Ed.), *Advanced Mathematical Thinking.* (pp. 231–243). Dordrecht, The Netherlands: Kluwer Academic Publishers.

Edwards, M. & Mercer, N. (1987). *Common knowledge: The development of understanding in the classroom.* London, UK: Methuen.

Ellerton, N.F. & Clarkson, P.C. (1996). Language factors in mathematics teaching and learning. In A.J. Bishop, K. Clements, C., Keitel, J. Kilpatrick, & C. Laborde (Eds.), *International handbook of mathematics education* pp. 987–1034. Dordrecht, The Netherlands: Kluwer Academic Publishers.

Ernest, P. (1991) *The Philosophy of mathematics education*: Studies in mathematics education, London, UK: Falmer.

Evans, J. (1999). Building bridges: Reflections on the problem of transfer of learning in mathematics. *Educational Studies in Mathematics 39,* 23–44.

Gorgorió, N. & Planas, N. (2000) Minority students adjusting mathematical meanings when not mastering the main language. In B. Barton (Ed.), *Communication and language in mathematics education: The pre-conference publication of working group 9, ICME 9,* MEU Series no. 2 (pp. 51–64). New Zealand: University of Auckland.

Hoyles, C. & Sutherland, R. (1989) *Logo mathematics in the classroom,* London, UK: Routledge.

Hughes, M. (1986). *Children & number: Difficulties in learning mathematics.* UK: Blackwell.

Hundeide, K. (1985). The tacit background of children's judgements, In J.V. Wertsch (Ed.), *Culture, communication, and cognition: Vygotskian perspectives* (pp. 323–347). Cambridge, UK: Cambridge University Press.

Janvier, C. (1987). Translation processes in mathematics education. In C. Janvier (Ed.), *Problems of representation in the teaching and learning of mathematics* (pp. 27–32). Hillsdale, NJ: Erlbaum.

Johnsen-Høines, M. (1987). *Begynneropplæringen: Fagdidaktikk 1–6 klasse.* Rådal, Norge: Caspar Forlag.

Khuzwayo, H.B. (1999). *Selected views and critical perspectives: An account of mathematics education in South Africa from 1948 to 1994.* Unpublished doctoral dissertation, Aalborg University, Denmark.

Kissane, B. (2002). Technology in secondary mathematics. In L. Grimison & J. Pegg (Eds.), *Teaching secondary school mathematics: Theory into practice.* (pp. 248–270). Nelson Thomson Learning. Australia: (First published 1995 by Harcourt Brace and Company).

Kitcher, P. (1984). *The nature of mathematical knowledge.* Oxford, UK: Oxford University Press.

Krummheuer, G. (1995). The ethnography of argumentation. In P. Cobb & H. Bauersfeld (Eds.), *The emergence of mathematical meaning: Interaction in classroom cultures.* (pp. 229–270). Hillsdale. NJ: Erlbaum.

Krummheuer, G. & Voigt, J. (1991). Interaktionsanalysen von Mathematikunterricht: Ein Überblick über einige Bielefelder Arbeiten. In H. Maier & J. Voigt (Eds.), *Interpretative Unterrichtsforschung, Untersuchungen zum Mathematikunterricht* 17. (pp. 13–32). Köln, Germany: Aulis Verlag Deubner & Co.

Labinowicz, E. (1985). *Learning from children.* Menlo Park, CA: Addison–Wesley.

Laborde, C. (1993). The computer as part of the learning environment: The case of geometry. In C. Keitel & K. Ruthven (Eds.), *Learning from computers: Mathematics education and technology.* (pp. 48–60). Germany: NATO ASI Series, Springer.

Lampert, M. & Cobb, P. (1998). *White paper on communication and language for standards 2000* (Interim report).

Lave, J. (1988). *Cognition in practice.* Cambridge, UK: Cambridge University Press.

Lave, J. & Wenger, E. (1991). *Situated learning: Legitimate peripheral participation.* Cambridge, UK: Cambridge University Press.

Lee, B. (1985). Intellectual origins of Vygotsky's semiotic analysis. In J.V. Wertsch (Ed.), *Culture, communication, and cognition: Vygotskian perspectives.* (pp. 66–94). Cambridge, UK: Cambridge University press.

Leont'ev, A.N. (1981). The problem of activity in psychology. In J.V. Wertsch (Ed.), *The concept of activity in Soviet psychology.* (pp. 37–71). Armonk, NY: Sharpe.

Lerman, S. (2000). Some problems of socio-cultural research in mathematics teaching and learning. *Nordic Studies in Mathematics Education, 8*(3), 55–72.

Lerman, S. & Zevenbergen, R. (in press). The socio-political context of the mathematical classroom: Using Bernstein's theoretical framework to understand classroom communications. In R. Zevenberger and P. Valero (Eds.), *Researching the socio-political dimensions of mathematics education: Issues of power in theory and methodology.* Dordrecht, The Netherlands: Kluwer Academic Publishers.

Lesh, R., Post, T. & Behr, M. (1987). Representations and translations among Representations in Mathematics Learning and Problem Solving. In C. Janvier (Ed.), *Problems of Representation in the Teaching and Learning of Mathematics.* (pp. 33–40). Hillsdale, NJ: Erlbaum.

McDermott, R.P. (1993). The acquisition of a child by a learning disability. In S. Chaiklin & J. Lave (Eds.), *Understanding practice: Perspectives on activity and context.* (pp. 269–305). Cambridge, UK: Cambridge University Press.

Mellin-Olsen, S. (1987). *The role of thinking tools in mathematics education.* Southampton, UK: CIEAEM, University of Southampton.

Mellin-Olsen, S. (1991). The double bind as a didactical trap. In A.J. Bishop, S. Mellin-Olsen & J. van Dormolen (Eds.), *Mathematical knowledge: Its growth through teaching.* Mathematics Education Library 10. Dordrecht, The Netherlands: Kluwer Academic Publishers.

Mellin-Olsen, S. (1993). *Kunnskapsformidling: Virksomhetsteoretiske perspektiver,* Bergen, Norge: Caspar Forlag.

Mercer, N. (1995). *The guided construction of knowledge: Talk among teachers and learners.* Clevedon, England: Multilingual Matters.

Morgan, C. (1998). *Writing mathematically: The discourse of investigation.* Bristol, UK: The Farmer Press.

Morgan, C. (2000). Review of [Language in use in mathematics classrooms: Developing approaches to a research domain]. *Educational Studies in Mathematics 41,* 93–99.

Mudaly, V. (2004). *The Role and function of sketchpad as a powerful modeling tool in secondary schools.* Unpublished doctoral dissertation, University of Durban-Westville, South Africa.

National Council of Teachers of Mathematics [NCTM]: (1989). *Curriculum and evaluation standards for school mathematics,* Reston, VA.: NCTM.

Niss, M. (1999). Kompetencer og Uddannelsesbeskrivelse. *Uddannelse, 9,* 21–29.

Núñes, R. (2000). Mathematical idea analysis: What embodied cognitive science can say about the human nature of mathematics, In T. Nakahara & M. Koyama (Eds.), *Proceedings of the 24th Conference of the International Group for the Psychology of Mathematics Education.* (pp. 3–22). Japan: Hiroshima University.

Nunes, T. Schliemann, A. & Carraher, D. (1993). *Street Mathematics and School Mathematics.* Cambridge, UK: Cambridge University Press.

Perret-Clermont, A.N., Perret, J.F., & Bell, N. (1991). The social construction of meaning and the cognitive activity in elementary school children. In L.B. Resnick, J.M. Levine & S.D. Teasley (Eds.), *Perspectives on socially shared cognition.* (pp. 41–62). Washington: American Psychological Society.

Pimm, D. (1987). *Speaking mathematically: Communication in mathematics classrooms.* London, UK: Routledge.

Pimm, D. (1995). *Symbols and meanings in school mathematics.* London, UK: Routledge.

Rotman, B. (1988). Toward a semiotics of mathematics. *Semiotica* 72(1/2), 1–35.

Saxe, G. (1991). *Culture and cognitive development: Studies in mathematical understanding.* Hillsdale, NJ: Erlbaum.

Sfard, A. (1996). On acquisition metaphor and participation metaphor for mathematics learning. In C. Alsina, J.M. Alvarez, B. Hodgson, C. Laborde & A. Pérez (Eds.), *8th International Congress on Mathematical Education: Selected Lectures* pp. 397 411. Sevilla. Spain: SAEM "Thales".

Sfard, A. (2001). Review of [Communicating to learn or learning to communicate? Mathematics education in quest for new answers to old questions]. *Zentralblatt für Didaktik der Mathematik/International Reviews on Mathematical education 33*(1), 1–9.

Sfard, A. & Kieran, C. (2001). Cognition as communication: Rethinking learning-by-talking through multi-faceted analysis of students' mathematical interactions. *Mind, Culture and Activity, 8*(1), 42–76.

Sfard, A., Nesher, P., Streefland, L., Cobb, P. & Mason, J. (1999). Learning mathematics through conversation: Is it as good as they say?, *For the Learning of Mathematics, 18*(1), 41–51.

Shirato, T. & Yell, S. (2000). *Communication and culture: An introduction.* London, UK: Sage Publications.

Sierpinska, A. (1998). Three epistemologies, three views of classroom communication: Constructivism, sociocultural approaches, interactionism. In H. Steinbring, M. Bartolini-Bussi, & A. Sierpinska (Eds.), *Language and communication in the mathematics classroom.* (pp. 30–63). Reston, VA: National Council of Teachers of Mathematics.

Slammert, L. (1993). Mathematical spontaneity: Innovative mathematical learning strategies for South Africa today. In C. Julie, D. Angelis & Z. Davis (Eds.), *Political Dimensions of mathematics education 2: Curriculum reconstruction for society in transition* pp. 115–121. Cape Town, South Africa: Maskew Miller Longman.

Solomon, Y. (1989). *The practice of mathematics.* London, UK: Routledge.

Steinbring, H. (2000). Mathematics as language: Epistemological particularities of mathematical signs in social communication, In B. Barton (Ed.), *Communication and language in mathematics education: The pre-conference publication of working group 9, ICME 9,* MEU Series no. 2. (pp. 81–92). New Zealand: University of Auckland.

Steinbring, H., Bartolini-Bussi, M. & Sierpinska, A. (Eds.) (1998). *Language and communication in the mathematics classroom.* Reston, VA: National Council of Teachers of Mathematics.

Undervisningsministeriet [Danish Ministry of Education] (2001). *Klare Mål - Matematik - Faghæfte 12.* København, Denmark: Author.

van Oers, B. (1998). From context to contextualising. *Learning and Instruction 8*(6), 473–488.

Vergnaud, G. (1987). Conclusion. In C. Janvier (Ed.), *Problems of representation in the teaching and learning of mathematics.* (pp. 227–232). Hillsdale, NJ: Erlbaum.

Vithal, R. (2000). *In Search of a pedagogy of conflict and dialogue for mathematics education.* Unpublished doctoral Thesis, Department of Mathematics and Computer Science, Institute for Electronic Systems, Aalborg University, Denmark.

Voigt, J. (1995). Thematic patterns of interaction and sociomathematical norms. In P. Cobb & H. Bauersfeld (Eds.), *The emergence of mathematical meaning: Interaction in classroom cultures.* (pp. 163–202). Hillsdale, NJ: Erlbaum.

vom Hofe, R. (1995). Grundvorstellungen mathematischer Inhalte, *Texte zur Didaktik der Mathematik,* Heidelberg, Germany: Verlag.

vom Hofe, R. (1998). On the generation of basic ideas and individual images: Normative, descriptive, and constructive aspects. In J. Kilpatrick & A. Sierpinska (Eds.), *Mathematics education as a research domain: A search for identity.* (pp. 317–331). Dordrecht, The Netherlands: Kluwer Academic Publishers.

von Glasersfeld, E. (1995). *Radical constructivism: A way of knowing and learning.* London, UK: Falmer.

Vygotsky, L.S. (1978). *Mind in Society: The Development of Higher Psychological Processes,* M. Cole, V. John-Steiner, S. Scribner & E. Souberman (Eds.), Cambridge, MA: Harvard University Press.

Vygotsky, L.S. (1986). *Thought and Language,* (A. Kozulin, Ed. and trans.), Cambridge, MA: MIT press.

Walkerdine, V. (1988). *The mastery of reason: Cognitive development and the production of rationality.* London, UK: Routledge.

Walkerdine, V. (1998). *Counting girls out: Girls and mathematics.* London, UK: Falmer.

Wenger, E. (1998). *Communities of practice: Learning, meaning, and identity.* New York: Cambridge University Press.

Wertsch, J.V. (Ed.) (1985) *Culture, communication and cognition: Vygotskian perspectives.* Cambridge, UK: Cambridge University Press.

Wertsch, J.V. & Stone, C.A. (1985). The concept of internalization in Vygotsky's account of the genesis of higher mental functions. In J.V. Wertsch (Ed.), *Culture, communication and cognition: Vygotskian perspectives.* (pp. 162–182). Cambridge, UK: Cambridge University Press.

William, D. (1998). A framework for thinking about research in mathematics and science education. In J.A. Malone, B. Atweh & J.R. Northfield (Eds.), *Research and supervision in mathematics and science education.* (pp. 1–18). London, UK: Erlbaum.

Wittgenstein, L. (1953). *Philosophical Investigations.* Oxford: Basil Blackwell.

Wood, T. (1995). An emerging practice of teaching. In P. Cobb & H. Bauersfeld (Eds.), *The emergence of mathematical meaning: Interaction in classroom cultures.* (pp. 203–229). Hillsdale, NJ: Erlbaum.

Worthington, M. & Carruthers, E. (2003). *Children's mathematics: Making marks, making meaning.* Paul Chapman, London, UK.

Wyndhamn, J. (1993). Problem-solving revisited: On school mathematics as a situated practice. Doctoral dissertation, *Linköping Studies in Arts and Science, 98,* Linköping University, Sweden.

Zack, V. (1999). Everyday and mathematical language in children's argumentation about proof, L. Burton (Ed.), [Special Issues]. The culture of the mathematics classroom. *Educational Review 51*(2), 129–146.

# part II

## CONTRIBUTED CHAPTERS

Part II of the book consists of nine contributed chapters which are organized in three themes; Theme I: Communication: register, representations, context(s), Theme II: Communication: social interactions, social setting, classroom activity, and Theme III: Communication: practice, identity, community, politics.

Although these themes, as already mentioned, can be seen as interrelated, this organization serves not only to highlight distinctions among them, but also to initiate reflection and dialogue about recent developments concerning the theory and practice of communication in mathematics classrooms. Its goal is to help us realize what the issues are that still need further attention, and what the research perspectives are that can potentially provide a challenging direction to the field.

THEME I

# COMMUNICATION: REGISTER, REPRESENTATIONS, CONTEXT(S)

This theme consists of three chapters. Chapter 2 by Marcelo Borba is called *Humans with Media: Transforming Communication in the Classroom.* Chapter 3 by Triandafillos Triandafillidis and Despina Potari: *Integrating Different Representational Media in Geometry Classrooms.* And Chapter 4 by Tamsin Meaney: *Mathematics as Text.*

We find that the research focus has increasingly moved away from seeing mathematical language register as separate from its use. There is now recognition that cultural context, natural language, and classroom discourse influence students' ways of "reading" symbols, register and representations. The challenging direction here is to consider register, representational media and mathematical language within the human and physical context(s) in which their use is realized and experienced, thereby leading to an increased emphasis on the relationship between communicative contexts and the development of learners' mathematical knowledge and competencies.

*Challenging Perspectives on Mathematics Classroom Communication*, page 49

CHAPTER 2

# HUMANS-WITH-MEDIA

## Transforming Communication in the Classroom

**Marcelo C. Borba**
***State University of São Paulo at Rio Claro, Brazil***

### ABSTRACT

In this chapter, I discuss how communication in the classroom can be viewed from a different perspective if cognitive processes are seen as partially generated by media. In particular, the notion of technologies of intelligence is introduced as a way of overcoming the dichotomy between humans and technology. An example from a mathematics classroom for biology majors at the State University of São Paulo at Rio Claro, Brazil, is presented. In this course, functions and derivative are the main concepts. The data presented are from an episode in which there is a transition from the work with functions to derivative. Such a transition is based, among other factors, on a rich debate in which the students use metaphors that are related to their work with graphing calculators. I claim that the use of technology, together with pedagogies that are in resonance with this new medium, reorganize thinking.

*Challenging Perspectives on Mathematics Classroom Communication*, pages 51–77

## INTRODUCTION

In this chapter, I present a notion regarding the relationship between media and humans which is relevant to the process of communication in the classroom. The main claim of the chapter is that media are central, not peripheral, to communication among members of an average classroom. I propose the notion of *humans-with-media* as the basic unit for production of knowledge in the classroom. From this perspective, knowledge is viewed as not solely a product of humans, but rather as a product of a humans-with-media unit. Technology and humans are seen as integrated instead of dichotomous. To develop such ideas, I first present a theoretical discussion about technology and cognition, and briefly discuss the history of media. An example from research is presented, followed by a discussion of the example in the light of the theoretical ideas developed.

## TECHNOLOGY AND COGNITION

My view of the relationship between technology and cognition is heavily based on the work of Oleg Tikhomirov (1981). Tikhomirov describes three theories which relate computers and human activity: substitution, supplementation, and reorganization. In the theory of substitution, as suggested by the name, computers are seen as substituting for humans, since they have the capability to solve problems which previously only humans could solve. Tikhomirov rejects this theory, arguing that the heuristic mechanisms used by computers and humans to solve problems differ significantly from one another. In other words, he claims that, although the output offered by a computer can at times be the same as that offered by humans, this does not mean that the computer software can be placed on the same level as human thinking, since the problem-solving skills are different. As a result of his research, Tikhomirov claims that "[a] large part of the control mechanisms of search in humans in general are not represented in existing heuristics programs for computers. When computer heuristics do resemble human ones, they are significantly simpler and are not comparable in some essential way." (p. 259).

The supplementation theory is described by Tikhomirov as viewing the computer as a complement to humans, increasing the capability and speed of human beings to perform given tasks. In this view, there is a juxtaposition of computer and human: computers and humans are seen side by side but with no *intershaping relationship*. Elsewhere I have argued that, in such a relationship, media (a computer with software, for example) shape humans, and humans shape the design of a given software for uses other than the ones planned by the software designers (Borba, 1995a; Borba &

Confrey, 1996; Gracias & Borba, 2000). It should be noted that, although this was a step forward in terms of emphasizing the role of media in cognition, it was a timid one, as I hope to make clear in this chapter. In any case, Tikhomirov drew on his view of cognitive psychology to disagree with the notion of only a juxtaposition between media and humans.

Tikhomirov cites information process theory, which is based on the idea that "complex processes of thought consist of elementary processes of symbol manipulation" (p. 260), as the basis for this argument against supplementation. He criticizes this theory, which views thinking as the activity of solving problems, arguing that thinking involves not only solving problems, but formulating them as well. Tikhomirov writes

> the formulation and the attainment of goals are among the most important manifestations of thinking activity. On the other hand, the conditions in which a goal is formulated are not always "defined" ... Consequently, thinking is not the simple solution of problems: it also involves formulating them. (1981, p. 261)

It is interesting to note that a paper written by a Russian psychologist in the 1970s and published in English in the 1980s, had already presented a view of cognitive psychology which supported this emphasis on including the formulation of problems among the legitimate activities to be developed in the classroom. It was only in the '80s and '90s that having students propose problems became more popular in mathematics education. (e.g., Skovsmose, 1994; Christiansen, 1997; Borba, 1995b; Borba, 1999a; Vithal et al., 1995). This movement in mathematics education has proposed that students have input, at various levels, in the choice and design of problems to be studied in the classroom. As will be briefly discussed later in this chapter, this pedagogical approach is one which is consonant with the use of the new technologies in mathematics education. But, at this point, I shall examine the option presented by Tikhomirov (1981).

Tikhomirov intensifies his criticism of information process theory, suggesting that it does not take into account meanings that are given to the manipulated symbols and other important characteristics of solving and formulating a problem. He is particularly concerned that such a theory does not take human values into account. In this case, I would add that such a view of cognition, which simply breaks down complex processes into parts, ignores the political ideology inherent in processes like defining and solving a given problem—a perspective that can be dangerous as it separates politics from cognition. Alternatively, he proposes that computers should be seen as reorganizing human activity, instead of substituting or supplementing it, and that emphasis should be given to human-computer systems, and to problems which can be solved by them. Tikhomirov proposes, based on Vygotskyan theory, that a tool is not just added to a human being, it actually reorganizes human activity: "as a result of using computers, a transformation of human activity occurs, and new forms of activity emerge" (1981, p. 271). In Tikhomirov's view, computers play a role similar to language in Vygotskyan theory, representing a different way of regulating human intellectual activity. Human-computer systems lead to new forms of teacher-student relationships, and can suggest new ways of legitimating and justifying findings in the classroom (see also Borba, 1994).

The research developed by our research group, GPIMEM, has indicated that the notion of the human-computer system should be expanded in two ways (Borba and Penteado, 2001). First, the notion of computer itself should be expanded to incorporate all the different kinds of interfaces which are interconnected. Thus, devices such as calculators, graphing calculators, printers, modems, video, etc., should be incorporated with the notion of computer. Second, these different interfaces, which are transforming our daily lives, no longer allow us to think in terms of a single computer as an isolated unit, as Tikhomirov did at the time, with good reason. I also believe that it should be emphasized that, for the most part, computers have invited interaction among humans; I propose that we think more in terms of *several* humans interacting with computers, as opposed to *individual* humans. I have therefore proposed the notion of *humans-with-media* (Borba, 1999b; Penteado & Borba, 2000; Borba & Penteado, 2001). I have also proposed that we think of this system as a unit that generates knowledge: neither humans alone nor media alone. From this perspective, computers cannot produce knowledge by themselves, as substitution theory suggests, nor can humans; both medium and human must be present in a given system. It is necessary to have an environment, as well as the different types of information technologies mentioned above, and humans. Both humans and technology produce the environment and are shaped by it.

I believe that holding such a view does not mean there is no need for pedagogy. The design of tasks for educational practices has to take into account discussions such as the one outlined above. For instance, if one sees the computer as merely a supplement, one may be inclined to design tasks which are similar to those designed to be solved without computers, restricting the use of computers (or portable computers, such as graphing calculators) to verification of results or illustration of a given topic. In the Research Group on Technology and Mathematics Education (GPIMEM) research program, there has been an attempt to use tasks which take advantage of these new resources. Elsewhere I have discussed practice in classrooms in which the graphing calculator has a central role in students' discussions and the reorganization of their thinking (Borba, 1997, 1999b). In these previous examples, the calculators were being used when a conjecture or a main idea was developed. In this chapter, I will present an example that suggests that, even when the graphing calculator is not being used by all the students (the students were not using it at the time of the debate from which the excerpts for this chapter were taken), the graphing calculators themselves and the tasks have led to such a reorganization of thinking. Before I present the example however, I will take the reader into a discussion about the nature of technology, and specifically, about how technologies of intelligence have been important for the transformation of knowledge over time. Media and intellectual technologies will be presented as very close concepts.

As a community of mathematics educators, we have underestimated the role of media in knowledge. The main debate during the 1990s in mathematics education was centered on whether knowledge was produced by a "lonely knower" or by a "social knower." But in both cases, the unit which produces knowledge includes only humans.

It is not worth getting into the details of such a debate, since there is a rich literature about it, including a lengthy debate in the *Journal of Research in Mathematics Education* toward the end of the 1990s, and because reviewing such a polemic issue would probably take another book. But it can be said, in a nutshell, that on one side of the debate were researchers who claimed that knowers, anchored in basic schemes, developed more complex schemes as they interact with the social world. Let me call this perspective A. On the other side, were those who supported the notion that cognition is a social process, in the sense that interpersonal activity precedes the development of intrapersonal cognition. Let me call this perspective B. There were also those who said that perspective A and B could not be reconciled and others who said they could. There are certainly many authors who would disagree with this description of perspectives A and B, but as I mentioned, the goal was to briefly summarize this debate. For the purpose of this chapter, I will situate the above dispute asking another question: "Who won this theoretical debate"? Perspective A is not proved by any set of data presented by a given author, nor is perspective B recognized by everyone as being the "right way" of seeing cognition, language, and so on. Empirical data and careful research are necessary means of persuasion, but sometimes historical and political arguments can be more persuasive than those based on empirical research.

I will present data from the classroom, not as a way of proving that an epistemological position that I support is correct (perspective C), but to contribute to the debate. In perspective C—and this was my second motivation to summarize the above debate—the role of media is paramount. In addition to the presentation of empirical data, I will also build on research developed by Pierre Lévy on the history of media which emphasizes the interplay between media and the production of knowledge (1993, 1999).

How such a view helps us to understand communication in the classroom is the focus of this chapter. Clearly, communication has been altered by the possibilities of information technology since the mid-1990s. For example, if we consider the new possibilities of "distance education" with the use of chats, teleconferences, e-mail lists, and other available options, it is quite clear that technology has dramatically changed communication possibilities and patterns (see one example in Borba & Penteado, 2001), and the very notion of dialogue and classroom. In this chapter, however, I want to argue that technology, used in "face-to-face education"—as opposed to distance education, in which the teacher may be miles away from students and connected via computer—also transforms the notion of communication among students and between the teacher and the students.

## MEDIA THROUGHOUT HISTORY

Many philosophers of technology, like those reviewed by André Lalande (1999), tend to dichotomize the relationship between humans and technology. Humans are creative while machines are repetitive. Humans are susceptible to aesthetics while technologies are not. If this discussion is extended to the issue of communication, especially "face-to-face communication," such as that which takes place in regular classrooms, we may think that technology and setting have little to do with how people communicate and relate to each other. I want to claim that this is not the case. In order to build one more argument, I will summarize some of Lévy's ideas regarding technology and humans (1993).

This author suggests that technology and humans have been completely intertwined, since technology is developed by humans and, in turn, shapes what humans can do, in a way similar to what I have labeled the "intershaping relationship." However, Lévy takes it a step further when he suggests that some technologies, called "technologies of intelligence," not only shape but are actual subjects of knowledge (1993, 1999). Such technologies are orality, writing and computers. The argument can be extended to other kinds of socio-cultural products, and to notions such as cities, libraries, houses, and so on. They can also be seen as active producers of a given type of knowledge. But in this chapter, I will concentrate only on the influence of the technologies of intelligence.

Lévy takes an excursion into history to argue that knowledge was strongly based on memory, before writing was pervasive. Culture rested strongly on orality, a technology of intelligence, and knowledge was produced and reproduced via propositions that needed to be repeated periodically so they would not disappear. A given culture, then, had its circular forms of generating and transmitting knowledge. In such cultures, the subjects of knowledge are humans with this almost invisible technology: orality. This includes not only the actual spoken words but also body language, mimics, gestures and intonation.

Once writing became more generally available, there were substantial changes in the way knowledge was generated. As memory was extended to paper, it was possible for theories to be born. In the case of mathematics, the opportunity emerged for long demonstrations to be developed and stored. In other words, memory had been extended. It is relevant to note that writing did not abolish orality. On the contrary, it created what Lévy (1993) labeled secondary orality, which would be an orality related to reading what has been written. In this case, the technology of intelligence, writing, became much more visible, even though it has not been sufficiently visible to those who claim that cognition is solely a mental/internal process.

Orality and writing shaped most classrooms up to the 1980s, and have continued to do so in spite of the introduction of computer technology in many classrooms around that time. Most communication between teachers and students occurs through orality; traditional learning seems to be based on copying with the purpose of memorizing, so that students can demonstrate in a written test that they can "repeat" in writing what was transmitted orally in class. Again, the classroom is one case where writing has not caused orality to disappear. However, while most communication is oral, I would claim that, because most tests are written, and "what counts" in society is often one's test grades, this technology of intelligence has been given more importance. This may explain why oral and body language based explanations are not given as much importance in schools. In other words, orality may be more present in everyday life at school, but what matters when it comes to rewards is writing.

This case may be similar to scholars who see scientific knowledge as superior to knowledge developed by socio-cultural groups whose main medium is orality. Many scientists in general, as well as mathematics educators in particular, tend to see knowledge as unrelated to media, and therefore they use measuring sticks based on a view of knowledge that attributes the highest value to a particular genre of writing to evaluate meaning produced by members of cultures where orality is (or was) the main medium. Ethnomathematics, a trend in mathematics education, has emphasized how different groups produce relevant knowledge (D'Ambrosio, 2001). One who sees media as peripheral to knowledge production may react to ethnomathematics as attributing scientific value to "common sense," to oral knowledge produced by common people on the streets, in contrast to "real, *written* scientific knowledge."

On the other hand, if one holds an epistemological position that knowledge is perceived to be produced by a collective of humans and nonhumans (in particular media), one may be inclined to support the political program developed by the Ethnomathematics movement. Ethnomathematics—which has emphasized the knowledge of groups which are, for the most part, discriminated socially (Knijnik, 1996), socially and culturally (Borba, 1995b) or culturally and historically (Ascher & Ascher, 1981)—could advance its program of strengthening diversity of knowledge (D'Ambrosio, 2001) if such a perspective were taken into account. In many of the cases analyzed in the Ethnomathematics research program, the oral tradition is very strong, and therefore knowledge should also be understood as a product of humans-with-media, in which one of the main—or *the* main—technology of intelligence is orality (Borba, 1995b). Therefore, criticisms of these different types of Ethnomathematics for their lack of scientific basis represent a prejudice based on the notion that knowledge is solely a result of humans-with-writing. Humans, together with available media, are the ones who produce science. It should always be seen, therefore, as a result of collectives which include at least humans-with-media. Thus, mathematics, which is produced by qualitatively different units of humans-with-media, should be assessed as such, and not with a common measuring stick based on some ahistorical view of the role of media. If one wants to analyze the basis of Ethnomathematics produced by cultures in which the main technology of intelligence is orality, then it is important to consider how the orality has shaped the knowledge, for instance making it more circular compared to the linearity of written texts.

It is relevant for a political discussion involving knowledge that we see different kinds of knowledge as based on the different media with which they are associated. If one accepts this view of knowledge, then it may be difficult, if not impossible, to compare knowledge produced by units composed of qualitatively different media. For example, one could view knowledge produced by the Incas (Ascher & Ascher, 1981) as incommensurable when compared to science as it has developed since the 1800s when the technological apparatus for writing became widely available.

I want to claim that seeing the classroom through an epistemological lens based on the construct humans-with-media may very well highlight aspects of classroom communication that may be overshadowed by other views in which the role of technologies of intelligence are not emphasized. In particular, this approach is powerful for classrooms in which computer technology has been used, if one accepts Lévy's view that computer technology is a medium that is qualitatively different from writing and orality, since it extends memory in different ways (1993). To this I would add that it associates orality, writing and images in ways that are completely different from previously available technologies of intelligence. Computers are qualitatively different media, whereas overhead projectors, for example, are not, because the former completely change systems of writing and the way pieces of information can be stored and memorized, while the latter is a simple adaptation of the traditional blackboard or poster. Overhead projectors change a given collective, formed by humans-with-media, but they do not represent a qualitative different change as computers do.

Besides the notion that computers are different media, I argue that the computer has dramatically changed communication in the classroom. Of course, if we consider the Internet, as mentioned before, such a transformation is obvious, but this is not the kind of change that is my concern in this chapter. I want to present an example in which face-to-face education is taking place where media, such as graphing calculators, help to transform what students and teachers consider to be a valid argument.

Before moving to a specific example, it should be noted that I make no claim that graphing calculators or any kind of media transform things by themselves. I believe that transformations are extremely complex, and no cause and effect relationship can be established. Instead, I present evidence that the type of communication that took place in the classroom would not have been possible without devices such as the graphing calculators. It should also be emphasized that I do not believe that media determine what knowledge is produced. My claim is that knowledge is shaped by technologies of intelligence and that such technologies are actual actors in producing knowledge and communication, even though there are many other factors that shape such processes. For example, investigating pedagogies that are consonant with these computer technologies is another facet of my research, albeit not the focus of this chapter. Establishing connections between the ways students use different technologies of intelligence (orality, writing, and computing) is another goal of the research program developed at GPIMEM and will be discussed following the presentation of the classroom communication example.

## THE RESEARCH IN THE CLASSROOM

For the past ten years, I have headed one of the GPIMEM research projects which investigates the use of technology with first-year biology majors. These biology students are required to take one mathematics course, scheduled during their first semester at the university. The course meets for four hours per week. There are approximately 40–45 students enrolled in the course each year. In 1998, the Biology program as a whole was expanded, and a new evening course was created to accommodate those students who work during the day. I have been teaching both courses, among other reasons, so that I can develop the fieldwork that served as the source of the data presented here. The excerpt presented here is from the 1997 class. The school year in Brazil runs from March through November, and this particular course was offered in the first half of the year. As well as the teacher and the students, there were other people present in the classroom: a doctoral student who was developing a different study with the students (Villarreal, 1999, 2000), a technician who filmed most of the classes, and an undergraduate research assistant who helped manage the data. Whenever a member of the research group involved in this project, or myself, detected a filmed episode that needed to be analyzed in greater depth with respect to our research questions, the whole "scene" was transcribed to provide a more comprehensive analysis by members of the research team. Although I was the teacher of the course in question, I will refer to myself in the third person so that it is easier for the reader to follow the critical reflections that I, as a researcher, have of my conduct as a teacher back in 1997.

Although it is a college-level course, it is actually more comparable to an advanced high school course in which functions are analyzed in greater depth, the derivative is introduced, and notions of integrals are outlined. Thirty percent of the student's grade is based on what we call the *modeling approach.* In such an approach, students work in groups studying a topic of their choice. The participation of the group in choosing a theme and delimiting a problem to be investigated is stressed as very important. Thus, the definition of the problem is formally treated as important for these students. The teacher collaborates with all the groups in defining the limits of the problem, and in proposing possible paths to solve it or generate another problem. This approach is similar to the one described by Renuka Vithal, Iben Christiansen and Ole Skovsmose (1995) or Rodney Bassanezi (1994), in that the problems worked on by the students are researched for four months, include both oral and written presentations, and include the participation of the teacher in helping them solve the problem. They differ, however, in that Vithal, Christiansen and Skovsmose's examples are much more embedded in a politically progressive agenda with respect to the choice of the themes, and Bassanezi's examples are more related to the context of teachers' continuing education. Students were also told that it would be desirable, but not necessary, for their investigation in some way to involve the main theme of the course: functions, derivative, and integrals. Students were expected to present partial and final written versions of their work and to make oral presentations to the class. Other examples from this part of the class, which raise other issues, can be found in Borba (1997, 1999a) but are beyond the scope of this chapter.

Parallel to this type of work, we have developed the *calculator-experimental approach,* in which the graphing calculator was used as a vehicle for experimentation by students in response to assigned tasks. Graphing calculators were chosen since they were easy to use in the classroom, a special room is not needed to use them as is the case with computers and, at the time, UNESP, the university where I teach, had very poor computer facilities.

Besides, graphing calculators were affordable and they were suitable to some of the topics which had to be addressed in the course. Twenty Casio fx-8700 were available during each class. During the last five years, a complete computer laboratory and a set of 27 Texas Instruments TI-83 and interfaces such as CBRs and CBLs have been available. A typical class using this approach would open with the students working in small groups (2–3 participants) on tasks given by the teacher. One of the students deals directly with the calculator, a second one writes a group report to be handed in, and a third, when there is one, monitors both activities. Periodically the students switch roles. The teacher is responsible for raising challenging questions and encouraging the students to pursue conjectures and share some of the questions that arise from a given group; he also facilitates a "summary section" in which the students' findings are debated. The class also includes lectures by the teacher and textbook work, but these account for less than a fifth of the entire class time.

The data collected included written examinations, written group reports about their modeling work, and written reports about their group activities with the calculators using the experimental approach. Beginning in the third year, the parts of the class involving group work with the calculators and group presentations about students' modeling projects were videotaped. Reflexive notes were taken by the teacher.

Different sources of data were used to improve the trustworthiness of the research. Toward this end, the data were "triangulated" by different sources and analysis was peer-debriefed (Lincoln & Guba, 1985; Denzin & Lincoln, 2000). For example, a conjecture which emerged from a piece of data would be compared to evidence from other types of data. The other members of the research group also looked at given pieces of data and considered my own interpretations, offering criticisms or alternative interpretations. Thus, research data, like that presented here, were shown to members of the research group and alternative interpretations of the data were tried by different members of the group, especially by Mónica Villarreal, who was conducting a different study in the classroom and could easily challenge my interpretations. Thus, the data were "triangulated" by different subjects and by different sources of data.

It is not enough to be careful about the data collection and analysis. Although it is relevant to develop ways of reducing the "opinion effect" of a given researcher and to increase the credibility of the research through procedures such as triangulation, it is also important to strive for consistency between research procedures (e.g., interviews, videotaping, and triangulation), the view of knowledge inherent in a research methodology, the research question, the object of the investigation, and, in this case, the pedagogical approaches being used (Penteado & Borba, 2000; Skovsmose & Borba, 2000).

In the classroom analyzed, both pedagogies—modeling and the calculator-experimental approach—emphasize students' ability to solve problems in different ways and arrive at different points. The teacher also allows for detours or different roads to be taken by students as they cope with learning, and values students' input as much as possible. In these approaches, students can choose the problem (modeling) or they can take their investigation in different directions (calculator-experimental approach). For example, in the modeling approach, students would choose a theme to study and it would be part of the teacher's responsibility to relate it to mathematics, or to discuss why such a connection could be artificial or not desirable. In the experimental approach, as will be made clear later, the students could take the problem in directions unexpected from those planned by the teacher.

There must be resonance between such pedagogies and the view of knowledge held by the researcher (Lincoln & Guba, 1985). Thus, I believe that the view of knowledge previously discussed, which emphasizes the role of humans and media in the production of knowledge, is consistent with a pedagogy that emphasizes process as opposed to product (correct answers). It would be very hard to develop such a study based on test results, since a view of knowledge based on outputs would be in conflict with pedagogies that stress processes. In the same way, the research questions driving the studies I have been developing are related to the role of technology in the classroom and not, for instance, to proving whether technology does or does not improve a given aspect of learning, the latter being more in resonance with quantitative research based on test outputs, for instance. Test outputs would contribute little with respect to questions regarding how collectives formed by humans and media are transformed when joined by new "non-human actors." In this way, the pedagogical perspective practiced in the classroom and the qualitative methods—such as interviews, videotaping and analysis of students' written work—are in resonance with the questions, and with the view of knowledge.

As I present the example of students-with-technology, I hope the coherence between the research methodology—both in the broad sense of epistemology and in the narrow sense of procedures—and the questions I have been pursuing and the educational proposal under scrutiny becomes even clearer. In the last section of the chapter, I discuss how the theoretical framework helps to shed light on the analysis itself.

## STUDENTS, FUNCTIONS, TANGENT LINE, AND GRAPHING CALCULATORS

As part of the experimental approach, we developed activities in which students used the calculator to consider the distance between two points, the equation of a straight line, and the relationship between coefficients and graphs of different families of functions, such as linear, quadratic, logarithmic, and exponential. In addition, the notion of function was stressed as a way of modeling data, when students analyzed a variation of a problem posed by Christopher Schaufele and Nancy Zumoff (1995) in which coal consumption is modeled with functions. It was expected that the tasks developed by students in the experimental approach would relate to the modeling part of the course, as the work with the transformation of functions (Borba, 1995c; Borba & Confrey, 1996) in the former could lead to adjustment of curves in the latter. At the end of this part of the course, the students had had approximately 15 hours of practice with the calculators in the classroom, so it could be said that they were comfortable with the calculator.

Over the past several years, I have used the following task, after the section on functions, as a means of introducing the notion of derivative: Is it possible to make the graph of $y = x^2$ using only straight lines? This task triggered the discussion which constitutes the data presented in this chapter. I intended with this task to take advantage of two of the main features of computer technology: visual facilities, and experimental possibilities. So, in contrast to the notion of introducing limits and derivative using a formal approach, I intended to have the students experiment with local linearization (Tall, 1991) through a visual and experimental approach. Students were expected (with guidance, if necessary) to realize that they could approximate the parabola with straight lines if they took the points belonging to the parabola with the shortest distance between the x-coordinates. If they then had a greater number of straight lines using points of the parabola which were closer, they would get an increasingly better approximation. After that, the teacher would introduce more formal notation for the notion of derivative.

For this task, the students were divided into small groups of 2–4 participants, and a lively discussion began that lasted about 10 minutes. It was not possible to know what each group was thinking about the task. However, following this period of discussion in small groups, a debate among the entire class began, led by some of the participants. As this discussion was occurring, many participants continued to work with their graphing calculators. The first major input was made by Iris and her group who explained their approach to the problem; it consisted of considering the side of $y = x^2$ with positive x-coordinates, taking the points (0,0), (2,4) and (3,9) and then, by trial and error, finding the straight lines which went through those points. They also claimed that they could draw parabolas—consisting of straight lines—more precisely if they had taken more points. Camila then entered the discussion and added that she had taken the points (2,4) and (3,9) and had calculated

$$\frac{\Delta y}{\Delta x}$$

to find "a" in $y = ax + b$; she used algebraic calculation with paper and pencil to calculate the value of "b." Camila and her group had not used the calculator to find the equation $y = 5x - 6$, except as a way of checking whether their algebra looked the way they thought it should on the graphing calculator. At this time, the following dialogue took place (as captured on videotape). The translation from Portuguese into English was made by Mónica Villarreal, Anne Kepple and me):

Fernanda: Teacher, I don't think we will get a parabola just right, because if it is a curve... it will always be pieced together, there will be a bump.

Teacher: There will be a bump, hum...?

Fernanda: It will not be complete, maybe in the calculator it looks like [it is smooth].... no, I don't think so, if you stop and think, even if there were very short straight lines, there would be little pieces to be put together... because it is straight line, do you understand? And the thing [a parabola] is a curve, and I don't think it will be possible.

It seems that Fernanda is uncomfortable with the method presented by both Iris and Camila, since they see that if we keep taking smaller $\Delta x$, they will eventually be able to answer yes to the question initially posed by the teacher ("Is it possible to make the graph of $y = x^2$ using only straight lines?"), while Fernanda thinks that this method will not lead to a satisfactory answer since there will always be some bumps left. This debate resembles a historical problem in mathematics regarding the possibility or not of $\Delta x$ turning into zero. The discussion continued:

Iris: But teacher, what if I put a straight line at the little bump.

Fernanda: But then you are going to smooth out the bumps, but there will be points, we are going to have only points on top of the curve.

Iris: But if you smooth them out with straight lines, you won't have bumps.

Fernanda: But then you will end up having dots, and this is not a straight line, then there are points to fill out...the parabola, I think....

Mayra: What I thought was the following, so ..., on each point of the parabola, we could find a straight line which passes by this point.

Fernanda: But then we are going to have points, not straight lines.

The above debate is interrupted by about ten different people speaking at the same time; it is impossible to understand from the recording what anyone was saying except for the teacher's attempt to get the students to speak one at a time. However, what is important to emphasize in the above excerpt is Fernanda's reaction to Iris' solution. Fernanda perceived Iris' solution to imply that the parabola will be made up of points, not of line (segments), and she is reluctant to accept this solution. Mayra's suggested solution is to "find a straight line which passes by this point." This "passing by" could refer to tangents. After the disruption of the debate, she expands:

Mayra: For example, if we get 30,000 points in the parabola... 30,000 straight lines which pass by the parabola...

Fernanda: Then there are many straight lines, but to fill out...the bumps, it is going to get to another point in which there will be only one point...

Iris: He [the teacher] said straight lines, he didn't say whole straight lines.

Fernanda: But what defines a straight line is not only a point.

There is a new "eruption" in the class as everybody speaks at the same time. It seems that Fernanda is working with line segments, or straight lines with a restricted interval (see Figure 2.1), while Iris seems to be thinking about straight lines with the domain $D = \mathbf{R}$. The interpretation of Fernanda's thinking is also corroborated by a drawing she made in her written report about her activities during the day (see Figure 2.1). By working with line segments, Fernanda seems to believe that, in the end, she will not have a line, but a point, while Iris thinks that she still has a line if it is a tangent to the parabola. Fernanda also seems to stick to reasoning developed in previous tasks that stressed the notion that two points are needed to generate a straight line. She wants, at this point, to have these two points on the parabola. It is possible to say, then, that there are two notions of straight line. Fernanda seems to put together the notion that it is necessary to have two points from the parabola to have a straight line, since it is necessary to have two points to define it, and the feature posed by the teacher in the proposed task: a straight line that approximates the parabola. The latter might have been understood as a need for the two points to belong

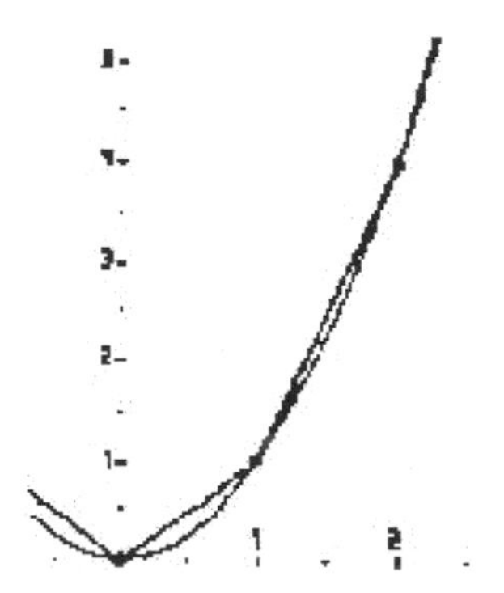

Figure 2.1. A figure which resembles the one F drew in her written report.

to $y = x^2$. Apparently, Mayra is bringing up the notion of a tangent line, even though there is no evidence that she was thinking about derivative as the slope for a straight line that is tangent to the parabola at a given point.

It is important to notice that this is the interpretation of the researcher reflecting back on the experience, since the teacher, at the time of the class, had no awareness of what we now believe was the root of the problem. It should also be noted that the role of the teacher was more one of just calling on different students so that the discussion would not come to a dead end. The teacher was fascinated by the discussion and had made a conscious decision to interfere as little as possible. Fernanda thinks that the other solution is a form of cheating, so to speak, since the task presented mentions lines and not single points, as the next interaction shows:

Fernanda: ... in order to have a straight line I need many points, right? and I believe that to fill out a parabola we would need a straight line with many points inside the parabola.... I don't understand the idea of having a straight line ... with just one point belonging to the parabola, do you understand?

Tonini: Professor, wouldn't it be the case that the points which go outside ... the parabola are so insignificant in relation to the scale of the graph ... that they would be negligible ... could they be negligible? ... If you magnify the scale of the graph and did a giant parabola [makes a parabola-like shape with his arms] so you could see "the millimeters" of all those points, of ... those bumps of the straight line, wouldn't it be so insignificant and therefore negligible ... ?

For the main purpose of this chapter, it is relevant to note that students have been using metaphors such as bumps throughout the discussion and that, at this point, Tonini explicitly brings "mathematical body language" into the debate as he uses his arms to think and to express himself.

The teacher, in turn, was trying to let the discussion flow, and thus his comments had a management tone, bringing the class back to organized discussion twice. He had also indicated that he was happy with the nature of the discussion about the task presented and the issues raised by Mayra related to continuity. His notes indicate that he was trying not to spoil the discussion and decided not to jump into the debate for fear of inhibiting or shutting down the discussion. He attempted to summarize the discussion up to that point, without great success, but he was able to get Mayra to be more explicit about her ideas, as she brings up the notion of tangent line in an explicit way: "then, this is what I thought: for each point of the parabola, you can have a tangent line...." Fernanda repeats her idea about points being different from lines, and Tonini gets back into the debate:

> Tonini: what she [Mayra] said about the 30,000 points that the straight line passes through...that a point of the straight line can pass by the parabola, correct?...and if a little bump shows up...it can be neglected...

The teacher was interested in exploring the idea of 30,000 points as a way of passing from secant lines to tangent lines and introducing derivative, even though he was aware that Mayra's notion resembled the idea of a continuous curve being similar to a necklace, as described by Paul Goldenberg & Marlene Kliman (1990). Mayra explicitly introduces the notion of tangent line for the first time, and another student characterized it as touching the parabola just at one point but, "staying outside the curve." Many students bring in arguments which have been presented before, and the discussion starts to go in circles. One exception is an idea brought up by one student related to zoom and microscopes, which may allow us to discuss reorganization of thinking, not only in terms of technologies of intelligence, but also other types of technology, as we will see in the next section.

The teacher notes that class time is almost up, and tries to summarize the issues raised by the students, using a graphing calculator on top of an overhead projector to give his interpretation of what had happened. He also asks the students to bring to the next class answers to the question: "Is there a more precise method of drawing a parabola with straight lines?" With this question, he intended to get back to the discussion about tangent lines. He instigated the students with the idea that, if Iris' and Mayra's ideas are used, one may arrive at a different way of seeing tangent line, and raised the question of how to calculate the slope of this straight line given just one point instead of two, as they had discussed in some of the earlier classes. The teacher intended, with this activity, to move students from a more static view of tangent line that they learned in high school to the more dynamic idea of tangent found in calculus. In the next class, he introduces derivative, without using the terminology yet, using the informal idea of limits and discussing the possibility of making $\Delta x$ arbitrarily small ($\Delta x \to 0$) in

$$\frac{\Delta y}{\Delta x}.$$

In the next section, I discuss how the data presented are related to the theoretical framework presented earlier.

## STUDENTS-WITH-MEDIA: THE REORGANIZATION OF CLASSROOM COMMUNICATION

In this chapter, I have claimed that communication in the classroom should also be seen from an epistemological perspective that emphasizes the role of media in generating knowledge in such a setting. As one possible way of seeing this, I presented the notion of reorganization of thinking. In such a view, media are not just peripherally important for production of knowledge; they have actually provided the possibilities for the production of the various kinds of knowledge throughout history or in a given classroom. I proposed, then, the metaphor of humans-with-media as subjects who produce knowledge.

Different metaphors have been formulated regarding knowledge: "construction" as in constructivism, "fields" as in the many different versions of semantic fields, "*tabula rasa*" as in behaviorism, and "computers" in the case of artificial intelligence. Metaphors have been used to express our ideas about cognition due to our difficulty or the impossibility of describing thinking, among other reasons. On the other hand, the metaphors chosen carry strong meaning which is culturally bounded. It is also possible that metaphors are necessary to express theoretical ideas, as we try to link some abstract ideas of a theory to our experience. In this sense, *humans-with-media* as a subject that produces knowledge should be seen as a metaphor which brings media to the center of knowledge production and, at the same time, builds on the idea developed by Lévy that we live in a cognitive ecology, and that technologies of intelligence in particular, such as computer technology, are actors in the process of knowledge production (1993).

But what do we gain with such an approach? How does it help us to see what we could not see using different glasses? Do the data validate or corroborate the theoretical framework presented? Is the technology "the cause" of the style of communication that took place in that classroom? The rest of this chapter is devoted to answering these questions as I bring together the ideas about the history of media, reorganization of thinking and how technology transformed the communication that took place in the episode presented.

The first issue that should be clarified is that I do not believe knowledge development can be described by a straight forward cause-and-effect. In such a view, there is a cause to be found for every effect. In what Yvonna Lincoln and Egon Guba (1985) and Norman Denzin and Yvonna Lincoln (2000) called positivistic and post-positivistic paradigms, social science research imported a model from the early physical sciences in which there is a cause for every effect. Thus, a reader who holds such a view, when presented with data such as that presented above, would be inclined to ask whether the technology was *the* cause, or whether the teacher, the pedagogy, or the climate were not *the* cause for such a view.

Alternatively, authors such as Lincoln and Guba (1985) propose a paradigm in which, rather than cause and effect, one thinks more in terms of events occurring synergistically; events mutually shaping each other. It is not possible to isolate the teacher as a "variable" and control it. Instead, it is necessary to understand all the events and factors as interconnected and webbed. This is why I do not claim that media in general, nor the graphing calculator in this specific case, were the cause of the communication that took place or even for the style of communication. In a more modest sense, I want to show signs of the role of technology in the communication process that took place in the classroom. To do so, I also use the notion of humans-with-media to emphasize the point that humans cannot produce knowledge without a medium, but also that the type of knowledge produced can change over time, shaped by media that are developed. Such a notion will be extended in the sense that the role of media used in the classroom is also paramount to the communication process. In this sense, humans-with-media produce knowledge and communicate it. Communication is, on the other hand, part of the process of generating knowledge in the classroom. The data do not validate the framework but help to extend it, as in the example of the microscope which is presented shortly. In the paradigm in which I have been working, there is neither refusal nor confirmation of a theory or theoretical proposition. There is instead negotiation of what is considered valid, and theories are "shadowed by clouds." Therefore, at the end of this chapter, you will not have clear sky or a rainy weather forecast, or a confirmation that humans-with-media is or is not a fine theoretical construct. It will add to other papers where similar arguments are made and will contribute to the debate about seeing learning and communication through its lens.

In this sense, as a qualitatively different medium begins to form part of the humans-with-media unit, a reorganization takes place at different levels. Elsewhere I have been dealing with issues such as the way that reorganization of thinking occurs with increasing use of visualization and experimentation; the change in the way that "mathematical discovery" takes place in the classroom; the transformation of the power relationship between student and teacher; and the physical reorganization of the classroom. Other authors have emphasized these aspects as well. For example, Anna Chronaki (2000) and Miriam Penteado (2001) have asked that the new roles of the teacher be discussed as technology enters the school scene. Jere Confrey (1993) has required listening to students' voices, while Goldenberg and Kliman (1990) have discussed students' interaction with graphing programs. All these aspects—involving teachers, students, design, etc.—could have been explored in the episode presented. But in this chapter, a different aspect is emphasized: reorganization of communication in the classroom as different media enter the scenario. Of course, as proposed before, the other issues mentioned above also shape the technology in the classroom, but I want to build the argument that there is some specific influence of technology. Therefore, technology is not "the cause" of communication in the classroom but is nonetheless feared by some.

Arguments that were common in the early 1990s that students would become attached to computers, turning them (and people in general) into uncommunicative individuals, are rarely heard nowadays. Since the mid-1990s, however, with the popularization of the Internet, a new wave of arguments have emerged: "people will spend hours in on-line chats and e-mail lists," raising concerns that people will not talk (face-to-face conversation) to each other anymore. In countries like Brazil where, due to economic reasons, the use of technology is less intensive compared to other countries, teachers still fear that the use of information and communication technologies in the classroom may cause such problems. I believe that part of this fear is a result of the theoretical dichotomy between humans and technology present in much of the literature on philosophy of technology.

It is important, therefore, for countries such as Brazil, that the metaphor humans-with-media be recognized as one possible way of thinking about knowledge, since it stresses that knowledge is always produced by humans and by at least one medium. Such a metaphor may shift the discussion from whether computers are bad or good to one about the transformation of knowledge in the classroom and how communication changes, particularly in the classroom. The dichotomy between technology and humans does not leave room for hope regarding the use of technology, while the unit proposed in this chapter centers the debate on how we want to transform knowledge, and how we want to integrate this new technology with other technologies of intelligence, such as orality and writing.

The metaphor of humans-with-media as knowers is integrated not only with the research question, research design, and the pedagogy in question, as discussed earlier in this chapter; it is also relevant to another aspect of methodology: data analysis. In the last section, the episode showed how communication took place in a mathematics classroom as the discussion started with the problem relating tangent lines and parabolas. I used the metaphor of humans-with-media in the analysis to understand specifically how different media, e.g., graphing calculators, are also actors in the communication that takes place. Technology was not the cause of it; but there is strong evidence suggesting that it was paramount for the discussion that took place, and that the communication process was qualitatively altered by the presence of information technology.

As mentioned earlier, I do not want to claim that the graphing calculator caused the communication process, nor the teacher, nor the pedagogy. I do, however, want to draw on the theoretical framework developed above to make the case that the technology was relevant in the communication that took place. But in what way? Based on the analysis of this episode, I suggest that the graphing calculator fosters the practice of conjecture. For the purpose of this chapter, the act of conjecturing, sharing with small groups and having the conjecture challenged by another group or the teacher, transformed the way communication took place in the classroom. Moreover, the discussion of visual aspects of the graphing calculator led many students to try metaphors and body language to describe the knowledge that was being produced by this collective of students-graphing-calculators-and-paper-and-pencil. Students using calculators, paper and pencil, and orality built arguments to discuss problems connected to local linearization. A theme which is not trivial in mathematical terms was debated by first-year biology students, and it is all pervaded by their experience with calculators.

It is in this context that the example presented should be viewed. The discussion led by Mayra, Fernanda, Iris, Camila, and Tonini, illustrates how a discussion about conceptual issues can take place in a mathematics classroom where an experimental approach with intense use of technology is practiced. The discussion that took place was based on the calculator-experimental pedagogy. This approach is closely connected to the computer technology used, since this medium, together with appropriate pedagogy, supports a process of investigation that includes trial and error, generation of conjectures, and assessment of conjectures. This experimentation provokes intense debate in the small groups as well as in the whole class. In this sense, the use of this medium in the classroom reorganized the communication process as different small groups, or using the terminology of this chapter, the different units of humans-with-media, had the chance to generate their conjectures and then test them in the larger group. Many of the transcribed passages presented earlier in this chapter can illustrate this reasoning, but the following is exemplary and also points to another issue:

> Fernanda: It will not be complete, maybe in the calculator it looks like [it is smooth] .... no, I don't think so, if you stop and think, even if there were very short straight lines, there would be little pieces to be put together... because it is a straight line, do you understand? And the thing [the parabola] is a curve, and I don't think it will be possible.

In this excerpt, Fernanda uses the calculator to support her argument. She says it is not possible to have local linearization, even if it looks that way in the calculator because of the limitation of its screen. The relevance of the calculator seems very strong in Fernanda's discourse, which leads me to affirm with confidence that the calculator was a part of the system that was producing knowledge. Moreover, she was reorganizing her thinking as she was communicating with colleagues and the teacher. Thus, Tikhomirov's notion of reorganization of thinking can be extended to one of reorganization of communication.

Although the graphing calculator was not being used at all times by the students, particularly at the time of the debate when it was being used little, it can be said that humans-graphing-calculator systems were in action during that class. For instance, without the graphing calculator, Iris could not have used her trial and error approach—and then tried to connect them to other approaches developed by her colleagues—to find straight lines which connect two points belonging to the parabola. And it is very unlikely that, without her experience with the graphing calculator, without the presence of this non-human actor, Mayra could have come up with the argument "...for example, if we get 30,000 points in the parabola, ... 30,000 straight lines which pass by the parabola...." When Mayra made this statement, the calculators were not being used in the classroom, showing that thinking and communication do not need the presence of such an artifact to be reorganized.

I also want to claim that the use of this medium did not suppress the use of other media in the classroom; orality and "paper and pencil" were some of the media used to structure the discussion that took place. In other words, the reorganization of thinking proposed by Tikhomirov can have many different facets, as in the case of Fernanda, who thought with the calculator when she came up with the idea of the impossibility of drawing a parabola with straight lines, but also used paper and pencil and orality in order to structure her arguments. Other students, Camila and her group, used the graphing calculator "just to check their result," while others used paper and pencil to find the equation for straight lines. But what matters for the main purpose of this chapter is that there is a reorganization of communication in the classroom. The style of communication, in which the teacher is the center, was substituted by one in which the students took the lead in mathematical arguments, with the calculators making it possible for such a collective to make several "experiments," conjectures, communication, and testing.

Orality and writing did not disappear despite all the incentives to use the calculator. This explains the plural form of "media" in "humans-with-media," because there is often more than one medium being used (Borba, 1995a; Borba & Confrey, 1996; Villarreal, 1999). A person or a group of students may use different media, and each representation (e.g., graphic representation) may gain a "life of its own" in a different medium. This example also illustrates that "humans" should be plural and not singular, in the sense that group work was essential in generating conjectures. Humans-with-media, then, is a theoretical construct that enables us to see communication being reorganized by different media.

Tonini, one of the students, seems to extend these ideas—and those of Tikhomirov and Lévy—which emphasize the role of media in general, and computers in particular. His comments about scale, and about bumps being negligible suggest that he had incorporated his previous use of the graphing calculator in other tasks, and that we can think of a human-computer system as being the actor of his argumentation as well. This student connected the idea of the zoom and scale of the calculator to another issue: the use of the microscope, which is a very familiar piece of equipment for biology majors. Calculators and microscopes were not present in the classroom at that time, but they were part of their experience, they were part of this collective humans-with-media unit which was producing and communicating knowledge. When he talked about bumps, he implicitly showed the idea of local straightness of a differentiable curve, as suggested by David Tall (1991), this is strongly linked to the media. As he uses the microscope as a subject of knowledge, too, he extends the metaphor of humans-with-media to incorporate the microscope. If we look closely at what Tonini said, it is possible to strengthen further the argument about the suitability of thinking about humans and media, both in the plural form:

> Tonini: teacher, wouldn't it be the case that the points which go outside the parabola are so insignificant in relation to the scale of the graph . . . that they would be negligible . . . could they be negligible? . . . If you magnify the scale of the

> graph and did a giant parabola [makes a parabola-like shape with his arms] so you could see "the millimetres" of all those points, of...those bumps of the straight line, wouldn't it be so insignificant and therefore negligible...?

This excerpt suggests that he was drawing heavily on his own and his colleagues' experience with the graphing calculator, showing the reason for the use of the plural in "humans," and also illustrating the reason for the plural in media. Not only were orality, written language, and computer language becoming actors in knowledge production: his discourse suggests that his body, in particular his arms and gestures, was thinking the mathematics he was producing. In this sense, this student-with-media, in conjunction with other students, illustrates how thinking of humans and technology as a dichotomy is not adequate. Computers, which are machines that were developed based on algorithms, help students to do trial and error, visualize, and use their body to think and communicate. I believe that overcoming such a dichotomy is fundamental for learning, for political debate, and for ways of developing communication in the classroom that are based on student input.

In this chapter, I tried to emphasize the role of technology in communication in the classroom. In doing so, it helped to develop further the notion of humans-with-media as a way of connecting the voice of the students (Confrey, 1993), the paramount role of the students and teacher, technologies of intelligence, open research questions, and pedagogies which emphasize students' input. Such networking of theoretical constructs and practice enabled me to illustrate how I see the reorganization of communication in the classroom. I highlighted how the graphing calculator was an actor that transforms the possibilities of students making conjectures, sharing with their peers, checking with colleagues and with the graphing calculator and with other technologies of intelligence such as writing. Based on their experience with the calculator, and being biology majors, they connected the thinking in both worlds with the graphing calculator and the microscope to understand the problems of continuity and tangent line involved in the issue they were pursuing. The calculator, albeit absent at that very moment, was an important actor in that collective that was making sense of a mathematics problem. Their patterns of communication were reorganized in the sense that the students communicated with peers, and with the machine or the software design of the graphing calculator. It was a dialogue in which some of the students had other humans as partners, and had the graphing calculators as partners as well. Software used in the calculator and writing were extensions of their memory that changed the nature of communication, since they could use the results they obtained (e.g., a graph) and the lack of results (is that line really a tangent?) to communicate with their colleagues. There is also another feature in which the graphing calculator could be considered as an actor: students could refer to graphs, points, zoom, and approximation, because they had experience with the calculator (and some with other graphing software). So they had this common ground, or this "channel," to disagree and agree about the different problems they were raising. In other words, they were a collective of humans-with-media producing knowledge as they communicate among the humans and non-human actors involved in such an endeavor.

## ACKNOWLEDGMENTS

This chapter is part of a research project named "*Pensamento Matemático, Funções, Computadores e Outros Meios de Comunicação II/III*" e "*Novas e Velhas Tecnologias da Informação em Educação Matemática*" *(Grant 520033/95–7),* which is sponsored by CNPq, a funding agency of the Brazilian Government, and it was developed by GPIMEM (Research Group on Technology and Mathematics Education). Although they are not responsible for the content of this chapter, I thank Miriam Godoy Penteado, professor of the Mathematics Department of UNESP-Rio Claro; Telma Gracias and Ana Paula Malheiros, members of the Research Group in Mathematics Education and Technology; Mónica Villarreal, State University of Córdoba, Argentina; Ole Skovsmose, Aalborg University, Denmark; and Anne Kepple, Brazil for their comments on earlier versions of this chapter. The example presented in this paper was also used in a joint paper with Mónica Villarreal (Borba & Villarreal, 1998) which was presented at PME 22, South Africa.

## REFERENCES

Ascher, M. & Ascher, R. (1981). *Code of the Quipu.* Michigan: The University of Michigan Press.

Bassanezi, R.C. (1994). Modelling as a teaching-learning strategy. *For the Learning of Mathematics 14*(2), 31–35.

Borba, M. (1994). A model for students' understanding in a multi-representational environment. In J. Ponte & J. Matos (Eds.), *Proceedings of 18th PME.* (pp. 104–111).

Borba, M. (1995a) *Students' understanding of transformations of functions using multi-representational software,* Associação de Professores de Matemática de Portugal, Portugal. (Doctoral dissertation in mathematics education. Cornell University, 1993).

Borba, M. (1995b). *Um estudo em etnomatemática: Sua incorporação na elaboração de uma proposta pedagógica para o "Núcleo-Escola" da Vila Nogueira-São Quirino,* Associação de Professores de Matemática de Portugal, Portugal. (Master Thesis in Mathematics Education. UNESP, 1987.)

Borba, M. (1995c). Overcoming limits of software tools: a student's solution for a problem involving transformations of functions In L. Meira, & D. Carraher (Eds.), *Proceedings of PME 19* (2), (pp. 248–255).

Borba, M. (1997). Graphing calculators, functions and reorganization of the classroom. In M. C. Borba, T. A. Souza, B. Hudson, & J. Fey (Eds.), *Proceedings of Working Group 16 at ICME 8: the role of technology in mathematics classroom* (pp. 53–60). Rio Claro, Brazil: UNESP.

Borba, M. (1999a). *Calculadoras gráficas e Educação Matemática.* Rio de Janeiro: Universidade Santa Úrsula.

Borba, M. (1999b). Tecnologias Informáticas na Educação Matemática e reorganização do pensamento. In M. A. V. Bicudo (Ed.), *Pesquisa em Educação Matemática: Concepções e Perspectivas.* (pp. 285–295). São Paulo: UNESP.

Borba, M. & Confrey, J. (1996). A student's construction of transformations of functions in a multiple representational environment. *Educational Studies in Mathematics, 31*, 319–337.

Borba, M. & Villarreal, M. (1998). Graphing calculators and reorganization of thinking: The transition from functions to derivative. In A. Olivier & K. Newstead (Eds.), *Proceedings of the 22nd PME Conference* 2. (pp. 136–143). Stellenbosch, South Africa.

Borba, M. & Penteado, M. G. (2001). *Informática e educação matemática.* Belo Horizonte, Brazil: Editora Autêntica.

Confrey, J. (1993). Voice and perspective: Hearing epistemological innovation in students' words. In N. Bednarz, M. Larochelle & J. Desautels (Eds.), *Revue des Sciences de l'education.* Manuscript, to be published.

Christiansen, I.M. (1997). When negotiation of meaning is also negotiation of task. *Educational Studies in Mathematics, 34*(1), 1–25.

Chronaki, A. (2000). Computers in classrooms: Learners and teachers in new roles. In B. Moon, S. Brown & M. Ben-Peretz (Eds.), *Routledge international companion for education.* (pp. 558– 572). London, UK: Routledge.

D'Ambrosio, U. (2001) *Etnomatemática—elo entre as tradições e a modernidade.* Belo Horizonte: Editora Autêntica,.

Denzin, N.K. & Lincoln, Y.S. (Eds.), (2000). *Handbook of qualitative research (2nd edition),* California: Sage.

Goldenberg, P.E. & Kliman, M. (1990). *What you See is What you See.* Unpublished manuscript, Educational Technology Center, Newton, MA.

Gracias, T.S. & Borba, M.C. (2000). Explorando possibilidades e potenciais limitações de calculadoras gráficas, *Revista Educação e Matemática da Associação de Professores de Matemática de Portugal, 56*, 35–39.

Knijnik, G. (1996). *Exclusão e Resistência: Educação Matemática e Legitimidade Cultural.* Porto Alegre: Artes Médicas.

Lalande, A. (1999). *Vocabulário técnico e crítico de filosofia (3rd edition).* São Paulo, Brazil: Livraria Martins Fontes Editora.

Lévy, P. (1999). *A inteligência coletiva: Por uma antropologia do ciberespaço (2nd edition).* São Paulo, Brazil: Edições Loyola.

Lévy, P. (1993). *As tecnologias da inteligência: O futuro do pensamento na era da informática.* Rio de Janeiro: Editora 34.

Lincoln, Y. & Guba, E. (1985). *Naturalistic Inquiry.* California: Sage Publications.

Penteado, M. (2001). Computer-based learning environments: Risks and uncertainties for teachers. *Ways of Knowing Journal I*(2), 22–35.

Penteado, M. & Borba, M. (2000). *Informática em ação: formação de professores, pesquisa e extensão.* São Paulo, Brazil: Olho d'água.

Schaufele, C. & Zumoff, N. (1995). *Earth algebra: College algebra with applications to environmental issues.* New York: Harper Collins College Publishers,.

Skovsmose, O. (1994). *Towards a philosophy of critical mathematics education.* Dordrecht, the Netherlands: Kluwer Academic Publishers.

Skovsmose, O. & Borba, M. (2000). Research methodology and critical mathematics education. *Center for Research in Learning Mathematics, 17.*

Tall, D. (1991). Intuition and rigour: The role of visualization in the calculus. In W. Zimmermann & S. Cunningham (Eds.), *Visualization in teaching and learning mathematics.* (pp. 105–119). Reston, VA: M.A.A. Notes.

Tikhomirov, O.K. (1981). The psychological consequences of computarization. In J. V. Wertsch, (Ed.), *The concept of activity in Soviet psychology.* (pp. 256–278). New York: M.E. Sharpe Inc.

Villarreal, M. (1999). *O pensamento matemático de estudantes universitários de cálculo e tecnologias informáticas,* Doctoral dissertation in Mathematics Education, Universidade Estadual Paulista, Rio Claro, Brazil.

Villarreal, M. (2000). Mathematical thinking and intellectual technologies: The visual and the algebraic. *For the Learning of Mathematics, 20*(2), 2–7.

Vithal, R., Christiansen, I. & Skovsmose, O. (1995). Project work in university mathematics education: A Danish experience—Aalborg University. *Educational Studies in Mathematics, 30,* 199–223.

CHAPTER 3

# INTEGRATING DIFFERENT REPRESENTATIONAL MEDIA IN GEOMETRY CLASSROOMS

**Triandafillos A. Triandafillidis**
***University of Thessaly, Greece***

**Despina Potari**
***University of Patras, Greece***

## ABSTRACT

The integration of different representational media is suggested by a number of researchers as an approach that leads to more unified cognitive constructs of mathematical ideas. In the present paper, we attempt to extend this view to accommodate the changes that the integration of representational media may bring to the discourses existing in the mathematics classroom. More specifically, we combined written and verbal forms of representation in a learning situation that included the haptic exploration and construction of models of three-dimensional objects. We present findings from the implementation of the activities in one fourth grade and one fifth grade class in a primary school in Patras, Greece. From our analysis it became evident that the integration of representational media helped students to become more conscious of the processes of organising coherent texts to communicate

*Challenging Perspectives on Mathematics Classroom Communication*, pages 79–108

mathematical understandings. Moreover, the integration of different modes of communication may be able to challenge existing ways of communication in mathematics classrooms.

## BACKGROUND

The importance of integrating different representational media in the mathematics classroom has been acknowledged by a number of researchers. The advantages of this approach have been especially demonstrated in computer environments (Kaput, 1992; Schwartz & Yerushalmy, 1995). Other researchers have studied pupils' mathematical performance and strategies in which they integrate visual and written forms of representation (Ben-Haim, Lappan & Huang, 1989, 1985; Gaulin, 1985). In these studies, analysis is focused on children's cognitive constructs, and the ways that these are shaped through the proposed interventions. Guillerault and Laborde for instance, have investigated difficulties that pairs of 11- to 13-year olds face when they "encode" visual information in order to produce a written message containing no figure, as well as when other pairs "decode" these messages in order to redraw the figure (1986, 1982). Even though the authors acknowledge that the classroom environment contributes to children's difficulties with the activities, they focus their analysis on the use of the representational medium. Furthermore, the pairs that decoded the written messages were not classmates of those who encoded them, thus creating a "clinical" artificiality. In another case, Brown (1994), in an in-service course for prospective teachers, has integrated haptic, verbal, and written forms of communication. His aim was to underline the fact that mathematics is bound up with linguistic practices, as well as to suggest the rich potential that lies in integrating various representational media to build mathematical understandings.

In this chapter, we address issues of communicating mathematical understandings in a classroom environment. Any discussion about communication in the mathematics classroom immediately raises issues concerning the medium of communication, the organization and interpretation of mathematical notions, as well as the animate and inanimate "participants," (i.e., teachers, pupils, researchers, and discourses prevailing in the classroom). Our aim is to investigate decisions made by children in the process of organizing and assigning meaning to texts, focusing on the potential value that the alternation of different representational media carries in communicating and building understanding of mathematical ideas in the classroom. In the process of organizing and assigning meaning to a text, the social environment in which the text is grounded is critical. Thus, we are equally interested in determinants and conventions referring to the environment in which texts are initially constructed, and later on interpreted. Examples of these determinants are the nature of the given task itself as an opportunity for producing a mathematical text, the directions given by the teacher/adult, the "culture" of the particular classroom, and the discourses which it engages, the background knowledge of the children and their personal engagement and direction of attention to the task at hand. These factors frame the production and interpretation of a text and affect the emergence of meaning in the mathematics classroom. In the present study, we examine verbal and written descriptions produced by Greek children between the ages of nine and eleven, during the course of haptic exploration of three-dimensional objects, and the construction of physical models based on these written descriptions.

## THEORETICAL FRAMEWORKS

In our experiment we consider all children's work, including written reports and constructions based on these reports as texts embodying their understandings and interpretations of the tasks at hand. We also view the objects given to them to explore haptically as texts. A text designates any coherent configuration of signs interpretable by a community of users (Hanks, 1989). This description raises issues of sign form and the representational media incorporated in a text. Texts then can be written, verbal, or material constructions (a painting, a sculpture, a photograph) or a combination of these, according to the media chosen by the user in order to communicate his or her emotions, ideas, perceptions, and intentions. As Ingarden suggests, all representation is inherently incomplete, full of blank spots of indeterminacy (1973). These blank spots contained in a text may relate to assumptions of background knowledge, omitted details, conceptual gaps, stylistic peculiarities of the text, and ineffective use of the medium. Textuality is a term that addresses the quality of coherence presented by the text, which may depend on the inherent properties of a text, the interpretative activities of a community of readers and viewers, or a combination of the two. Concretization is seen as a process whereby the blank spots are partially filled in during the process of building an understanding of the text. The level of the achieved concretization of a text is determined, then, by the overall structure of the text and the interpretative activities of a community of users (Hanks, 1989).

Geometry is a field where concepts traditionally relate or even come to coincide with a visual image. In teaching geometric concepts, vision is customarily considered dominant over other senses. Researchers, however, have increasingly realised that this prioritization has hindered the development of children's geometrical understanding (Clements & Battista, 1992;

Fischbein1993; Hershkowitz, 1989; Parzysz, 1988; Vinner, 1983). Prototype phenomena in the teaching and learning of geometrical concepts, for instance, can be seen as a result of such prioritization. A number of studies have suggested ways of building geometrical understanding to overcome prototype phenomena (Clements & Battista, 1990, 1989; Fischbein, 1993; Markopoulos & Potari, 1996; Marrioti, 1989), mostly through the use of computers. All these cases, however, employ visual practices. In the present study, we make an effort to extend research beyond ocular boundaries.

The ocular emphasis in the teaching and learning of geometrical concepts cannot be "seen" in isolation from the domination of vision in other cultural and social practices in the history of Western civilization. Indicative, for instance, is the pervasiveness of visual metaphors in language (Jay, 1993; Pimm, 1995). This prioritization should also be viewed as a cultural structure detected in ideas, opinions, ways of thinking and behaving, systematically built in the process of teaching geometry in the classroom. In that sense, the prioritization of vision in geometry classrooms is culturally constructed constituting certain practices as more "appropriate" in systematically forming the object of which they speak (Foucault, 1993). The reliance on a type of reasoning that emphasizes a close reference to a static figure, and a focus on its visible characteristics, are examples of such practices that dominate the choice of material and the modes of communication in geometry classrooms.

One way to overcome prototype phenomena is to accompany the teaching process of geometrical concepts with the "view" from another "sense." We discuss below the significance of such alternative and complementary to vision approaches. Until the 18th century at least, touch was one of the master senses that "checked and confirmed what sight could only bring to one's notice" (Mandrou, 1975, p. 53). The role of sight was clearly secondary to that of hearing and touch, providing impressions that were not as reliable. Nowadays, in a culture so wrapped up in ocular practices, the haptic exploration for example, particularly in the absence of vision, denotes a state of impairment. In the mathematics classroom haptics are found to have a value in promoting children's learning (Triadafillidis, 1995). On a macro-perspective level, touch opens up the possibilities to restore the proximity of self and the other, of self and objects in the environment, of actually situating the self in a lived world (Jay, 1993). The decision to use haptics in the teaching of geometry could be, then, a step taken against the hegemony of spectator learning practices in geometry classrooms.

In the haptic exploration of a geometric shape the positioning of the object in the real world is achieved through the sense of touch. By haptic exploration, we refer to the process of intentional and conscious movement of one's hands about an object without being able to see the object. The slow nature of this process, compared to the swiftness of vision, opens up the possibilities for a dynamic mental "reading" and reconstruction of the shape of the object. In learning environments that rely on vision the term "dynamic" characterizes any presentation of images that involves movement. Nicolet's films are examples of dynamic presentation of images. Geometer's Sketchpad and Cabri Geometre's environments are also dynamic, providing the learner with the opportunity to transform the image or even alter one or more of its geometrical characteristics. Changes of the image's geometry are accomplished indirectly on the screen of the computer through the actions of a virtual hand (for a "hands-on" environment where changes are achieved directly on the object see Markopoulos & Potari, 1999). There are fundamental differences, though, between these environments and those that rely on haptics. With haptics there is a "hands-on" sense of achieving a "physical" immediacy with the object's geometry, joining at the same time visualization with action (Piaget & Inhelder, 1956). Furthermore, the step-by-step recognition, discrimination, and "animation" of the object's qualities and relationships among these qualities facilitates performance in geometry, and further provides support for the development of reasoning (Triadafillidis, 1995).

The perspectival deformation of objects, so evident and determining in activities that rely on the modality of vision, is closely related to perceptual and logical fallacies in geometric thought. However, this phenomenon appears also in activities that involve the haptic exploration of geometric shapes, though not to the same extent and form. In the haptic exploration of an object when we aim to disclose the characteristics and/or the shape of that object, we may direct attention to the physicality of the object, (i.e., characteristics concerning material, texture, form, indentations, protuberances or other anomalies of the surface, and weight). Children develop haptic exploration strategies that supplement the non-specific identification of form and physical characteristics. Gross or fine exploration of geometric features and relations with these strategies lead, then, to a comprehensive percept about the object at hand. Exploration strategies used by children, it is suggested, are regulated by changes that follow the developmental trends that the van Hieles suggested for geometric thinking (Triadafillidis, 1995).

Organizing a written text around haptic exploration, we suggest, not only reflects a geometrical understanding, but also plays a critical role in its shaping. Written texts are often perceived as descending from words spoken to oneself or to others, a way of inscribing our thoughts or prior speech (Rotman, 1994). Such a theorization sees the production of a written text as a result, in a linear fashion, of some prior firm thinking. On the contrary, written texts can be perceived as produced or organized by human beings, but not actually possessed by either their writers or readers. According to this view, any one speaker or writer cannot control words, written or spoken. Half of every word said or written already belongs to someone else, Bakhtin (1981) informs us. This is because "each word tastes of the context and contexts in which it has lived its socially charged life" (p. 293) since every written text or utterance is always situated in a certain socio-cultural, historical moment. "Words," then, written or spoken, "seek not to report rules thought to be 'discoverable in nature' but constitute the very rules that create the society's sense of nature" (Connolly, 1989, p. 2). In this sense there is more to writing than a stylistic presentation of thoughts since writing as a practice partakes in the process of constructing knowledge in the first place. To become fully concretized, a text demands commitment both from the writer's side in order to communicate his or her ideas, and from the reader in order to assign meaning to the text. The roles of the writer and the reader, besides being equally important in "reading" a text, are also alternated. The writer is also the reader of the text since he or she organizes the text with an intention to communicate with a certain group of readers. In this sense the intended readers become co-writers of the text. Meaning-making then is co-constructive, socially-situated, and not created in isolation by individuals (Spivey, 1995; Tedlock & Mannheim, 1995). Furthermore, writing in the mathematics classroom can be a way of heightening awareness of conceptual relations and thought processes (Connolly, 1989; Shepard, 1993).

The process of building concrete representations of geometric objects can be seen as another way to complement vision in teaching geometry. In addition to an engaged reader, building physical models of three-dimensional objects based on a written report about an object requires a sound mental representation of the form of the object with an identification and analysis of its different components (Potari & Spiliotopoulou, 1992). As Fischbein (1993) suggests, "unfolding" objects presents an excellent opportunity for training the handling of figural concepts in geometrical reasoning; however research studies on children's work of folding and unfolding three dimensional objects are few. The process of "building" solids requires the child to "see" the objects and recognize their characteristics, as well as combine the latter information in a transformed position taking also into account the reverse process. Through the process of building a solid, the child can develop an understanding of various concepts that are related to the shape of the object. In discussing the designing process, Alan Bishop notes "What is important for us in mathematics education is the plan, the structure, the imagined shape, the perceived spatial relationship between object and purpose, the abstracted form and the abstracted process" (1988, p. 39). Constructing models is a similar process to the designing process in the sense that it incorporates the same attributes.

In the present study, we consider the haptic exploration and the production of a written report, as well as the interpretation and construction of a visual embodiment of the report. The order in which these approaches are forming the phases of the experiment does not imply the imposition of a specific hierarchy of the different representational media. On the contrary, we attempt to examine how the suggested alternation of media is realized by the children as they try to incorporate their experiences from the different phases. The decision to encourage the incorporation of these representational media accords with the view that communicative acts within a given medium are normally co-occurent with acts in other modalities, and very rarely we can isolate a medium of communication from other media (Preziosi, 1986).

## METHODOLOGY—DESCRIPTION OF THE EXPERIMENT

The study is a classroom teaching experiment that was conducted in a primary school in Patras, Greece. The educational system in Greece is centrally managed. Mathematics syllabi are prescriptive, in the sense that hours spent in each area are predetermined. In the classroom mathematics is taught formally by talking to the students from the front, with activities engaging the whole class. Instruction is restricted to the textbook provided by the state. The study falls within an ethnographic research tradition (Wolcott, 1988) as it addresses the classroom environment in which children worked on the activities. We consider not only the ways in which pupils themselves assigned meaning to the activities, but also the ways in which these understandings and intuitions were communicated among pupils, in group- and whole-classroom arrangements, and between pupils and the teacher. We encouraged teachers to be actively involved in the study, introducing the activities to their class and assisting in the organization of the work throughout all the phases. This was not part of an attempt to situate us as external observers. Non-participant observation research paradigms reflect an illusion of seizing and reporting moments from the field exactly as they happened. Bystanders, eavesdroppers, and overhearers, all jointly construct even the simplest conversation in a cultural scene (Tedlock & Mannheim, 1995). Leaving room for teachers' initiatives, then, was not an attempt to assume the roles of bystanders. On the contrary we were aiming to investigate the dialogic relationships among the meanings that teachers, students, and researchers assigned to the activities, keeping in mind the existing hierarchies and discourses in these classrooms.

With its 300,000 inhabitants, Patras is the third largest city and the second largest port in Greece. The primary school where we conducted this study is located near the university campus in the outskirts of the city. Children of university professors and university employees, as well as of families from a nearby village attend the school. As a result, the school brings together pupils from a range of socio-economic backgrounds. In order to consider the role of classroom culture, as influenced by differences in age, school mathematics experience, and teaching, we worked with two classes, a 4th grade of 20 pupils (8–9 years of age) and a 5th grade of 22 pupils (9–10 years of age). Both teachers and the children had recent experiences with working with other researchers and prospective teachers. In particular, the 5th grade teacher had been seconded to the University in the Department of Education. The classes were informally visited for two weeks before the main study to get to know the classroom environment and familiarize the teachers with the materials we planned to use. The children worked mainly in groups of four, which consisted of two pairs that worked separately to produce the texts for each activity. All sessions were videotaped and the discussion in all groups was tape-recorded. In addition to the authors, three postgraduate students from the University of Patras participated as group observers.

The study consisted of four phases each lasting for 90 minutes. Phase one acted as an introduction to the "rules" of the game. An object was drawn randomly from a bag full of geometrical solids made out of cardboard, and other objects from everyday contexts. A volunteer was assigned the task of haptically exploring the object, and responding to various questions asked by the rest of the class about its features. To prevent pupils from seeing the objects while exploring them we used a "feely box" (Giles, 1981). The feely box is a cardboard cubic box with two holes cut on opposite sides. Seeing the object while exploring it, is considered "cheating." The rest of the class could see the explored object. In a variation of the game, the object was not visible to the class. Pupils then had to ask the volunteer questions in order to guess the type of object that was in the feely box.

In the second phase, each pair within a group had to explore haptically a plastic vinegar bottle or a glass bottle of soda, and produce a written description of it. The objects consisted of cylindrical and conical surfaces (see Figure 3.1). The written reports were exchanged, and used by the other pair of the group to construct the object out of cardboard. Therefore, we cautioned pupils to include in their descriptions as many features of the object as possible. Disclosing the name of the object was not an appropriate clue. In the third phase, pupils worked in the same manner on different objects made out of multi-link cubes in different arrangements. These constructions formed cubes made of three layers with a side of three multi-link cubes but with some pieces missing (see Figure 3.2). In this phase, the children were given multi-link cubes to build the object described in the written report. At the end, in both phases, the initial shapes were shown to the children. Each pair, then, had to evaluate their own constructions, and comment verbally and in writing on the other

Figure 3.1. The everyday objects used in the second phase.

Figure 3.2. The multi-link cubes constructions used in the third phase.

pair's written description. In the fourth phase, we discussed with the whole class two written descriptions, one for the vinegar bottle and one for a cubic arrangement, employing clues that pupils had used in the two preceding phases. This encouraged children to reflect on their own work and voice personal opinions concerning their choices.

In the analysis, data from different sources, transcriptions of video- and audio-recorded episodes from group- and whole-class discussion, and children's texts, were compared and contrasted to provide triangulation across the data. In this process we followed the phases of the experiment, analyzing each phase in reference to the previous. The second and third phases are jointly analyzed. We followed the common elements that characterize both phases, namely the texts produced by the children. We feel that we followed a bottom-up approach, based on issues that emerged from, and were supported by, our data. These data facilitated our attempt to understand the framework within which participants interpret their thoughts, feelings and actions (Chamberlin, 1974).

## FRAMING THE ACTIVITY: DOMINANT DISCOURSES IN THE CLASSROOM

In the first phase, children had the time to become familiar with the rules of the activity itself. At the same time a "consensus" about expected interaction for the forthcoming phases was laid out and implicitly formed. The actual framing of the activity ended up being shaped by the teacher's interpretation of the activity and the classroom culture in general. The classroom culture is characterized by a number of regularly occurring patterns of interaction, and is the product of what all participants bring to the classroom in terms of ideologies, presuppositions, feelings, aspirations, and knowledge (Nickson, 1992). Culture does not remain stable in the course of learning in a mathematics classroom. On the contrary, it evolves as it is continuously produced and reproduced within and outside the spatiotemporal boundaries of the classroom. To a certain degree, this classroom culture shapes and at the same time is, sometimes, shaped by the activity. The teacher's role in this first phase was crucial, since he or she was voicing the dominant discourse. Even within an assertive environment, though, there is always room for other voices to emerge.

In 5th grade the teacher, Nikitas, appeared to be following a type of interaction where the children participated mainly by asking questions and evaluating responses. He used frequent changes in the tone of his voice and body language in an attempt to attract the children's attention. On the surface it seemed as if children were shaping the "rules" in the classroom. In reality though, the children followed the norms of a classroom culture, in which the teacher is the initiator of the activity. In his desire to engage children and increase their interest, Nikitas builds on creating a competitive climate. In this climate a negotiation of meaning becomes a process of checking the correctness of an answer without necessarily justifying judgment: "You have to ask questions which put him [a child] in a difficult position, to see in the end how many of the questions he manages to answer correctly" (teacher is addressing the whole class).

During the first haptic exploration, Katerina had to identify the number of angles of a tetrahedron. The question was confusing, as angles at this age are discussed only in the context of two-dimensional shapes. In this process though, the teacher demands a quick answer and appear to be inattentive to the child's needs for time to understand the question and to develop her strategy. The class senses this message and acts accordingly:

Teacher: Come on, you have to answer! We will set a time limit.
Child: Yes ... two minutes to reply!
...
Teacher: Just say a number, any, to check it later.
Katerina: ...
Teacher: The question was very clear. Angles ...
Class: Ehh ... say a number ... any, say 10 ... 100.
Teacher: Say a number!
Katerina: Five.

This extract is characteristic of the type of interaction the teacher had with the children in the class during this first phase. It appears that a consensus is being formed about what is getting done in this scene, in this first phase of the experiment. This consensus is aligned with the teacher's interpretation of the activity. His interventions guide children to a possible way to interpret and experience the activity. As soon as the object was revealed to the child that performed the haptic exploration, a process of accounting started in which right and wrong answers were checked. Children participate in this accounting process eagerly, colluding with the teacher's approach and implementation of the activity.

On the whole, the teacher's questioning was of a closed type, emphasizing the "what" ("What is the number of faces"?) instead the "how" and "why" ("How did you find it"?, "Why is this so"?). As a result, children's responses were only announced and not further explored. There were times though, where this pattern was not followed. In those instances, the teacher attempted to explore and extend children's thinking, a decision that revealed a convergence between the aims of the teacher and those of the researchers. In these instances the teacher looked for the researchers' consent to the way that he was dealing with the experiment. So, the researchers' participation and the teacher's attempts to satisfy the researchers changes, even temporarily, the classroom culture and as a result the way that the activity is actually implemented.

Although, the organization of the familiarization phase of the experiment passed through the same stages in both classrooms, its actual implementation was very different. Lena, the 4th grade teacher, was not ready at first to take a leading role in this initiatory phase. Her hesitation in accepting such a role was probably due to her lack of confidence with mathematics and her experience of mathematics as a "subject difficult to teach." In contrast to Nikitas' confidence, Lena was not particularly pleased with her teaching and she perceived her participation in the project as an opportunity to learn. As a result, the researchers presented the "rules" of the activity while later Lena started to ask questions, along with the researchers, aiming to encourage children's thinking and understanding of the emerging concepts and processes. Lena's welcoming of co-operation, then, influenced the positioning of the researchers in the classroom's environment and, therefore, the implementation of the activity. In contrast to Nikitas' class, here the researchers shared an active role with the teacher, as questioning alternated between researchers and the teacher.

In the following extract, Lena encourages the children to resolve a conflict that they experienced by folding back to more concrete experiences (Pirie & Kieren, 1992). A girl called Despina haptically explored a metal mold for baking cakes that had the shape of a cone's frustum without top and bottom. In the process of verifying Despina's responses, the number of surfaces became an issue for the class:

Maria: It does not have any sides.
Researcher: Why?
Kostas: It has two sides ... one on the inside and one on the outside.
Researcher: Do all of you agree that it has two sides, and also why?
Giorgos: Only one, the outside!... because the inside holds the outside.
Researcher: Is there anyone who thinks that it has no sides?
Teacher: Can you make this shape with the piece of paper you have in front of you?
Class: No ... yes ... we can...
Teacher: Don't cut the paper ... don't bend it.
[children make a construction that approximates the shape of the object]
Teacher: If we open it, how many surfaces does this shape have?
Class: One, two ... [pointing to their constructs]

By this type of interaction, children come to realize that the object has sides. Children, though, view the objects as physical and not as idealized geometrical objects.

Even if we cannot claim that the teacher adopted a researcher's role, we can suggest that the researchers' goals and expectations, to some extent, shape her actions. Lena's effort to establish an environment where mathematical meanings are negotiated is also evident on the whole. She approaches a role of an apprentice of listening rather than a master of discourse in the classroom (Corradi-Fiumara, 1990). She did not remain within the logic of collecting right and wrong answers, but she perceived questioning as a way of deepening her knowledge about children's mathematical thinking, and as an opportunity to build on this thinking. Questions asked by the teacher and the researchers focused on children's thinking processes behind their responses. "How did you find it?", "What do you mean?", "In what ways is this object different from Angeliki's?",

"What else could you have said?", "What does it remind you of from your everyday life?", "Do you all agree?" are examples of the type of questions Lena was asking.

During the familiarization phase children in both classes asked questions about the physical and geometrical characteristics of the objects, as well as their possible function. In the 5th grade, children's questions initially focused on the physical characteristics of the objects: "What is it made of"?, "What is its color"?, "What is it used for"?, "What does it look like"?, "Does it have a label"?, "What do you think we put in it"? are some examples. These were followed by questions concerning geometrical aspects such as shape and size: "How many surfaces does it have"?, "How many angles"?, "How many vertices"?, "How many equal parts does it have"?, "How big is it"? The transition from contextual to geometrical questions at a stage where the children appeared to have exhausted their repertoire of questions was encouraged by the teacher's implicit interventions. In the 4th grade, children's questions concerning physical and geometrical characteristics of the objects were alternated. Children also used everyday or familiar objects to describe their perception of the objects' shape: "it is like a gift box," "it is like a robot" (rectangular box), "it is like a glass with a broken bottom" (metal mold for baking cakes). The integration of these elements, along with the communication in the classroom, acted as a base for the children to build intuitions and develop their geometric appreciation of the objects. In both grades, questions of a geometric nature focused mainly on the recognition of direct properties, overlooking relations between these properties. For instance, neither the children nor the teachers addressed questions concerning the exact size. In cases where the researchers prompted such questions, children gave approximate answers like "small," "big," "average," without employing a personal measuring unit.

## WORKING IN GROUPS: OTHER VOICES GETTING STRONGER

In the second and third phase, the experiment developed into a group-arrangement. After setting the rules for these phases, the teachers addressed the whole class only when they wanted to regulate time or direct attention to a certain activity (e.g., from exploring to writing to constructing to evaluating). They influenced group-work when, invited or not by the children, they made specific comments on the content of their work or the quality of cooperation in each pair. In this sense, the teachers assumed a role different from the one in the initial phase. More space was left for the children to assign their own meaning to the activities and for researchers' interventions to acquire a stronger voice. In the first part of the second and third phase of the activity, each pair's task was to communicate their perception about the objects to the other pair through the lines of a written report. In order to do so, children had to communicate their representations of the objects at hand within each pair. Children's cooperation, then, was of great importance. Through their joint work of haptically exploring and discussing the objects, they developed an appreciation of the overall shape and characteristics of the objects and wrote a report that represented their perceptions.

## Setting the Tasks

Following the whole class experience, teachers in both classes organized the group work and posed the tasks. They defined the rules:

> ...[W]ithout revealing the object, you have to write the object's characteristics ... Even if you have realized which is the object, you must not write what it is. You will try to see whether the other pair will manage to construct the object from your description. You will have to cooperate with your partner.

Nikitas at this stage builds on a competitive approach once more: "You should make your descriptions in a way so to make it tough for the other pair to understand what the object is." These rules were slightly altered in the third phase in the task of exploring the multi-link constructions. In this phase, in the light of the experiences from the second phase, instructions emphasized the need to take the reader into account. Lena, for example, gave the following instructions:

> Having the previous experience, try to describe it so the person that will read it will understand how to build it. Some descriptions were not complete in my opinion. Because you have the previous experience, you know what you need for making it, ... what you want to read for making it, this is what you have to write ... Not the material but the shape, the angles, the sides ... OK?

Both teachers emphasized the need to describe the shape analytically. Nikitas attempted to encourage children to be more precise: "Look carefully for the object's characteristics. If you say 'it reminds me...' it may remind you of something but it does not mean that it reminds the same thing to somebody else as well." In a way, though, he undermined the writing process by announcing "...[I]n the end we will look for the best report." Neither teacher encouraged the inclusion of physical characteristics in the reports, or the use of descriptions referring to everyday objects, as they did not seem to them connected to the shape. Teachers' decision to abolish expressions of similarity reflects a traditional resistance to thinking of mathematics as bound up with linguistic practices (Rotman, 1993; Walkerdine, 1988). On the other hand the researchers in the groups emphasized a different type of communication. For example, they encouraged children openly to express their thoughts, to use similes, to take into account the other children, to work cooperatively towards common goals. Especially in the classroom of Nikitas, this seemed to be in contrast with the whole classroom culture and sometimes created confusion for the children.

## Developing an Appreciation of the Objects: Cooperating Towards Writing a Report

The pairs of children cooperated in different ways to produce the report. It appeared that the process of exchanging notions about the objects' characteristics developed around their haptic exploration of the objects. Children's haptic exploration strategies can be characterized as global or local. In the global, children mostly embraced the object with their hands and/or got a sense of its content by raising and rocking the bottles in the feely box. In the local exploration children used their fingertips to feel the details of the surface, looked for a specific property, and in the case of the cubic arrangements attempted to keep a reference point in order to count the multi-link cubes.

In most pairs, one of the children was haptically exploring the object, announcing the specific characteristics to the other child, who was responsible for writing them down. In some cases, both children in turns haptically explored the object, discussing it, and verifying what they had found. The following example indicates a process of verification between two boys of the 4th grade:

> Yiannis: Can I write that its shape is a square with a hole in the middle?
> Alexandros: Can I see it? . . . I mean touch it . . . [he explores haptically]
> Yes, write it down.

In some pairs, one child posed questions and the other responded while exploring the object, imitating the exchange that took place in the initial familiarization phase. During their discussion, children attempted to apprehend an overall appreciation of the shape of the object creating a mental reconstruction: "We built it in our mind." Then, in their written report, they either started to describe the features of that shape, or they drew a representation of the overall shape of the object. In some cases, children represented the image of the shape by "drawing" it with their hands in the air. Drawing a representation reflects children's need to "see" the objects and the strength of visual practices in geometry classrooms. This process acted also as a medium for passing from the haptic to the written text. It also seemed to be a strategy, that helped children communicate their perceptions of the objects helping them to retain and agree on its assumed shape.

In this first part of the second and third phase, the children used a lot of similes, especially at the beginning of the exploration, which also contributed to getting a sense of the shape of the objects. Similes are forms of metaphorical language, and together with metaphors and metonymies are ways of expressing similarity judgments and analogical reasoning. The similes that the children used referred mostly to the physical characteristics of the objects. The children came to believe that such expressions and descriptions of the physical characteristics of the object were not acceptable: "We don't say 'look like'." Teachers had not encouraged their use since the beginning of the experiment. Nevertheless, they used a variety of similes resisting in a sense the teachers' approach. For example: "it looks like Walkman," "it looks like a well," "it looks like a table." Some of these expressions, though, did not eventually appear in their report. The similes acted as a basis for them to move to a more descriptive use of language, and not as a way to hinder understanding about the object.

Competitiveness mainly appeared between pairs in the 5th grade. The children did not want to use information that would help others to discover the object. "We often use it at home," Nikolia suggested. "Don't! Don't say that because they will guess it," Mirsini promptly responded. Children were probably acting according to the rules established in this class since the initial phase. In some groups, the researchers attempted to help children reconsider the goal of the task, encouraging them to reflect on their own needs as interpreters and "constructors:" "Would this information be enough if you had to build it or would you have asked for more"? Silence or affirmative responses usually followed this question, as children could not easily sense the needs of a future reader.

## ANALYZING THE CHILDREN'S WRITTEN TEXTS

In our analysis of the children's written texts we focus on the kind of information that the children provide about the objects. This information indicates the characteristics of the objects that children considered important in order to communicate their understandings. In general children had difficulties in composing a written text. These difficulties were more evident than those that they had in gathering information about the objects with the use of haptics. Revealing is Katerina's comment in the 5th grade:

"We all know what it is . . . but how do we write it down." This was supported by our observations of the classes before the implementation of the activity.

In some cases children's reports expressed a "dynamic recreation" of the object. This contributed to a visualization of the shape and to the quality of the coherence of the written text: "It is cylindrical and as it goes up it gets narrower and narrower, it looks like a long cylinder." The majority of the children of the fourth grade used similes to describe the shape of parts or of the whole object. This was not the case with the older children. To a certain degree the use of similes expresses an intuitive appreciation of the shape. Furthermore it acts as a medium for visualizing the shape, encouraging and complementing the use of more analytic strategies. As a result it leads to a high degree of textuality for the written texts. Examples of similes for the multi-link constructions are the following: "it is like a table," "it is like a small staircase," "it is like a square bracelet," "it is like a tube," "it is like a box." On the contrary, very few similes appeared in the descriptions of the bottles.

The first sense characteristics of the objects were intuitive evaluations of the shape of the object. The use of similes and a dynamic creation of the objects were the most common choices for presenting these characteristics. Children's expressions about the texture of the surface pointed to the protrusions, anomalies, dents, and lumps of the objects. The geometrical characteristics were based in the description on a familiar shape. The size acquired importance to children during the construction phase, particularly in the reconstruction of cubic arrangements. The geometrical characteristics of the objects also referred to the sides, edges, vertices, and angles. Physical characteristics like material, weight, and content appeared in all reports. Children seemed to appreciate their importance in building a sense of the hidden objects. Examples are: "it is made of glass," "it is plastic," "it is heavy," "it has liquid in it," "it has a tin cap." Children, in their attempts to help others guess the hidden object, included referents known to the children from their everyday experiences. Examples are: "it has a label," "it is known from TV," "we drink with this when we are thirsty." In most reports, the geometrical and physical characteristics of the objects coexisted.

## INTERPRETING THE WRITTEN REPORTS: FILLING THE BLANK SPOTS OF INDETERMINACY

In the second part of phases two and three of the experiment, children had to assign meaning to the other pair's report and, based on their interpretation of it, to reconstruct the object. In any text there is always a space of indeterminacy (Ingarden, 1973). In the case of the reports, then, children had to achieve a balance between what was said in the text and what was actually missing in order to assign meaning to the written text. In achieving such a balance their experience of having organised a similar report proved critical. We cannot suggest, though, that children's reconstructions of the objects actually mirrored their interpretations of the texts. Especially in the case of the everyday objects other elements interfered with the reconstruction process. Towards the end of phases two and three, after the objects were revealed to the groups, each pair had to evaluate the other pair's report. This was an opportunity for children to reflect on the processes of writing and reconstructing the objects.

## The Construction Phase

The children had difficulty in proceeding with the reconstruction of the object. They started to realize that the information they had in the written reports did not allow them to reach an accurate reconstruction of the object. "We can make an object which will not be exactly the same ... but it will look like it," they often said. The limitation imposed by the material used for the reconstruction was also realized at this point. The children needed more specific information for the construction, and asked for more details or explanations from the other pair in the group. The information that they needed concerned mostly the geometrical characteristics of the objects, especially in the case of the multi-link cubes where reconstruction required more analytic data. On the other hand, in these first steps the children generally disapproved of similes and approximations of size, especially in the case of the multi-link constructions, as they could not find a direct use for them. In the case of similes, however, the younger children were more successful in their attempts to utilize them at a later stage in developing an overall sense of the shape of the objects. In addition, they used new similes during the construction phase in their effort to assign meaning to the report. "It is like a cone … like a small hill. If we tape it on this side and squeeze it from the other it will eventually become what we have in mind," Despina and Irini said while reconstructing the vinegar bottle.

In the process of lessening the degree of indeterminacy in each report, children's experience in the previous phases of the experiment turned out to be critical. So, they assigned personal meaning to the written texts in rendering them more concrete. With the younger children, constructions were the least accurate and resulted from their attempts to give a form to their construction similar to the form of the object that they had explored. Younger children were also more easily satisfied with their constructions. Almost all the 5th grade children attempted to make a drawing of the object before constructing it, in contrast to the younger children who used the written report to form a mental three-dimensional image of the object. In Figure 3.3, we summarize the most common construction process that appeared among the 5th year children during the reconstruction of the bottles:

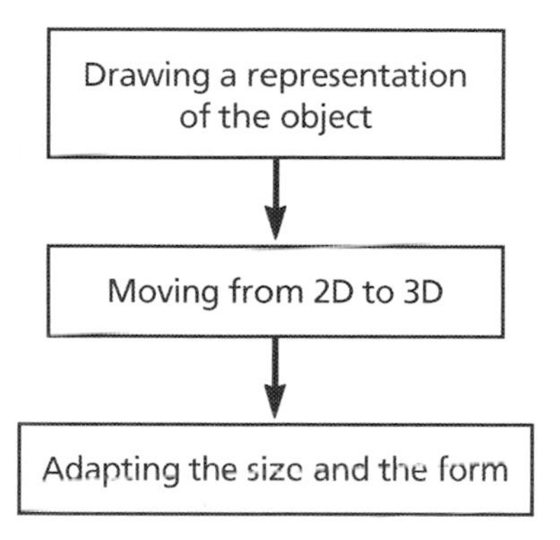


Figure 3.3. Most common construction process.

In most of the pairs, children first cut a rectangle to make the cylindrical part of the bottles. Then they built the object around this part. Their attempts were not systematic though, as adjustments concerning the size of the parts mainly were done after the parts had been constructed. In the case of the soda bottle, for example, the dimensions of the rectangle were arbitrary with no relation to those of the circular base. So, children either drew a smaller circle or adjusted the cylinder to match the size of the circle. In very few cases did the children use the cylindrical surface as a template to make the circle. In making the vinegar bottle, Yiannis and Alexandros first constructed three cylindrical surfaces. Next, they squeezed one end of the two cylindrical surfaces to give them the appearance of a truncated cone. Finally, they inserted the squeezed ends of these two cylindrical surfaces in each of the two ends of the third surface and taped them.

On the whole children used a variety of sources during the construction process. The written reports were acquiring a personal meaning for them as they were filtering the given cues through their own experience from the previous stages. On the one hand, the reports justified their initial representations of the objects. On the other hand, they acted as a base for verifying and altering these initial representations, as they were assigning meaning to more cues in the reports. The processes of concretizing the reports and building the objects, then, were interrelated as the latter represented a visual embodiment of the degree or quality of the concretization of the report. For example, Alkiviadis, a fourth grade boy, drew the following net to represent the way the bottle became thinner at the top. "If we fold these points towards the inside, it will become smaller, like the top

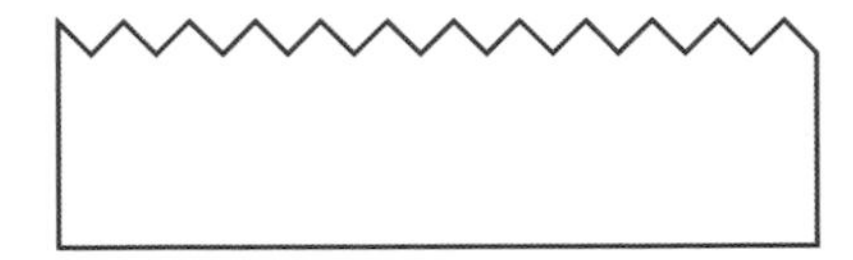

Figure 3.4. Konstantinos and Alkiviadis' net of the vinegar bottle.

part of the bottle," he explained. Konstandinos interrupting him said: "Alkiviadis wake up! They tell us that it is also fatter at the bottom." This comment made Alkiviadis alter his drawing of the net, cutting similar pieces at the bottom to make it "fatter." (see Figure 3.4)

The reports about the multi-link constructions were not very helpful for the children as they realized that they required more accurate descriptions. As a result a number of children admitted that they "made it from their imagination" while others stated that what helped them was the number of cubes that comprised each side. Moreover, the similes that were included in the reports did not prove useful for the interpreters. "I do not know what is the shape of a table," Annita said. In describing the everyday objects children used similes of a geometrical origin ("it looks like a cone"), and similes to supplement geometrical and physical characteristics of the objects. We can suggest that the regular and at the same time unfamiliar shape of the multi-link arrangements, rendered the use of similes highly idiosyncratic and therefore not particularly helpful. Even though the need for the size of the objects was quite apparent, only one construction had layers with side of four multi-link cubes. It appeared, then, that in this phase children based their constructions on the perceptions of the object that they had explored in the previous stage. Concerning children's specific strategies, these were very difficult to follow. In general, though, children chose a specific characteristic of the report and built the rest of the object "around" it. Therefore we could characterize children's building strategies as following, in a way, the categorization of their approaches in organizing the written reports on these objects.

## Evaluating the Written Reports

Children's experience during the previous phases framed the criteria that they used to evaluate the written reports. Sometimes they became critical of others' reports, especially when the whole climate of cooperation between the pairs was competitive. In almost half of the cases, children evaluated the written texts judging on whether the other pair had included in their report elements that they had included in theirs. In most of these cases, though, children included in their evaluation comments concerning missing elements both of theirs and the other pair's report. This observation clearly suggests that the evaluation process was another opportunity for children to reflect on their own work.

Concerning the completeness of the report, children's evaluative comments were general and unspecified: "You didn't describe it as well as you should," "They should have told us more things and explained it more clearly," or about specific elements which they had not included. The children either mentioned what these missing elements were, or added new elements supplementing the report. These missing elements were characteristics of the shape: "You should have told us that on the top becomes narrow;" the object's geometrical characteristics: "they did not tell us about the faces," "you didn't write that the neck is the cone;" the number of geometrical elements: "all the cubes are 24, all the angles are 7;" the shape's orientation: "they did not tell us which is the side on the top," "they did not say where the five pieces are missing;" size: "they did not tell us that the cover was small and that the label was big;" and material: "they did not say that it was plastic." Concerning the correctness of the reports, the mistakes that the children mentioned, related to geometrical characteristics like size, shape, number of sides, to physical characteristics, and to the use of the objects: "it is not exactly a cylinder," "it was not correct that they wrote that the object is full of angles," "it is not known from TV." The "irrelevant" elements that the children mentioned concerned the non-geometrical elements, the physical characteristics of the object like the content of the bottles, the material substance, the label, and its use.

Finally, the children considered the use of language in their evaluation of the reports. More specifically they commented on the accuracy of the mathematical language used: "They should have said cylindrical and not round." They also characterized the clarity of expressions: "We had a problem where it said that half of it is like an Indian tent with a neck;" "I did not like the expression 'On the top it has a rectangle and beside it has a smaller one';" "we did not understand anything."

## EVALUATING AN IMAGINARY REPORT: CHALLENGING DOMINANT DISCOURSES IN THE MATHEMATICS CLASSROOM

In this phase, the class discussed in a whole-class setting two imaginary reports, one for the vinegar bottle and another for a multi-link construction. In the case of the multi-link construction the class, in a collective effort, also had to reconstruct the object. The imaginary reports consisted of statements that appeared in the children's written texts in both classes (e.g., it is made out of plastic, it has curves, its two sides are squares, the object is wrapped by my two hands, etc). In selecting these statements we were mindful to include a variety of the elements and strategies that children used in their written texts. Our aim in this phase was to provide children with an opportunity to reflect on their experience of working in the activity as a whole. It provided us also with an opportunity to investigate our research questions further: to evaluate the impact of the activity on children's appreciation of the geometry of three dimensional objects, and to describe how the activity affected, and was affected by the classroom culture.

In comparing the two classes we noticed a difference in the degree and the form of participation. In the 5th grade, children expressed themselves without hesitation, often becoming "over-critical" in their desire to say something. It was clear in some instances that they were adopting the role of the teacher. Although the younger children on the whole made the same comments as the older ones, they had real difficulty in expressing their position verbally. So, both the teacher and the researchers had to guide the process of the discussion closely, addressing specific questions concerning the elements of the reports.

In the 5th grade the teacher implicitly followed the first phase's "accounting" process by asking the children to evaluate each statement in the reports separately. The fact that the children had the experience of working on the previous phases of the activity gave them the confidence to assume the leading role in the discussion. Children showed an appreciation for the connectivity of the reports, evaluating each statement in relation to the rest. For example, in discussing the statement "it has curves," Nikolia refers to the previous statement in the report: "a cylinder has always curves ... I mean if we relate it to the previous..." The following extract is another example of children seeking for connectivity in the reports, as well as moving towards the idea of the minimal characteristics in determining the shape of the object. Children discussed the relationship of the two sentences "it has curves," "it does not have sides and angles:"

Nicolia: Sir, a cylinder has curves, always a cylinder has curves, if we combine it with the rest.

Teacher: Do you think that he has already said it?

Nicolia: Yes, he has said that it is a cylinder, is it possible not to have curves?

Michalis: As it says that the bottom face is circular, how can it be ... circle first and then have angles and then to be a cylinder.

Katerina: I agree that the last sentence is right, but I believe that if we had used the other sentences correctly [she means in the construction] I think that this information would not be necessary.

In this phase, children's comments drew on their personal experience from the other phases. They reflected on the difficulties they had as writers and readers, and showed an appreciation of the process, of its production, and its implementation. In a way the children themselves integrated their experiences from writing, reading and constructing, and refined their understanding about the shape of the object. For example, in discussing the statement "it is made of plastic" the children used different arguments to support or criticize this information. We give below examples of these arguments in the order that they were given:

> It wouldn't help... because it does not say anything about the shape or... nothing for the object."
>
> "It wouldn't help because we made it from cardboard."
>
> "To understand its shape you must see it first [the object], you must imagine what it is."
>
> "My opinion is that this does not help us at all ... it can help you in a way but the point is not to say what the object is but to construct it.

These arguments appear as more specific to their experience during the construction phase. It seems that the children started to distinguish the differences between the object and its shape, and at the same time they recognized the contribution of non-geometrical information in developing a feeling for the object itself. Despite their appreciation of importance of the non-geometrical information, children's preference in both classes remained with the more precise, formal, and analytic statements. They eventually concluded that the statement "the top part looks like a cone cut in the middle" was the most accurate and helpful information in the report for the vinegar bottle. At first, though, they viewed it as inadequate since, as a 5th grade boy said:

> Sir, first of all I believe, as we had said, that similes do not help us at all, because someone believes this while someone else thinks otherwise. Here, it says like a cone cut in the middle, it doesn't say anything ... I mean ... does she mean the top part of the cone or the bottom part?

This comment indicates the effect of teacher's negative attitude in the initial phase concerning the use of similes in describing an object, on children's opinions.

In the case of the multi-link construction children considered the information analytically. Similarly to the construction phase, their directions to one of the researchers for reconstructing the object from the report were based on the strategy of building the object around a certain characteristic like the square hole, the L shape, or the stair. The problem of calculating the number of the small cubes came to the discussion in the 5th grade class, while the younger children did not face the issue of the size. The children used multiplication to count the number of small cubes in the imaginary cube and then subtracted the number of missing multi-link cubes. Concerning the size, the younger children did not find the information "all the object can be closed in my two palms" appropriate. Yiannis said that "a palm can be bigger than the other" but the class could not find another way to describe the size.

On the whole, the choice of more analytical and geometrical information reflected the elements that were missing from children's reports, and their difficulties during the construction phase. The analysis of this phase reveals that the children continued the process of reflection that they demonstrated during the construction phase, and developed further their understanding of the three dimensional objects. Concerning classroom culture, children's initiative in this final phase eventually frames the activity around their interpretation of it.

## CONCLUDING REMARKS: STUDYING GEOMETRY CLASSROOMS AS CULTURAL SITES

Our aim in this study was to challenge any hierarchization imposed on the choice of materials and teaching paradigms in geometry classrooms. We suggested that the integration of different representational media might provide valuable experiences in understanding the processes involved in communicating and building geometrical ideas in the classroom. As the children moved through the different phases of the experiment, they became more conscious of the processes of organizing coherent texts. Especially in the activities of evaluating the other pair's reports and the imaginary ones in the final phase, children appeared to reflect on their own decisions in organizing the reports. It is clear that children's experiencing of each phase resided in their reflection on the previous. We can suggest then, that the integration of different media provided children with the opportunity to present their ideas, think about the "others," and to reflect on their own and others' choices in process of the production and interpretation of texts.

Efraim Fischbein's (1993) notion of figural concepts was an attempt to characterize the depth of properties, qualities, and experiences "hidden" behind a geometrical concept. In this sense, a number of intertextual relations (Hanks, 1989) support the development of a geometric concept within and across the various textual representations of the concept. The meaning that the children attributed to the geometry of the objects accommodated physical, everyday, and geometrical characteristics. Even though the need for an intuitive representation of the shape of the object was never really abandoned, children moved to a geometrical and analytical awareness of the shape, and properties of a three-dimensional object. In this process of building an awareness of the geometry of a three-dimensional object, the role of haptics was critical. Besides suggesting a different voice in the teaching of geometry, haptic exploration as a process "demanded" reflection and an intuitive appreciation of the geometry of the objects, emphasizing at the same time their geometric characteristics.

On a different level, the incorporation of representational media constituted a way of challenging the norms of communicating in geometry classrooms. Culture does not remain stable in the course of learning in a mathematics classroom. On the contrary, it evolves as it is continuously produced and reproduced within and beyond the spatiotemporal boundaries of the classroom. In this process of creation and recreation of the culture of the mathematics classroom, the discourses voiced by the participants play a critical role. In the present study the 5th grade teacher assumed a storyteller's style in his teaching. In that sense, the telling of stories of mathematical concepts were non-negotiable as they are "naturally" considered as pre-existing (Triadafillidis, 1998). Even in cases where children appeared to be in control, the teacher was dominating the learning process. During the experiment the teacher clearly offered his own interpretation as the only one, and expected children to act accordingly. More evidently in the initial phase, power was centralized in him and rarely diffused to other participants. Other voices in this initial phase were either silenced or subsumed by the teacher's confident opinion. This is not surprising for, as Foucault has suggested, discourses are usually organized around such practices of exclusion (1981). As a result, tension was built in this phase supported by the teacher's imposition of his interpretation of the activity and by the uncommon nature of the activity. On the other hand, the 4th grade teacher's willingness to listen, both to the children and the researchers, contributed in creating a climate where meaning was negotiated.

As the experiment proceeded other voices started to emerge. As a result, other interpretations of the activity started to develop. Teachers appeared as voicing the dominant discourses in mathematics education that advocate the divorce of mathematics from the spheres of the senses, feelings, intuition, and non-exact practices. For example, analogical thinking was not approved by either of the teachers as a "proper" way to describe the shape of the objects. These dominant discourses marked the organization of work and production of texts during the experiment. Each phase was characterized by instances of intertextuality, where children brought together, directly or indirectly, consciously or unconsciously, the dominant discourses existing in the particular spatiotemporal situation. However, the children's need to establish their perception of the explored objects around the physical and first sense characteristics of those objects, as well as their everyday experiences with them, was a form of resistance to the dominant discourses. In particular, similes played an important role in developing this resistance towards the presented "naturalness" of the position advocated by the teachers. The researchers assisted in the strengthening of children's voice, by supporting and encouraging analogical thinking throughout the experiment. More clearly in the 4th grade, similes expressed a unified perception about the shape of the objects against the countless reports that could possibly be written about these shapes. In the final phase of the experiment the 5th grade children developed an even stronger resistance towards their teacher's interpretation of the activity. In this phase children spoke with the confidence that their first-hand experience with the activity had invested in them. Not only did they point to the need for an intuitive appreciation of an object's shape. They also superseded their teacher's "accounting" process in evaluating the imaginary reports. Similar were the 4th graders' reactions in this phase, despite their difficulties to express verbally on their experience. Moreover, the changes in the classroom's climate were less dramatic in this grade as the teacher appeared in general to be more attentive to the children's cognitive and emotional needs.

From the previous remarks we can conclude that one's position within discourse is never fully achieved, and cannot be taken for granted. On the contrary, individuals are always weighing their perceptions of their own position within the discourse against what they assume others perceive of their position (Mills, 1997). The researchers welcomed a more "active" positioning for the children in the classroom. This new positioning was shaped in the different phases of the experiment, each phase being marked by the affordances and limitations of a certain representational medium. Furthermore, these representational media and the type of communication they suggested resisted the dominant discourses in geometry classrooms. The ocular paradigms in the teaching of geometrical concepts were challenged by the use of haptics. Writing, reading and constructing were advanced as ways of reflecting, and eventually shaping, one's understanding of geometrical concepts and thinking processes. Lena, the 4th grade teacher, in an interview with the authors after the conclusion of the experiment, agreed to the previous argument, mentioning that despite the time-consuming element of the activities they could help in "developing a way of thinking about geometric shapes and communication in the classroom." We have no evidence on Nikitas', the 6th grade teachers, view.

The "anthropological turn" in mathematics education developed as an attempt to direct attention to the human element of mathematics, by looking into the cultures of production of mathematical meaning. Relating mathematics to practices such as communication, mediation, semiosis, interpretation, representation, signification, and discourse (Rotman, 1993, Walkerdine, 1988), underscores the view that mathematical constructions are the emergent "products" of the work of all participants located in socio-historically grounded scenes of human interaction. Transferred into the classroom, the "anthropological turn" in mathematics education reflects a willingness to study mathematics classrooms as cultural sites. As the present study suggests, in these sites mathematical meaning emerges through the contributions of all participants as they are playing together, "pushing and pulling each other toward a strong sense of what is probable or possible" (McDermott & Tylbor, 1983, p. 278), as well as of what is required, expected or anticipated. In this study we investigated such instances of "playing togetherness" among contending discourses in the classroom. More research is required in order to describe the processes that shape the organization and production of meaning in the mathematics classroom.

## ACKNOWLEDGEMENTS

The authors are indebted to Penelope C. Papailias for the discussions concerning theoretical points that appear in the paper, as well as for her editing of the written text. We would also like to thank the teachers of the primary school that we visited, and the postgraduate students that participated in the research.

## REFERENCES

Bakhtin, M.M. (1981). *The dialogic imagination.* (C. Emerson & M. Holquist Trans.). Austin, TX: University of Texas Press.

Ben-Haim, D., Lappan, G. & Houang, R.T. (1985). Visualising rectangular solids of small cubes: analysing and effecting students' performance. *Educational Studies in Mathematics, 16,* 389–409.

Ben-Haim, D., Lappan, G. & Houang, R.T. (1989). Adolescents' ability to communicate spatial information: Analysing and effecting students' performance. *Educational Studies in Mathematics, 20,* 121–146.

Bishop, A. (1988). *Mathematical enculturation: A cultural perspective on mathematics education.* Dordrecht, The Netherlands: Kluwer Academic Publishers.

Bliss, J., Monk, M. & Ogborn, J. (1983). *Qualitative data analysis for educational research.* London, UK: Croom Helm.

Brown, T. (1994). Describing the mathematics you are part of: A poststructuralist account of mathematical meaning, In P. Ernest (Ed.), *Mathematics education and philosophy: An international perspective.* (pp. 154–161). London, UK: Falmer.

Chamberlin, J.G. (1974). Phenomenological methodology and understanding education. In D. Denton (Ed.), *Existensialism & phenomenology in education: collected Essays.* (pp. 119–138). New York: Teachers College Press.

Clements, D. H. & Battista, M. T. (1989). Learning of geometric concepts in a Logo environment. *Journal for Research in Mathematics Education, 21,* 450–467.

Clements, D. H. & Battista, M. T. (1990). The effects of logo on children's conceptualisations of angle and polygons. *Journal for Research in Mathematics Education, 21,* 356–371.

Clements, D. H. & Battista, M. T. (1992). Geometry and spatial reasoning. In D.A. Grouws (Ed.), *Handbook of research on mathematics teaching and learning.* (pp. 420–464). New York: MacMillan,

Connolly, P. (1989). Writing and the ecology of learning. In P. Connolly & T. Vilardi (Eds.), *Writing to Learn Mathematics and Science.* (pp. 1–14). New York: Columbia University.

Corradi-Fiumara, G. (1990). *The other side of language: A philosophy of listening.* London, UK: Routledge.

Fischbein, E. (1993). The theory of figural concepts. *Educational Studies in Mathematics, 24,* 139–162.

Foucault, M. (1981). The order of discours. In R. Young (Ed.), *Untying the text: A poststructuralist reader.* London, UK: Routledge & Kegan Paul.

Foucault, M. (1993) *The archaeology of knowledge and the discourse on language,* New York: Barnes and Noble.

Gaulin, C. (1985). The need for emphasising various graphical representations of 3-dimensional shapes and relations. In L. Streefland (Ed.), *Proceedings of Psychology of Mathematics Education Conference 9. Vol.2.* Netherlands. (pp. 53–71).

Giles, G. (1981). *Factors affecting the learning of mathematics. DIME projects.* University of Stirling, Scotland: Dept. of Education.

Guillerault, M. & Laborde, C. (1982) Ambiguities in the description of a geometrical figure. In F. Lowenthal, F. Vandame & J. Cordier (Eds.), *Language and language acquisition.* (pp. 151–156). New York: Plenum.

Guillerault, M. & Laborde, C. (1986). A study of pupils reading geometry. In F. Lowenthal & F. Vandame (Eds.), *Pragmatics and education.* (pp. 223–238). New York: Plenum.

Hanks, W. F. (1989). Text and textuality. *Annual Review of Anthropology, 18,* 95–127.

Hershkowitz, R. (1989). Visualisation in geometry: Two sides of the coin. *Focus on Learning Problems in Mathematics, 11,* 61–76.

Ingarden, R. (1973). *The literary work of art: An investigation on the borderlines of ontology, logic and theory of literature* (G. Grabowicz trans.). Evanston, IL: Northwestern University Press.

Jay, M. (1993). *Downcast eyes: The denigration of vision in twentieth-century french thought.* Berkeley, CA: University of California Press.

Kaput, J.J. (1992). Technology and mathematics education. In D. Grouws (Ed.), *Handbook of research on mathematics teaching and learning.* New York: Mac Millan.

Lowenthal, F. (1986). Non-verbal communication devices: Their relevance, their use and the mental processes involved. In F. Lowenthal & F. Vandame (Eds.), *Pragmatics and education.* (pp. 29–46). New York: Plenum.

Mandrou, R. (1975). *Introduction to modern France 1500–1610: An essay in historical anthropology* (R.E. Hallmark trans.). London, UK: Edward Arnold.

Markopoulos, C. & Potari, D. (1996). Thinking about geometrical shapes in a computer based environment. In L. Puig & A. Gutierez (Eds.), *Proceedings of the Psychology of Mathematics Education Conference 20. Vol. 3.* (pp. 337–340). Valencia, Spain.

Markopoulos, C. & Potari, D. (1999). Forming relationships in three-dimensional geometry through dynamic environments. In O. Zaslavsky (Ed.), *Proceedings of the Psychology of Mathematics Education Conference 23. Vol. 3.* (pp. 273–280). Haifa, Israel.

Marrioti, M.A. (1989). Mental images: some problems related to the development of solids, *Actes de la 13e Conference Internationale, Psychology of Mathematics Education. Vol. 2.* Paris.

McDermott R.P. & Tylbor, H., (1983). On the necessity of collusion in conversation. Mouton De Gruyter, *Text 3*, 277–297.

Mills, S. (1997). *Discourse.* London, UK: Routledge.

Nickson, M. (1992). The culture of a mathematics classroom: An unknown quantity. In D. A. Grouws (Ed.), *Handbook of research on mathematics teaching and learning.* (pp.101–114). New York: MacMillan.

Parzysz, B.: 1988, "Knowing" vs. "seeing:" Problems on the plane representation of space geometry figures. *Educational Studies in Mathematics, 11,* 79–92.

Piaget, J. & Inhelder, B. (1956). *The child's conception of space.* London, UK: Routledge & Kegan Paul.

Pimm, D. (1995). *Symbols and meanings in school mathematics.* London, UK: Routledge.

Pirie, S.E.B. & Kieren, T.E. (1992). Creating constructivist environments and constructing creative mathematics. *Educational Studies in Mathematics, 23,* 505–528.

Potari, D. & Spiliotopoulou, V. (1992). Children's representations of the development of solids. *For the Learning of Mathematics 12,* 38–46.

Preziosi, D. (1986). The multimodality of communicative events. In J. Deely, B. Williams & F.E. Kruse (Eds.), *Frontiers In Semiotics.* Bloomington, IN: Indiana University Press.

Rotman, B. (1993). *Ad infinitum: The ghost in Turing's Machine—Taking the God out of mathematics and putting the body back in.* Stanford, CA: Stanford University.Press,.

Rotman, B. (1994). Mathematical Writing, Thinking, and Virtual Reality. In P. Ernest (Ed.), *Mathematics education and philosophy: An international perspective.* (pp. 76–86). London, UK: Falmer.

Schwartz, J. & Yerushalmy, M. (1995). Using microcomputers to restore invention to the learning of mathematics. In I. Wirzup & R. Streit (Eds.), *Developments in School Mathematics Around the World.* (pp. 623–636). Reston, VA: NCTM.

Shepard, R. G. (1993). Writing for conceptual development. *Journal of Mathematical Behaviour, 12,* 287–293.

Spivey, N. N. (1995). Written discourse: A constructivist perspective. In L.P. Steffe & J. Gale (Eds.), *Constructivism in Education.* (pp. 313–329). Hillsdale, NJ: Erlbaum.

Tedlock, D. & Mannheim, B. Eds. (1995). *The Dialogic Emergence of Culture.* (pp. 1–32). Chicago, IL: University of Illinois Press.

Triadafillidis, T.A. (1995). Circumventing visual limitations in teaching the geometry of shapes, *Educational Studies in Mathematics, 29*, 225–235.

Triadafillidis, T.A. (1998). Dominant epistemologies in mathematics education. *For the Learning of Mathematics, 18*, 21–27.

Vinner, S. (1983). On concept formation. *ZDM 15*, 20–25.

Walkerdine, V. (1988). *The mastery of reason: Cognitive development and the production of rationality*. London, UK: Routledge.

Wolcott, H.F. (1988). Ethnographic research in education. In R.M. Jaeger (Ed.), *Complementary methods for research in education.* (pp. 187–206). Washington, DC: AERA.

CHAPTER 4

# MATHEMATICS AS TEXT

**Tamsin Meaney**
***University of Otago, New Zealand***

## ABSTRACT

Language is considered by many to be important in the development of mathematical ideas, whether as a result of new deliberations by mathematicians or as the reproduction of existing ideas by new learners. In this chapter, mathematics is considered as a text which can be analyzed. Using ideas from Michael Halliday's Functional Grammar, a model is outlined to illustrate how the linguistic choices made by the speaker or writer as a result of their contexts of situation and culture are intricately connected to the mathematics which is being developed and interpreted. In a similar way, mathematical understanding can also influence the contexts of culture and situation and thus the linguistic choices which are perceived as being appropriate in the production of a new text. This model is then used to consider aspects of the development of texts of mathematics within classrooms.

## BACKGROUND

For many years, my teaching has involved working with students with whom I did not share a first language or culture. The difficulties that these students had in learning mathematics appeared to be similar to the difficul-

*Challenging Perspectives on Mathematics Classroom Communication*, pages 109–141

ties that I had when learning some of the knowledge from their communities. A connection appeared to exist between the cultures in which we had been raised and the ease with which we were able to learn particular knowledge systems. However, neither these cultures nor our understandings are things which can be seen and easily analyzed. Coming from a background of teaching English as a second language, I was drawn to exploring the ways that language acts as a viaduct for interactions between culture and understanding. As it can be recorded, language also allows these interactions to be investigated.

There is a symbiotic relationship between our linguistic choices, our culture and our understanding, as each one can influence the others (Halliday, 1978, p. 126). For example, information is passed in mathematics lessons not just through the content contained in our messages, but also in the way that we encode it. Interpreters of the message use their cultural background and previous understanding in making sense of it (Burton and Morgan, 2000, p. 434). Knowing whether the typical ways for encoding the messages are followed enables both the producer and the interpreter of the message to add further layers of meaning (p. 431). The understanding gained can then expand previously held knowledge and alter our perception of the possibilities of the action, interaction and linguistic choices available. These ideas are not new. Amir Alexander showed how the Elizabethan mathematician Thomas Hariot's language in describing his ideas was heavily influenced by the cultural milieu surrounding the voyages to the New World (1995). What Halliday's Functional Grammar enables us to do is to look at the individual words as a means for understanding how the text has been molded by the context of culture and the immediate situation. In this chapter, such an analysis is used as a starting point for considering how culture affects our development of mathematical understanding.

## ANALYZING MATHEMATICS AS TEXT

For this analysis, mathematics is considered as a text or as a piece of communication through which ideas are expressed and interpreted. Halliday stated that:

> [w]e need to see the text as product and the text as process and keep both these aspects in focus. The text is a product in the sense that it is an output, something that can be recorded and studied, having a certain construction that can be represented in systematic terms. It is a process in the sense of a continuous process of semantic choice, a movement through the network of meaning potential, with which each set of choices constituting the environment for a further set. (Halliday and Hasan, 1985, p. 10)

By thinking of it as a process, it is possible, in investigating why choices were made, to include considerations of unused alternatives. A text of mathematics can be spoken or written and can be of any length. This is a broader definition than that usually given to "mathematics texts" which tends to focus exclusively on written material (see for example Morgan, 1998) and so the terms "mathematics as text" and "texts of mathematics" are used in this chapter.

Identification of the ways that the linguistic choices of the writer connect to the cultural background of the reader provides insights into how human experience comes to be understood (Rosenblatt, 1995, p. 7). However, an interpretation of texts can only provide clues to the producer's culture and understanding as any interpretation is also influenced by the culture and understandings held by the interpreter. The first part of the chapter describes a model which is used to illustrate a possible way of visualizing how, as the text arises from the participants' perceptions, mathematical understandings are developed. An e-mail exchange between two mathematicians is used as an example to highlight different aspects of the model.

The second part of the chapter uses the model to illustrate why only some interactions within mathematics classrooms are successful. The relationship between mathematics and language is used to explore, in particular, creativity in mathematics classrooms, and how the use of more than one language could contribute to an understanding of mathematical concepts.

## CULTURE, UNDERSTANDING AND THE MATHEMATICS REGISTER

Halliday puts forth the theory that texts are developed through the context of situation which in itself is surrounded by the context of culture (Halliday and Hasan, 1985, p. 46). When we live in a culture and participate in familiar situations, we make, unconsciously usually, the appropriate linguistic choices so that we have the greatest chance of being understood by our listeners or readers. When particular situations occur continually, sets of terms and grammatical constructions come to be seen as the most appropriate ones to use (Halliday, 1988, p. 162). These terms and constructions are known as the register. Halliday stated that

> [L]anguage also varies according to the function it is being made to serve: what people are actually doing, in the course of which there is talking or writing involved; who the people that are taking part in whatever is going on (in what statuses and roles they are appearing); and what exactly the language is achieving, or being used to achieve, in the process. These three variables (what is going on; who is taking part; and what role the language is playing) are referred to as FIELD, TENOR and MODE; and they collectively deter-

> mine the functional variety, or register, of the language that is being used. (1985b, p. 44)

When ideas are being discussed as a way to gain or improve understandings, then it seems to me that it is possible to identify how these ideas are likely to be developing by looking at the language the producers use. In regard to mathematics, the ideas which we expect to be discussed would be those of quantity, space, and relationships (Barton and Frank, 2001, p. 137). Although there are many definitions for mathematics, they all contain some or all of these elements. Davis and Hersh began their book with a definition of mathematics as "the science of quantity and space" (1981, p. 6). Others have considered anything to be part of mathematics as long as it is generalized and the truth about it deduced (Thomas, 1996, p. 14). Thomas gave his own description as:

> Mathematical objects are the grammatically (and psychologically) necessary posits that allow us to speak of the relationships that they have. Relationships, including how things behave together (not just in what are normally called structures) offer, I believe, a sufficiently reasonable way to describe what mathematics is about. (1996, p. 15)

The model in Figure 4.1 shows how I am envisaging the context of culture and context of situation to be reflected in the text through the linguistic choices made within the field, tenor, and mode, and then through the language into the mathematical understandings which are developed. This section of the chapter goes on to look at definitions of culture and understanding before outlining some of the typical features of the mathematics register (or the typical linguistic choices used to discuss mathematics). To aid this discussion, I use an e-mail exchange between two mathematicians. From knowing something about these mathematicians' backgrounds, I make assertions not only about the linguistic choices that they did make, but also about other linguistic choices which may have been open to them but which they chose not to use. Of course, as stated above, I am constrained by my own cultural background in interpreting what they have said. However, if any understanding of the relationship between culture and understandings is to be reached, some interpretation is required. Therefore, it is worth considering my interpretation as one which enables me to raise particular points. In the e-mail, K and M are discussing an idea of K's. The relationship between the mathematicians is long standing and they describe their relationship as one between colleagues of equal status. They are both of European origin and although M is a first language speaker of English, K is not. He would use another European language in most of his day-to-day interactions.

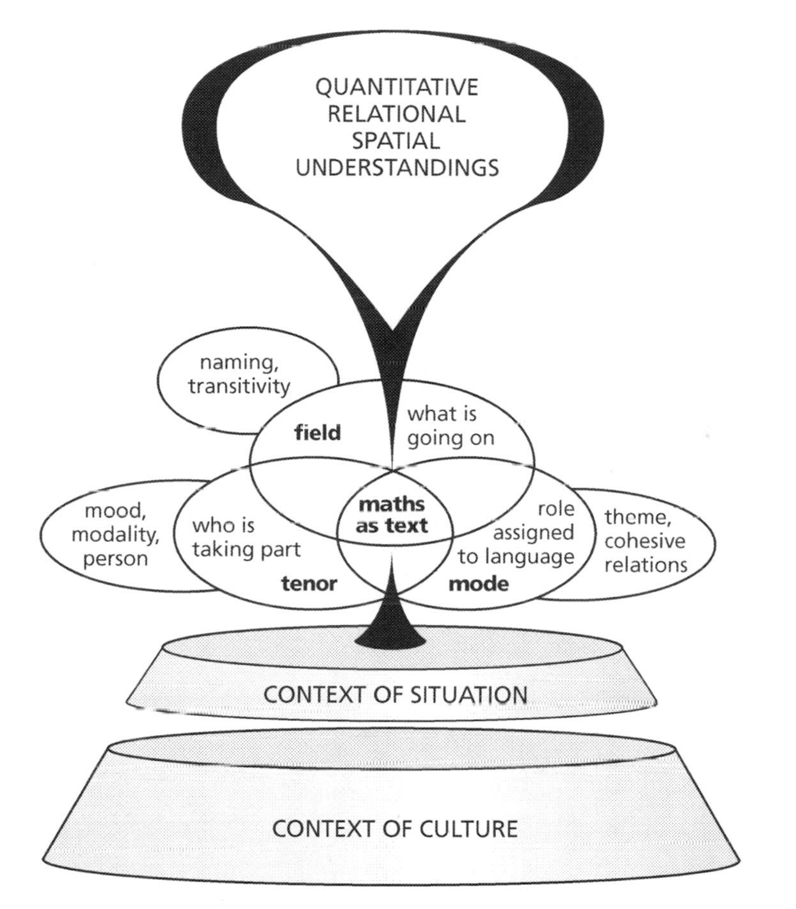


Figure 4.1. A model showing how language acts as a viaduct between culture and understanding.

Subject: Re: A specific problem on closed classes

Date: Thu., 19 Feb. 1998

>Dear K
> I think you need to add some qualifiers to your general hope (that, when
> there is a finite basis, the infeasible permutations are bounded away from
> the maximum by an appreciable amount). Suppose the basis were 21. Then
> the only feasible permutation of each length is 123 ... n so the number of
> infeasible ones is n!-1.
>
> Nevertheless, I see where you get the general idea. With a finite basis there
> is a limit on how many restrictions you can put on the feasible

> permutations, so there must be a lot of feasible permutations.
>Best wishes
> M

Thanks M, I jumped the gun a bit. In fact I had looked at the example you give but this was good evidence for the conjecture I had in mind (honest!) though of course not the version on counting. The original idea was that for large enough n, the sequence that gives the number of infeasible permutations of length n must be regular in some sense (which is how Hilbert functions came in as an analogy). The trouble is I'm not sure that this would be useable hence the hasty resort to counting. Still it seems to me that this might be interesting; do you know if there has been any work in this direction?
Regards
K

## Culture

In considering what is meant by context of culture, there is a need to define what is a fluid concept. Geertz stated that:

culture is best seen not as complexes of concrete behavior patterns—customs, usages, traditions, habit clusters—as has, by and large, been the case up till now, but as a set of control mechanisms—plans, recipes, rules, instructions (what computer engineers call "programs")—for the governing of behavior. (1975, p. 44)

The "control mechanism" view of culture begins with the assumption that human thought is basically both social and public—that its natural habitat is the house yard, the market place, and the town square. Thinking consists not of "happenings in the head" (though happenings there and elsewhere are necessary for it to occur) but of traffic in what have been called, by G. H. Mead and others, significant symbols—words for the most part, but also gestures, drawings, musical sounds, mechanical devices like clocks, or natural objects like jewels—anything, in fact, that is disengaged from its mere actuality and used to impose meaning upon experience. (Geertz, 1975, p. 45)

If culture controls a person's behavior by providing symbols which give meaning to personal experiences, then it must also be accepted that one person's culture can never be the same as another's. The repertoire of symbols available to an individual arises from that person's previous experiences. For example, as a daughter, a person will have learned not only how to act in a family but also the verbal expressions which accompany and describe those actions. In situations with people of similar age to the parents, it may that daughter-like actions and words are chosen for use in interactions and the meaning given to the situation is therefore restricted as a result. The context of culture for the mathematicians in the above extract is made up of their everyday culture from their family and societal roles as well as their culture from being mathematicians. Although these aspects can be discussed as though they are separate entities, it is unlikely that the mathematicians, K and M, would themselves perceive such a distinction. They are, at the same time, mathematicians interacting with other mathematicians, using mathematical terms to describe their ideas; and members of a wider society who operate within its boundaries of acceptable behavior, such as interacting politely when they do not agree with suggestions. The symbols chosen to give meaning to a new personal experience would be drawn, sometimes unconsciously, from, at times, the culture of the wider society and, at other times, from the culture of the community of mathematicians. When Kepler determined that planets speeded up when their elliptical paths came close to the sun, he felt that he was unable to express this adequately with the mathematical language he had at the time. So he expressed the paths of the planets musically. It was not, however, a composition that he intended to play but rather it allowed him to show how the movements of the planets interacted with each other and the sun. Other mathematicians found this musical notation limited and so new ways were developed to express these ideas mathematically.

Although M and K do not accept the ideas offered by the other, in the above extract it can be said that they have chosen very gentle ways to disagree. Within their everyday culture, there would be a range of ways of disagreeing but they may have chosen this one in order to keep the exchange going. There are lots of "I statements" to lessen the demand for changes to the other's ideas, and acknowledgment that they have agreement about some issues such as M's statement that "[n]evertheless, I see where you get the general idea" and K's "[t]hanks M, I jumped the gun a bit." Other people, including other mathematicians, may have felt more comfortable using stronger confrontation when they disagreed. As a result of using this softer disagreement strategy, the discussion revolved around the need for adaptations and the ensuing mathematics was a modification of an idea. If the original idea was rejected then the discussion, if it continued at all, may have resulted in the development of a completely new one.

Although the politeness that the mathematicians expressed was a result of following the rules from their wider culture, they also conformed to the rules pertaining to the culture of the community of mathematicians. In perceiving themselves as mathematicians, they use a similar set of symbols to give meaning to their experiences. Burton and Morgan stated that "[t]he language used in mathematical practices, both in and out of school, shapes the ways of being a mathematician and the conceptions of the nature of mathematical knowledge and learning that are possible within those practices" (2000, p. 445). Although there would be significant differences in beliefs about mathematics between mathematicians working in different areas (p. 445), in this extract there is no discussion of what might constitute good new mathematics. Both mathematicians operate as though they know what is necessary to determine that "when there is a finite basis, the infeasible permutations are bounded away from the maximum by an appreciable amount." As mathematicians, their role is to produce mathematics. It will be others who determine whether what they developed is interesting enough to be published. Van Bendegem in his paper on the development of proofs and the relevance of style in defining proofs stated, "the proof as a mathematically accepted proof, exists only on a social level. Hence the basic unit to consider is not the individual mathematician but the mathematical community" (1993, p. 32, emphasis in the original).

In order to develop their new mathematics, these mathematicians used terminology such as '"infeasible permutations" which has a particular meaning in regard to this area of mathematics (closed classes). Levi-Strauss described how people both inside and outside a society know if they belong to that society by the level and amount of communication (1963, p. 296). This shared terminology therefore reinforces their membership of the community of mathematicians, and excludes others who do not have access and use it.

Both the language in which it is developed and the mathematics itself are parts of the way in which cultural control patterns are enforced (see Zevenbergen, 2001, p. 204 for a description of this in the classroom situation). They control, by coloring the perceptions of participants in regard to what they are doing, who is involved and the way that they believe language should be used in these situations. By considering the cultural environment as the ways in which interactions between people are constrained, we can also consider the understandings which arise from the development of texts of mathematics as both individual and social.

## Understanding

In adding understanding to the model in Figure 4.1, I have moved beyond Halliday's functional grammar which looks exclusively at language. However, for me in trying to understand the difficulties faced by my students, it was necessary to look at the outcomes from developing texts of mathematics. Others such as Whorf have used features of a language to make comments about thought processes (1956). For example, Vygotsky stated that "each word is ... already a generalization. Generalization is a verbal act of thought and reflects reality in quite another way than sensation and perception reflect it" (1962, p. 5) and this model fits into this tradition. Yet Figure 4.1 could be considered misleading in that shared mathematical understandings would also be part of the mathematicians' cultures. New ideas would arise out of discussions which draw upon these shared understandings. As these ideas are stabilized into new understandings, they too become part of the culture of mathematicians who work in this area.

Like culture, defining mathematical understanding is not easy (see for example Love and Tahta's [1991, pp. 266–268] description of the word understanding). A mathematical understanding such as "permutations" should not be considered as something that a person has or has not. Labeling it as an understanding about "permutations" does not cause it to exist as an entity, but rather can enable a person to make connections between ideas. The act of labeling something mathematical also illustrates that there is a social purpose in connecting these ideas (Confrey, 1990, p. 110). This is discussed further in the section on the mathematics register.

In this chapter, an understanding is taken as information gained from giving meaning to an experience either as a result of individually participating in an activity or from jointly negotiating that meaning with others. Language acts both as a tool to analyze perceptions and as a means by which others can be included in the process of coming to understand (Mercer, 2000, pp. 8–9). Wittgenstein believed that mathematical truth did not arise from a discovery about the environment but rather as acceptance of this truth as a rule by a society (Bloor, 1983, p. 92).

However, an understanding needs to be considered as more than knowledge gained by adding meaning to our perceptions. Perkins & Blythe suggested that knowing involved being able to reproduce facts or skills, but that "understanding is a matter of being able to do a variety of thought-demanding things with a topic—like explaining, finding evidence and examples, generalizing, applying, analogizing, and representing the topic in a new way" (1994, p. 5). This kind of understanding can be seen in the discussion between the two mathematicians when one identified a problem but in discussing it, they redefined what it was that they were seeking to solve.

In the extract, it is the relationship between numbers (permutations, feasible and infeasible) rather than the numbers themselves which are of interest. The use of "n" by both mathematicians would also suggest that they are looking for a general case. For all of the mathematicians in Burton's study, connections with other mathematics or real-world data were very important (1999, p. 135).

In the model, a new mathematical understanding is therefore the added meaning given to our perceptions which enables us to do things mathematically. As we make use of this new understanding it becomes part of the culture upon which we draw in doing mathematics.

## Language

Another person's mathematical understanding can only be guessed at by interpreting how it is used in particular situations. Analyzing the linguistic choices that they make as they use their mathematical understanding provides clues about what they know. In showing how K's main point (his "general hope") needs to be qualified, M restates it before summarizing what he considers to be the useful part of this idea. K replies by describing how the idea was originally conceived and then giving his concerns about it so that its development into idea sent in a previous e-mail to M can be understood. K then restates what he feels is the most interesting point and asks for suggestions about other mathematicians who may have looked at a similar area. In this way, it can be seen that the "general hope" becomes the idea around which discussion is centred. This enables the mathematicians to make links between different mathematical understandings. As was suggested in describing the ways that they disagreed, the manner in which mathematical understandings are encoded can affect how they are perceived and therefore their usefulness in resolving problems.

Mathematical language consists of everyday terms as well as those used only when discussing mathematics. What language is used is decided upon as a result of the producer's awareness (conscious or unconscious) of the context of culture and context of situation (Butt et al., 2000, p. 3). Although I do not understand the mathematics being discussed in the e-mail exchange, I can recognize it as a text of mathematics because of the use of terms such as "permutations," "n!-1" and "functions" and the grammatical constructions such as "suppose ... then". Halliday stated

> [w]e can refer to a "mathematics register," in the sense of the meanings that belong to the language of mathematics (the mathematical use of natural language that is: not mathematics itself), and that a language must express if it is being used for mathematical purposes. (1978, p. 195)

I would extend this to say that the mathematics symbols and expressions are an integral part of the mathematics register as they mold and constrain the way that a text of mathematics is developed in the same way that the use of "if ... then" or "because" do which are also parts of the mathematics register.

Within the contexts of situation in which mathematics is produced, there will be differences in the texts which arise. Speakers and writers react to changes in the field, tenor, and mode of the context of situation by altering the language used (Butt et al., 2000, p. 4). In determining what are the elements of the field, tenor, and mode, participants are more easily able to choose what they feel are the appropriate ways to interact by drawing upon previous experiences (Halliday in Halliday & Hasan, 1985, p. 28). Mastery of the mathematics register allows mathematicians not only to present their ideas in an appropriate manner but also to develop and manipulate them. The acceptance of specific features of the mathematics register by mathematicians occurred at the same time as mathematics, itself, changed and evolved (see Halliday and Martin, 1993 for a description of how Western science and the grammatical resources of Western languages developed concurrently). These features have been accepted because of the efficiency gained from describing ideas in these ways. However, it is suggested in the model that the choice of particular expressions and grammatical constructions can constrain the possible meaning conveyed about an idea. In the e-mail exchange, K stated that he had "good evidence for the conjecture he had in mind." By referring to "good evidence" and a "conjecture" K has clued M into giving more value to the subsequent description of this "evidence" than if they were described as "random thoughts." In describing it as "good evidence," M is restricted in how he proceeds with the discussion. If on reading them, he felt that they were random thoughts with no logic, he cannot treat them as such but must instead provide rational arguments to K for discounting them before moving onto what he might feel was a more promising topic.

In the model in Figure 4.1, understandings are joined to the context of culture and the context of situation through the linguistic choices of the producer(s) of the texts. The next sections describe some of the linguistic choices which are related to perceptions of field, tenor, and mode of the mathematics register. By understanding these features more fully, it is possible to investigate how texts of mathematics enable mathematical understandings to be developed in particular ways.

***Field***

Within the field, choices are made about how to describe who did what to whom under what circumstances (Butt et al., 2000, p. 66). In mathematics, the ways in which mathematical objects are named are important as these are most often the doers or receivers of the actions in the texts (see Figure 4.2). Specific mathematical expressions, such as radius, are used, but everyday terms also gain more specific mathematical meanings. For example, "half" has a more precise definition in mathematics than it does in conversational language. (See for example, Pimm's discussion of the term diagonal, 1987, pp. 83–86.) By sharing a set of restricted terms, mathematicians are able to explain their ideas succinctly. In the e-mail exchange, many terms were used but not defined as both mathematicians appeared to feel that they had the same understanding of them, even

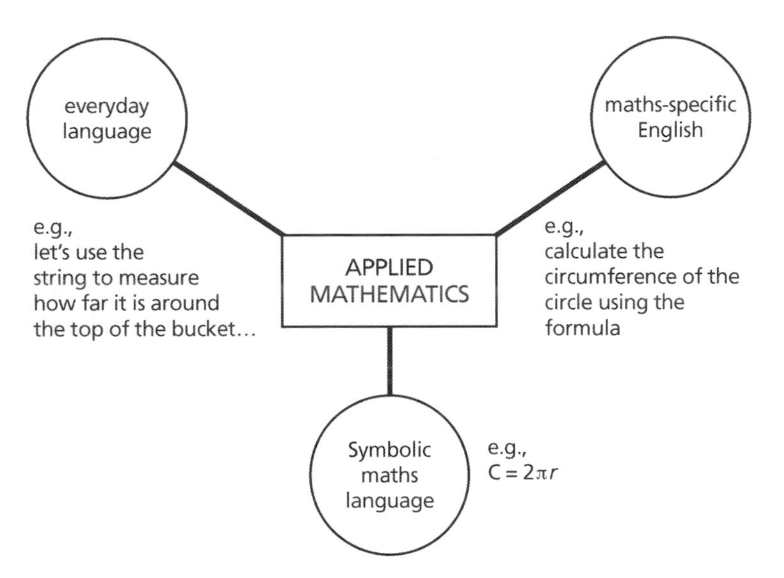


Figure 4.2. Diagram showing types of mathematical language (adapted by Bubb, 1994 from Ballard and Moore, 1987).

though they were trying to narrow down a conjecture. This is unlike situations such as in mathematics education where, for example, several pages of this chapter were devoted to defining the terms "culture" and "mathematical understanding." The culture of the community of mathematicians, which emphasizes the elegance and simplicity of proofs, uses and reinforces the need for precision.

Mathematicians also use nominal groups to express groups of ideas or chunks of information. These nominal groups contain not just information about who takes part in the action but also the circumstances under which the action occurs. They are groups of words, often clauses (or clauses within clauses) which have noun-like qualities. When a verb has been replaced by a noun, this is known as a nominalization. If

$$\frac{d}{dx}(x^2 + 3x + 5)$$

is verbalized, it becomes "the derivative of *x* squared plus three *x* plus five." "Derivative" is a nominalization as it is replacing the action of differentiating. "*x* squared" and "three *x*" are nominal groups but with implicit actions (multiplications) contained within them and so can be considered to share some of the characteristics of nominalizations. As single entity,

$$\frac{d}{dx}(x^2 + 3x + 5)$$

can be discussed and manipulated. However, there would be other times when it is more useful to be able to work with the separate parts. This includes being able to recognize the implicit actions and knowing how undoing them would impact upon the manipulations.

Nominalizations are grammatical metaphors. "Derivative," "*x* squared," and "three *x*" all belong in this category as there has been a shift of meaning as "the event becomes an object and the language is no longer congruent with our experience" (Butt et al., 2000, p. 75). Morgan stated that using grammatical metaphor hides human action but allows for more flexibility in the mathematics because the mathematical objects are also processes (1998, p. 15–16). By choosing to use nominalizations, mathematicians reinforce membership of their community but they also develop mathematical ideas as though they are entities in their own right.

This is reinforced when we consider what role verbs have in mathematics if they no longer represent actions. With actions often being expressed within nominalizations, the main verbs in texts of mathematics tend to be passive and relational. For example, "equals" or "=" is very common in texts of mathematics but describes the relationship between items. When an action is mentioned, it is often inanimate objects which are made active. In "when there is a finite basis, the infeasible permutations are bounded away from the maximum by an appreciable amount," the verbs "is" and "are bounded" make the number sequences sound as though they have a life of their own rather than being something created by people (see Morgan, 1998, p. 14).

By using nominal groups for labeling of ideas and having verbs marking relationships, mathematicians endow mathematical ideas with further meanings. By thus treating mathematical objects as entities in their own right, a perception arises that mathematics is discovered not made. This description of ideas results in facilitating connections to some understandings while disguizing connections to others.

### *Tenor*

The relationship between participants, their roles and their status, is given in the interpersonal meanings expressed in the text (Wells, 1999, p. 174). In the extract, two mathematicians of equal status share information in a way that facilitates the maintenance of communication so that eventually a problem can be resolved. This is the tenor of the text and is reflected in the choice of mood and the way that participants refer to themselves.

The mood is reflected in the type of sentences used. In situations in which information is being given as in the extract, the mood of the texts is declarative, as the participants use statements. These emphasize the certainty of the logic or validity of the argument being presented by incorporating words such as "clearly," "it is obvious that" and "trivial" (Burton & Morgan, 2000, p. 438). In the extract, the only question was about knowledge of related work. The certainty of an idea was queried when a specific example disproved what had been suggested.

Within texts of mathematics, there are often few references to the participants. Burton and Morgan found that many of the mathematicians interviewed felt that "academic writing was … neutral and that its authority stemmed from the facts and arguments contained within it rather than the personality or position of the author" (2000, p. 436). However, in examining examples of mathematicians' papers, they found that the authors infused their identities into their writing in other ways (pp. 439–445). In the given extract, the mathematicians do refer to themselves, as a way of lessening the insistence that changes need to be made to the other's ideas. For example, M suggested that K needed to use qualifiers for the idea to be valid. On the other hand, the mathematics itself is discussed as though independent of humans (as was discussed in the previous section). The removal of people, as the doers of the action, gives the impression that mathematics can be valued in its own right as an objective truth and not because of its relationship with people. One mathematician explained to me that it was easier to do mathematics if it was considered to be discovered rather than created (W. Solomon, personal communication).

### *Mode*

The mode refers to the role that language has in developing the function of text within the context (Halliday & Hasan, 1985, p. 12) and ensures that the text is a coherent whole (Butt et al., 2000, p. 134). For example, the theme as the first element in a clause before the main verb, provides the starting point for developing an idea. New information is usually given after the theme in what is called the rheme (Halliday, 1985a, p. 38). In the text above, the role of the theme as providing background to the clause can be seen in the sentence beginning "With a finite basis there …."

As mathematics has its own orthography or set of written symbols, the mode is extremely important. Mathematical symbols enable very dense sets of ideas to be expressed concisely. For example,

$$y = 150\cos\left(300t - \frac{\pi}{6}\right)$$

contains a number of interconnected ideas. As was discussed in regard to nominal groups, some of these ideas can be considered as a single chunk of information such as

$$\left(300t - \frac{\pi}{6}\right)$$

which allows for easy manipulation. The complexity of mathematics is a result of a series of clauses built or embedded into each other. This allows many lexical or content words to be incorporated together. Halliday stated that written language has a high lexical density which allows for an enormous elasticity in the way that the language can be exploited (1985b, p. 75). Symbolic mathematics language must be considered one of the extreme examples of this. Density of information is paralleled by the reading skills needed to interpret many mathematical equations. A reader of a mathematical equation must not only read from left to right but from right to left and from top to bottom. Even oral discussions of mathematics often make reference to written symbolic mathematics. The denseness of the mathematical ideas means that it is not always possible to ensure that all the meanings can be communicated or remembered without the prompt of the symbols.

These chunks of information provide lexical cohesion throughout texts of mathematics. As logical steps are built up, the chunks are modified but remain recognizable. They act as links to show that the same idea is being discussed, in the same way that after a character has first been introduced in a story his name and the pronoun "he" will be used to continue to show that "he" is what the story is about. In a similar way, mathematics relies on this chunking strategy to keep reminding the interlocutors of the main idea being discussed.

The logic within mathematics is illustrated in the relationship between ideas. This is reinforced by the use of relational clauses and logical connectives. In the extract, the first "that" in the third sentence in K's part of the e-mail signals the use of a relational clause which expands the meaning of what was written previously. "That for large enough n, the sequence... must be regular in some sense" gives more information about "the original idea" and the embedded clause "that gives the number of infeasible permutations of length n" elaborates on the kind of sequence. The clause in brackets beginning "which" provides an elaboration of the sense of the regulation. As well, logic in mathematics is reflected in the way clauses are connected. The mathematical register uses specific grammatical constructions to emphasize particular relationships between ideas. The use of what linguists call logical operators (Carston, 2002) and mathematics educators call logical connectives, words such as "and," "or," "if ... then," determine how particular ideas or concepts are related (Dawe, 1983, p. 216). In mathematics, the type of connection instills the cohesive relations between ideas. This can be seen in the extract where at the end of the first paragraph of M's communication there is a sentence beginning with "Suppose" followed by a sentence beginning with "Then" with a "so" within it. However, if the ideas within the e-mail were to be written up for a refereed journal where the simplicity of the argument will be valued (see Burton & Morgan, 2000, p. 438), then the logic would need to be tighter with stricter controls included to describe when and how the mathematics could be used.

## CULTURE, UNDERSTANDING AND THE MATHEMATICS REGISTER REVISITED

Figure 4.1 summarized how the context of culture and the context of situation could be reflected in the linguistic choices made in developing texts of mathematics. In analyzing the e-mail exchange, this model suggested that the choices made, as well as the alternatives which were not used, reflected the fact that the mathematicians belonged to everyday cultures, in this case late 20th century Western European, and this was exhibited in the soft disagreements that they employed in discussing their ideas. However, by including many features of the mathematics register, they also confirmed that they were members of the community of mathematicians and that the context of situation was one in which they were discussing mathematical ideas. By using the mathematics register to talk about these ideas, they gained a great ability to manipulate and combine different ideas together. This was through the use of nominal groups which often included the actions the objects had been engaged in and through the use of verbs to show the relationship between ideas. However, it also restricted how these ideas were described by removing any reference to human creation thus constraining the connections which could be made to other ideas.

From the model, participants' perceptions and the linguistic choices they make as a consequence can be assessed as being appropriate if the communication is successful and elicits an expected response. In the e-mail exchange, both parties were able to respond and ensure the exchange continued even though they were not in agreement. In this way, the problem was redefined and suggested strategies to solve it were discussed. By restating aspects of the problem, the mathematicians also effectively trialed different linguistic choices. These descriptions could have resulted in connections to other ideas being triggered.

The model separated what is a dynamic, intertwining of the threads of culture, language, and understandings. By describing each in more detail using the e-mail exchange between two mathematicians as an example, I hoped to have shown how the linguistic choices that are made can provide information not only about the producers' contexts of culture and context of situation but also the mathematical understandings which are developed. In the next section, by looking at an exchange between students in a mathematics class, I hope to show how changes in the context of culture and context of situation result in differences in the mathematical understandings which can develop.

## LANGUAGE AND THE LEARNING OF MATHEMATICS

The remaining sections of this chapter adopt the model to show how perceptions of culture and understandings are reflected in texts of mathematics produced by students. By analyzing their linguistic choices, it is possible to make suggestions about why only some texts developed by students are successful. This section therefore examines some of the issues involved in students learning mathematics, including perceptions of the context of situations, the role of creativity within mathematics classrooms and bilingual and bicultural students learning mathematics.

The context of culture for mathematicians is that they learn, use, and produce mathematical understandings with the emphasis on the using and producing. On the other hand, the context of culture for students is that they are expected to learn and use mathematical understandings with the emphasis on learning. Mathematical understanding produced by students as part of a class activity is different from the mathematics produced by a mathematician in explaining a new theorem in a refereed journal. A student's mathematical understanding is acceptable if it matches that of the teacher. The mathematical understanding described by a mathematician is acceptable if it resolves a problem either for the first time or in a new way. This judgment is made by peers, whereas that about student understanding is done by an authority (such as a teacher). The texts that they produce reflect the different roles of participants in the text (tenor), the different experiences expressed in the text (field), and the different purposes that the language is fulfilling in the text (mode). Although there may be some overlap in the linguistic choices perceived as being available to mathematicians and to students, there are also considerable differences.

As in the first part of the chapter, a text of mathematics is used to illustrate points made in subsequent sections. This extract comes from a much wider study which was done over ten years ago in a College of Technical and Further Education in Sydney (Roberts, 1990). The students were trying to solve a puzzle about the age of Diophantus when he died. Each student in the group was given a clue and then asked to discuss the clues and as a group determine the answer. They had been studying linear equations and had previously completed other group activities. The extract begins after they have already spent several minutes talking about the clues. Although the activity was based around finding the correct answer to the problem, as group activity it was used to help students develop their understandings about algebra.

Student 1: Miss, what's the question? There's got to be another question. What ya gotta do?
Student 2: There's the question.
Student 3: The problem is to determine how long
Student 1: So that's 19 years after his youth
Student 2: Wait, Diophantus married 19 years after his youth
Teacher: Why don't you say let, let "x" equal
Student 2: I know, that's what I'm gonna do
Teacher: the length of his life.
Student 2: And let "x" equal the sum
Student 2: I couldn't be bothered to write all that down and I just write 19 stuff and that what's that?
Student 4: But you gotta work it out
Student 1: What's that? When he grew a beard was quarter of
Student 3: But we don't even know how old he is
Student 1: He grew after quarter of his life
Student 4: Okay, put all that down and we'll see
Student 2: Wait there "x's" equals
Student 3: Give'm all to me. I'll work it out. Watch this.
Student 3: Yeah, you just say "x" equals Diophantus
Student 2: nineteen
Student 3: Diophantus
Student 1: Five years after Diophantus married his son was born. Doesn't matter
Student 3: So you gonna write five "x" equals
Student 2: *No*
Student 3: Why his son
Student 2: It's plus. It's a quarter
Student 3: A quarter
Student 1: Diophantus lived twice as long as his son
Student 2: It's plus a quarter
Student 1: Diophantus' youth lasted fourteen years. Alright.
Student 3: *Oh*
Student 1: Fourteen. Diophantus married 12 years
Student 3: After the youth, yeah
Student 2: Wait
Student 3: Nineteen years after the, ... Diophantus married nineteen years after his youth.
Student 2: Yeah
Student 1: Fourteen years
Student 3: Nineteen plus fourteen
Student 2: Yeah, I got that and plus a quarter
Student 1: For what quarter
Student 2: Don't worry ...

## Using and learning mathematical language

By examining the choices made by students as a result of perceptions about field, tenor, and mode, it is possible to tease out some of the differences between the texts that they produce and those of mathematicians. Through interpreting the students' linguistic choices, I make suggestions about the students' beliefs about the context of situation and their mathematical understanding.

### *Field*

The students' text has similarities with the text produced by the mathematicians, but without the advantages gained from being fluent in the mathematics register. For example, nominalizations are an important part of the mathematics register, as they facilitate the linking of mathematical ideas through the chunking of information. However, apart from their repetition of the clues, the students' utterances only refer to actions which they are about to undertake. The high proportion of references to actions would suggest that the students perceived the situation as a conversation (see Halliday, 1985b, p. 93) rather than a mathematical problem solving activity. By misreading the context of situation, it appears that they concentrated more on what they would do individually than the mathematical ideas they are discussing. In so doing, the mathematical ideas are not linked, and possibilities for solving the problem are reduced. Similarly, in other research, I have found that a student who wrote explanations for himself as opposed to his teacher or marker did not produce a fluent, cohesive text (Meaney, 2002, p. 487). This also seemed to inhibit his ability to make connections in mathematics.

The use of nominal groups to represent actions, although not unique to the mathematics register, rarely occurs with the same frequency in conversational language. If the students in the above extract were familiar with using nominal groups, they chose not to take advantage of the benefits they provided. In fact, the analysis of the semester's classes suggested that the students lacked fluency in this area of the mathematics register (Roberts, 1990). The next section considers some of the issues involved in developing fluency in the mathematics register.

Students also appeared to have difficulty with the metaphor in algebra, in which the letters, or variables, are used as though they were numbers. In the extract, the teacher set up the activity so that the students would need to relate the letter value to a numerical value. Although the students incorporated $x$ in their dialogue, there is no discussion let alone consensus about what $x$ should represent. As a result, the redefinition of the problem using a set of shared terms which occurred in the extract between the mathematicians does not occur in the students' text.

***Tenor***

The tenor of many mathematics lessons predominantly reflects the relationship between teacher and student as one of expert to novice. In the extract, the majority of interactions were between students. However, the interaction was heavily influenced by what the students believed were the expectations of the teacher about the purpose of the activity. The students seemed to believe that they would learn by discussing the clues. Finding an answer which made sense in the real world (they finally arrived at Diophantus' age as being 180 years) was not what they considered the purpose of the activity to be. Michael Christie's research compared Aboriginal children and non-Aboriginal children's beliefs about schooling (1985, p. 47). He found that non-Aboriginal children felt that success at school was achieved by working hard which would result in good learning and making the teacher happy. I would suggest that many students in mathematics classrooms partake in activities because they have faith that in doing what the teacher requests they will learn, even if it is only to learn to pass exams. Therefore, they rarely question what they are supposed to do and achieve. As the student/teacher relationship dominates any activity (Zevenbergen, 2001, p. 206), the teacher needs to be explicit in describing not only what students are expected to do, but what they are expected to achieve.

In the extract above, the teacher is called because although students had spent several minutes reading through the clues they were uncertain how to proceed. They listen and try to carry out the teacher's suggestion to incorporate $x$, but still remain unsure of how to do this successfully. This can be seen by the numerous suggestions that they make on what they should do next. Student comments included disparaging suggestions made by other students and assertions of their own ability—"Give'm all to me. I'll work it out. Watch this," probably reflecting the acceptable behavior from their everyday culture of interacting with peers. Although there were occasions where mathematical objects had actions (such as "x equals" and "nineteen plus fourteen"), the actions are limited and rarely completed. Apart from the question asked of the teacher, there was only one other query made by the students. There is no restating of others' ideas as there was in the e-mail exchange, rather each utterance is often independent of those before or after it.

***Mode***

In classroom situations, the mode reflects the fact that students are learning the ways of discussing mathematical ideas concurrently with clarifying and improving their understanding. This is significantly different to the mathematicians who are on the whole confident and fluent users of the mathematics. By giving written clues which did not include mathematical symbols, the teacher had felt that students were more likely to learn by using algebra to integrate the information. However, from the pages of transcript from which this extract came, it appears that the students on the other hand, seemed to believe that by repeating the clues the answer would appear. The language that they used therefore rarely facilitated the connecting of ideas. When bits of information were put together ("It's plus. It's a quarter, ... it's plus a quarter."), they are juxtaposed rather than being joined in the typical ways of the mathematics register through the use of logical connectives. As a result, the text lacks the cohesion found in that of the mathematicians. In their exchange, they rephrased their understanding using both mathematical terms and everyday language, including joining ideas in different ways. In the first part of the e-mail, M could have been discussing anything which needed qualification. It is only when the hope is described in brackets, that it is clear the discussion is about something mathematical and the mathematics register is evident. M then goes on to give an example of what he was describing.

As students learn mathematics, a tension develops for the teacher between providing knowledge about the mathematics register and encouraging fluency in using mathematics. In their review of *Language factors in mathematics teaching and learning*, Ellerton and Clarkson identified the need for mathematical fluency as an issue of concern within mathematics education since at least 1953 (1996, p. 987). As students learn, they are encouraged to discuss mathematics using their own language and then to move towards using specific mathematical terms. In this way, they are expected eventually to become fluent users of the mathematical register and also develop rich, mathematical understandings. As the extract illustrates, without this fluency, students are restricted in the ways that they can develop or redefine their mathematical understandings. Many students, such as these who do not come from middle class background, go through school without learning how to use classroom interactions to their advantage (Zevenbergen, 2001, p. 212).

However, too much concentration on the format of mathematics can restrict fluency in developing mathematical understanding. Paul Cobb recalled an interview in which a student added $16 + 9 =$ correctly by counting on and using her fingers but gained 15 when the algorithm was presented vertically (1988, p. 96). When queried about the fact that she had reached two different answers, the student suggested that for the vertical algorithm 15 was always the correct answer.

## Developing Fluency in the Mathematics Register

There are considerable differences between the field, tenor, and mode of the two texts of mathematics. The way the students described their ideas through concentrating on their actions rather than the mathematical objects did not aid their understanding. In the e-mail exchange, the mathematicians expressed their ideas as suggestions and examples, facilitating a shared understanding of a problem and its possible solution. The purpose of that exchange was apparent to both parties and was more straightforward than the situation in which the students tried to do the "right thing" by the teacher without being fully aware of what the "right thing" was. In trying to understand my own students' difficulties, it seems important that teachers are explicit not only about the objectives for doing an activity, but also about what makes mathematical communication successful, including information about asking questions, rephrasing information, and relating what they do in other situations both inside and outside the classroom. This description needs to include considering the advantages for expressing some ideas one way rather than another.

This development of language to talk about mathematics is not easy. For example, rephrasing actions as nominal groups may be in conflict with rules usually drawn upon to develop conversational language and thus cause difficulties for students (see Zevenbergen, 2001). The metaphorical nature of nominal groups can also add to this difficulty as students expect actions to be expressed as verbs. Explicit discussion of the advantages of using these linguistic choices while discussing mathematics needs to occur so that students can understand the advantages gained from doing this. Gadanides suggested that the type of activities which would encourage students to "explain and justify their understandings" are those which are open-ended and where the teacher does not necessarily know the answer (1994, p. 95). In order to operate successfully in these types of classrooms, students require a wider range of linguistic options. If the teacher works with students to determine answers, they can model how known information is built upon and rephrased in order for a solution to be reached. (See for example, Chapman, 1997.) If students always work on problems to which the teacher already knows the answer, it is unlikely that they would ever see examples of the type of exchange that occurred between the mathematicians. Students need models not just of using the mathematics register correctly but also of how to interact in problem solving situations. Scaffolding by the teacher which builds on the language the student knows can provide a link between conversational language and the mathematics register (Bickmore-Brand and Gawned, 1990, p. 43).

The analysis of the two interactions suggests that teachers need to be constantly aware of the linguistic choices they are modeling. Graham (1986) in her study of learning addition by Aboriginal children in remote communities, found that whether in the child's first language or in English, most discourse revolved around making judgements about the size of sets.

> They had no opportunity to represent in language their physical actions (e.g., putting together) or to refer to the mathematical processes (e.g., adding, counting) in which they were involved. Nor did they refer to the fact that solving such mathematical problems involved finding out and thinking. (p. 106)

For these students, learning was problematic because addition was described by the teachers, guiding them to talk about what they were doing, as an end result rather than a process.

I would also suggest that it is important for students to have skills in ascertaining changes in the field, tenor, and mode. Moving from using everyday language to using the mathematics register could be seen as changing the focus from the field to the mode, as the importance of what is being discussed is overshadowed by how it is discussed. It is useful for students to be aware of some of the ramifications of their linguistic choices on their learning.

## Context of Situation in Learning Mathematics

When the students finally determined that Diophantus' age was 180 years when he died, they did not consider whether this may have been an unreasonable length for a person's life. As others have found (Clarke, 1995), students often feel that the typical context of situation of a mathematics classroom means that the logic that they use in their everyday lives is no longer appropriate—as when the students accept that a person can live to be 180 years old. Using the mathematical register before they have fluency may limit how they represent logic in their texts of mathematics and contribute to their belief that a different logic operates in mathematics. In the extract, the students' perceptions of the context of situation seemed to suggest that choices within the field did not have to include terms which would make links to "outside-the-classroom" knowledge. One of the problems, it seems to me, is that students felt compelled to use what they do know of the mathematics register in discussing mathematics problems. However, without being fluent, they are limited in what they can do with language to help their understanding. If students in the extract had used their everyday language rather than the algebra terms they felt the teacher expected, they might well have been able to resolve the problem realistically. At this stage, they could then be asked to represent what had been done algebraically, which could have supported their gaining of fluency in the mathematics register. By assuming that their everyday language is not appropriate, these students lost alternative ways to describe what they are learning in their everyday language in which they are fluent.

In the Carraher, Carraher and Schliemann (1985) study of children selling produce in the street in Brazil, children displayed successful use of mental arithmetic abilities but were unable to perform the same tasks using pencil and paper techniques in a school setting. Their understanding of what they did outside of the classroom was unavailable to help them make sense of what they did within the classroom. If there is too much emphasis on how the mathematics register is used in school, then the arithmetic skills gained outside of the classroom will remain there. Sometimes links can be made between different texts on similar things by participants perceiving that sets of linguistic choices are appropriate in both texts. The amount of linkage depends on how the participant perceives the similarities between the field, tenor, and mode of the different texts. Strong links can be made only if both texts are described in similar ways. Therefore, students need to know how to describe the street vending text using the mathematics register and vice-versa. However, there are concerns that some cultural activities would then only be considered valuable if they could be discussed as mathematics (see Roberts, 1997 for a longer discussion of this).

The language used in mathematical textbooks often reinforces the view that the context of situation of mathematics is one of being able to do the mechanics of mathematics. This influences the type of understandings which develop. Gerot stated that mathematics textbooks did not explain concepts or make mathematical connections, but rather expected learners to do this while they were solving problems (1992). In her research, many students were unable to explain why solutions to particular problems were correct. Miller and Kandl expressed reservations about how the typical teaching of mathematics seemed to emphasize the getting of the right answer, but with little value being placed on knowing why (1991, p. 5). In other research, I have found that some students are very resistant to written explanations being part of mathematics as they feel that it spoils their ability to get on and do the calculations (Meaney, 2002, p. 487). Yet it is important that students realize that learning the language of explaining is part of learning mathematics. Without it, they are restricting their ability to connect different understandings together and so be creative mathematically. As can be seen from the extract, students were happy to suggest what should be done but did not provide reasons why.

Further complications seem to be added when the students come from backgrounds where schooling is a recent phenomenon. When their teachers do not understand the differences between their and their students' beliefs about schooling, the teacher's ability to provide meaningful learning experiences is limited. The following extract is from an exchange between me and one of my Aboriginal students from a community in a remote area several years ago.

Student: (reading) Twenty-four girls bought two eggs each, how many dozen is this?
Teacher: Now, can you work it out.
Student: Forty-eight
Teacher: That's how many eggs they had? ... Now
Student: Twenty-four
Teacher: How many eggs in a dozen?
Student: Twelve
Teacher: If you had to..., how did you get forty-eight eggs. You guessed or did you work it out some how?
Student: I guessed
Teacher: Oh, well, how do you think, how would you work it out, how many eggs you get.
Student: Six each ...

The student eventually gave the complete answer but never said more than two words at a time about what he was doing. He was obviously able to do more mathematics than he was prepared to explain. This was in stark contrast to when we were outside of school and were discussing activities such as football, which he would do in great detail. Yet without his contribution to the mathematical discussion, it was difficult to talk about the word problem with him. The interactions in mathematics classrooms are much closer to those of middle-class families than to those of Aboriginal families. Zevenbergen relates similar difficulties that working-class students have with language in mathematics classrooms (2001). In Yolngu Aboriginal culture, "[S]tories are passed on for the concrete knowledge they contain, and the pleasure they give, thus making verbal instruction on them redundant" (Harris, 1980, p. 80). This is in contrast to Western school culture where it is expected that students will actively discuss the knowledge being learned. "Aboriginal children are much better at talking to establish personal relationships with their teachers than they are at talking to transact knowledge inside the classroom" (Graham, 1988, p. 128). Answering questions with any response was acceptable, as it enabled the discussion to continue but was not necessarily seen as a way to ensure that learning occurred as learning is mostly done through observation. Knowledge of different possible linguistic choices is valuable but so is an understanding of the importance of discussion in mathematics classrooms. As a teacher, it was essential that I learned about the differences in how the beliefs about knowledge had an impact on our development of a text of mathematics.

The students in the previous extract are involved in reproducing existing knowledge. They are expected to use mathematical understanding that the school system considered necessary for their education. To be successful at this, the model suggests that they had to learn to make appropriate linguistic choices so that the mathematics that they produced was that valued in the mathematics curriculum (Zevenbergen, 1996, p. 110). Yet creativity is often referred to within curricula as an aim for mathematics education (NZ Ministry of Education, 1992, pp. 8–9). A teacher's or curriculum's interpretation of mathematics can be the defining limit of what becomes acceptable as knowledge within the classroom and the need for creativity limited to being creative users of mathematics.

The model presented in Figure 4.1 suggests that mathematical understanding is developed through language and is connected to the student's background and experiences. A mathematical understanding becomes richer if it can be related to understandings from other areas of knowledge. However, for an understanding to be considered useful in the mathematics classroom, it needs to be valued as being mathematical. There is then a dilemma for teachers (and lecturers) who must teach existing knowledge but in a way which does not devalue students' creativity. Too often by the time students reach the stage when they are suppose to produce new mathematics (when they can become mathematicians), many have lost the interest and ability to begin as they have spent so long replicating the results of others. By encouraging students to express their mathematical ideas in their own language as well as that of the mathematical register, it is possible that students would have more linguistic options to draw upon as producers. As in the e-mail extract, it is more likely that at the inductive stage of producing mathematics, mathematicians would use a combination of natural language and the mathematics register. In the student extract, the exchange revolves around a repetition of the words of the clues interspersed with a few mathematical terms. Students who come to see the mathematics register as the only legitimate form of language in the mathematics classroom could restrict the type of mathematical understandings that they can develop.

The context of situations found in the mathematics classroom will be influenced by the culture of the teacher and the students. Their beliefs about what mathematics is and how a mathematics classroom operates influence the perceptions about what mathematics can be produced. As the facilitator of the learning, it is the responsibility of the teacher to ensure that students are aware of changes to the context of situation. This could involve a discussion about expectations about learning mathematics.

## BILINGUAL MATHEMATICS

How has this discussion about language as a connection between culture and mathematical understanding helped me to come to grips with the problems that my students were facing? I believe that, viewing mathematics as a text which is produced from linguistic choices, does help to show how culture influences mathematical understandings.

Such a view means that it is through language that a teacher can describe aspects of their understanding and a student can clarify and develop their own understandings. Different languages will contribute to different understanding being developed. How different will depend on how the mathematical register operates within each language (see, for example, Mellin-Olsen, 1987, p. 124). Rudder, in describing the expressions used for length in the Yolngu languages of North East Arnhemland, found that "it is the quality of items, and not their lengths which is being compared in Yolngu evaluations" (1983, p. 23). Depending upon the situation, very long and very short could be equated with being a bad or good attribute of something. Although a similar evaluation is possible in English, it is rarely the main purpose for doing an evaluation of length. The relationship in Yolngu languages which is considered valuable is therefore not one of size but of usefulness. It is expressed as a set of paired comparisons rather than a continuum of different sizes (Rudder, 1987, p. 205). There are terms for "long" and "short" but not for "longer," "longest," "shorter" or "shortest." It must be assumed that understanding about quantities of length would be distinctly different to those developed in English because of the difference in the attributes valued and therefore highlighted in the different languages.

This has implications for students who come from different cultural backgrounds and who are expected to learn Western mathematics. For example, if the language in which the mathematics is produced does not have an extensive list of number names then it will be difficult to gain Western ideas about number. These problems are not necessarily going to ease if number names are imported into the language. Number names label a quantity and reflect a way of perceiving the world. Adding these labels to a language does not necessarily mean that these perceptions of the world will also be gained (see for example Berry, 1985, p. 20). The inter-relationship between the context of culture, the context of situation, and the choice within languages means that linguistic differences cannot alone explain differences in understanding. Differences at any or all of these levels will affect the mathematical understanding that an individual develops.

Similar understandings may fail to be connected if they develop in different contexts of situation. This can be exacerbated if the contexts of situation occur in different languages, such as would be the case if the language of instruction was not the same as that of the community. Cummins (1979) described two types of learning environment for the development of communicative competence in a bilingual person. These were: a *subtractive* environment where a complete linguistic repertoire in either language has not been achieved and so there are some ideas which a person cannot discuss in either language; and an *additive* environment where a person is fluent in both languages, allowing connections between understandings to occur which would not be available to speakers of only one of the languages. When students try to develop understandings in a language in which they are not fluent, it may be that few connections will be made to similar understandings which were developed through the students' first languages and so a subtractive bilingual model is most likely to arise.

In *Twice as Less,* Eleanor Wilson Orr provided many examples of where the dialect of English spoken by her African American students was interfering with their ability to express themselves in the mathematics classrooms (1987). Although the expressions that they used were taken as being similar to those of standard American English, they often had meanings which were substantially different. In another situation, Harris suggested that those teaching mathematics in English to Aboriginal children living in remote areas would find that

> in many instances, ways of expressing concepts in children's first language will be quite different from ways in which they are expressed in English, thus causing confusion with vocabulary and terminology. In some case, where concepts are totally foreign to the children's cultures, there will be no concise ways of explaining them in the children's own languages. Thus the children will be required to learn new vocabulary and new concepts simultaneously. (1987, p. 75)

There may be, however, cases where it is more appropriate to keep understandings about something separate from similar understandings from another context (Graham, 1988, p. 130). Harris in his book, *Two-way schooling,* put up a very compelling case for the separation of domains of knowledge and consequently the languages used to teach them (1990). This was because of the possibility of the dominant culture submerging understandings within the other culture (Harris, 1990). There is concern that the use of indigenous languages for discussing Western mathematics may result in profound changes to these languages. Barton, Fairhall, and Trinick, in examining the development of a Maori mathematics register, gave the following warning:

> Are the rationalism, the objectivism, the tendency to control, and the propensity to technological progress, all of which are inherent in mathematics, being felt within the Maori language, and subsequently in Maori culture? It is almost certainly too early to tell, and may be impossible to distinguish from the contemporary changes of a living language, however it will be interesting to watch as school-room mathematics discourse tips into the playground and then into everyday language. (1998, p. 7)

On the other hand, fluency in two quite distinctive languages, such as a Yolngu language and English, could result in quite complex understanding of quantifying being developed. If linearity is considered both as a graded continuum of lengths which allows evaluation of any number of items regardless of their purposes as is done in English, and as a matrix based on two-paired evaluations, usefulness (good/bad) and length (long/short) as is done in Yolngu, a much deeper understanding of the quantifying of lengths could be possible. However, Barton, Fairhall, and Trinick suggested that for mathematics to be developed in a Maori way, then there would need to be "an openness to changes in mathematics itself, and a high-level of linguistic awareness, i.e., a sophisticated understanding of mathematics and its symbolism, and how this relates to the English and Maori registers" (1998, p. 8).

Further research needs to be conducted into the mathematical understanding which can be developed by bilingual students and/or mathematicians. This needs to be done in coordination with language communities so that ramifications can be discussed. The model suggests that if a bilingual person is able to make connections between the different contexts of situation in which they operate, then not only could the range of appropriate linguistic choices be increased but their understanding would become richer. Connections could then be made in new and innovative ways which could result in a person's context of culture being reshaped. By recognizing how changes in the context of situations affect perceptions of the field, tenor, and mode, students could perceive a larger range of linguistic choices as being appropriate. Bilingual students already have skills in critiquing appropriate linguistic choices from situations where a literal translation is not possible, and so may not find it as difficult as predicted to identify the effect of differences in contexts of situations.

## MATHEMATICS: A PATHWAY TO WHERE?

By viewing mathematics as a text, insights can be gleaned into the dynamic relationship between culture and mathematical understandings and how these are affected by linguistic choices. In the model in Figure 4.1, I suggested that the linguistic choices that a person sees available to them are determined by their perceptions of the context of culture and the context of situation. These linguistic choices can then influence the way that mathematical understandings are developed. However, once mathematical understandings become stabilized they then become part of the person's culture.

In the development of texts of mathematics, mathematical culture often moulds the pathway from a participant's background and experiences to the mathematical understandings. As was suggested in the section on the "context of situation in learning mathematics," students are more likely to be creative if they draw upon their everyday language and their developing mathematical register to describe mathematical understandings. If this does not occur students are at risk of not making links to other understandings that they already possess. By acknowledging that participants' culture can contribute to the mathematics that they develop, participants are encouraged to be creative in the production of their texts and their understandings.

From the analysis done in this chapter, it is suggested that teachers have an obligation to teach students the mathematics register, but need also to make explicit how appropriate linguistic choices are made. Students need an awareness of the ramifications for their learning of making different linguistic choices, as each set of terms and expressions will enable or impede students' abilities to connect to other understandings that they have. With this awareness, students can make use of the mathematical register, actively. These insights would also allow students to critique their understanding so that understandings from other contexts of situation can be combined with them in new and innovative ways. For example, in the section on "bilingual mathematics," it was suggested that bilingual students may be at an advantage, as using two languages to describe a mathematical idea could give them two different sets of connotations for the mathematical ideas being described. If these connotations are both considered valid, then our perceptions of what constitutes mathematics can be broadened.

The model presented in Figure 4.1 suggests areas for further research, particularly in regard to how students learning mathematics (either through their first or second language) make their linguistic choices, and how mathematicians integrate their cultural knowledge into the mathematics texts that they produce. For me in trying to understand how my students learn, this is a beginning, but there is still much that as a teacher I need to understand in order to find efficient ways to teach my students.

## ACKNOWLEDGMENTS

I would like to acknowledge the support provided by Dr. Bill Barton in putting these ideas together. I would also like to thank the mathematicians in the Department of Mathematics, University of Auckland and elsewhere for putting up with my endless questions about what they did, especially my husband, Mike Atkinson. Sue Spinks at Macquarie University in Sydney provided much needed advice about the use of functional grammar.

## REFERENCES


Alexander, A. (1995). The imperialist space of Elizabethan mathematics. *Studies in History and Philosophy of Science, 26*(4), 559–592.

Barton, B., Fairhall, U., & Trinick, T. (1998). Tikanga reo tatai: Issues in the development of a Maori mathematics register. *For the Learning of Mathematics, 18*(1), 1–17.

Barton, B. & Frank, R. (2001). Mathematical ideas and indigenous languages. In B. Atweh, H. Forgasz & B. Nebres (Eds.), *Sociocultural research on mathematics education: An international perspective.* (pp. 135–149). Mahwah, NJ: Erlbaum.

Berry, J. (1985). Learning mathematics in a second language: some cross-cultural issues. *For the Learning of Mathematics, 5*(2), 18–21.

Bloor, D. (1983). *Wittgenstein: A social theory of knowledge.* London, UK: MacMillan.

Bickmore-Brand, J. & Gawned, S. (1990). Scaffolding for improved mathematical understanding. In J. Bickmore-Brand (Ed.), *Language in mathematics.* (pp. 43–51). Melbourne, AU: Australian Reading Association.

Bubb, P. (1994). *Maths in context.* Darwin, AU: Northern Territory Department of Education.

Burton, L. (1999). The practices of mathematicians: What do they tell us about coming to know mathematics? *Educational Studies in Mathematics, 37,* 121–143.

Burton, L. & Morgan, C. (2000). Mathematicians writing. *Journal for Research in Mathematics Education, 31*(4), 429–453.

Butt, D., Fahey, R., Feez, S., Spinks, S. & Yallop, C. (2000). *Using functional grammar: An explorer's guide, 2nd ed.* Sydney, AU: Macquarie University.

Carraher, T. N., Carraher, D. W., & Schliemann, A. D. (1985). Mathematics in the streets and in the schools. *The British Journal of Developmental Psychology, 3*(1), 21–29.

Carston, R. (2002). *Thoughts and utterances: The pragmatics of explicit communication.* Oxford, UK: Blackwell.

Chapman, A. (1997). Towards a model of language shifts in mathematics learning. *Mathematics Education Research Journal, 9*(2), 152–172.

Christie, M. (1985). *Aboriginal perspectives on experience and learning: The role of language in aboriginal education.* Geelong, AU: Deakin University Press.

Clarke, D. (1995). Quality mathematics: how can we tell? *The Mathematics Teacher, 88*(4), 326–328.

Cobb, P. (1988). The tension between theories of learning and instruction in mathematics education. *Educational Psychology, 23,* 87–103.

Confrey, J. (1990). What constructivism implies for teaching. In R. B. Davis, C. A. Maher & N. Noddings (Eds.), *Constructivist views on the teaching and learning of mathematics* (JRME Monograph #4). (pp. 107–122). Reston, VA: National Teachers of Mathematics.

Cummins, J. (1979). Linguistic interdependence and the educational development of bilingual children. *Review of Educational Research, 49*(2), 222–251.

Davis, P. J. & Hersh, R. (1981). *The mathematical experience.* Harmondsworth, England: Penguin.

Dawe, L. (1983). The development of inquiry skills in mathematics and science: Problems for the bilingual child. In D. Blane (Ed.) *Essentials of mathematics education.* Melbourne, AU: Mathematics Association of Victoria.

Ellerton, N. & Clarkson, P. (1996). Language factors in mathematics teaching and learning. In A. Bishop, M. A. Clements, C. Keitel, J. Kilpatrick & C. Laborde (Eds.), *International handbook of mathematics education.* (pp. 987–1023). Dordrecht, The Netherlands: Kluwer Academic Publishers.

Gadanides, G. (1994). Deconstructing constructivism. *The Mathematics Teacher, 87*(2), 91–96.

Geertz, C. (1975). *The Interpretation of cultures.* London, UK: Hutchinson & Co.

Gerot, L. (1992). Explaining mathematics, keynote paper presented at the *Mathematics Teachers' Association of the Northern Territory Maths Meet '92.*, Darwin.

Graham, B. (1986). *Language and mathematics in the aboriginal context: A study of classroom interactions about additions in the early years.* Unpublished Master of Education thesis, School of Education, Deakin University.

Graham, B. (1988). Mathematical education and Aboriginal children. *Educational Studies in Mathematics, 19,* 119–135.

Halliday, M. A. K. (1978). *Language as social semiotic.* London: Edward Arnold.

Halliday, M. A. K. (1985a). *An introduction to functional grammar.* London, UK: Edward Arnold.

Halliday, M. A. K. (1985b). *Spoken and written language.* Geelong, AU: Deakin University Press.

Halliday, M. A. K. (1988). On the language of physical science. In M. Ghadessy (Ed.) *Registers of written english: Situational factors and linguistic features.* (pp. 162–178). London, UK: Pinter.

Halliday, M. & Hasan, R. (1985). *Language, context and text: Aspects of language in a social-semiotic perspective.* Melbourne, AU: Deakin University Press.

Halliday, M. A. K. & Martin, J. R. (1993). *Writing science: Literacy and discursive power.* London, UK: Falmer.

Harris, P. (1987). *Measurement in tribal aboriginal communities.* Darwin: Northern Territory Department of Education.

Harris, S. (1980). *Culture and learning: Tradition and education in Northeast Arnhemland.* Darwin, AU: Northern Territory Department of Education.

Harris, S. (1990). *Two-way Schooling: Western education and the survival of a small culture.* Canberra, AU: Aboriginal Studies Press.

Levi-Strauss, C. (1963). *Structural anthropology.* C. Jacobson & B. Schoepf, (Trans.) New York: Basic Books.

Love, E. & Tahta, D. (1991). Reflections on some words used in mathematics education. In D. Pimm & E. Love (Eds.), *Teaching and learning school mathematics.* (pp. 252–272). London, UK: Hodder & Stoughton.

Meaney, T. (2002). Aspects of written performance in mathematics learning. In K. Irwin, B. Barton, M. Pfannkuch & M. Thomas (Eds.), *Mathematics in the South Pacific: Proceedings of the 25th Mathematics Education Research Group of Australasia Conference.* (pp. 481–488). Auckland, NZ: University of Auckland.

Mellin-Olsen, S. (1987). *The politics of mathematics education.* Dordrecht, The Netherlands: Kluwer Academic Publishers.

Mercer, E. (2000). *Words and minds: how we use language to think together.* London, UK: Routledge.

Miller, D., & Kandl, T. (1991). Knowing ... what, how, why. *The Australian Mathematics Teacher, 47*(3), 4–9.

Morgan, C. (1998). *Writing mathematically: The discourse of investigation.* London, UK: Falmer.

NZ Ministry of Education (1992). *Mathematics in the New Zealand curriculum.* Wellington, NZ: Learning Media.

Orr, E. W., (1987). *Twice as less.* New York: W. W. Norton & Company.

Perkins, D. & Blythe, T. (1994). Putting understanding up front. *Educational Leadership, 51,* 4–7.

Pimm, D. (1987). *Speaking mathematically: Communication in mathematics classroom.* London, UK:.Routledge.

Roberts, T. (1990). *Mathematics classrooms: A spectators' sport.* Unpublished special topic for a masters in applied linguistics. Macquarie University, Sydney, AU.

Roberts, T. (1997). Aboriginal maths: Can we use it in school? In N. Scott & H. Hollingsworth (Eds.) *Mathematics: Creating the future.* Proceedings of the 16th Biennial Conference of the Australian Association of Mathematics Teachers. (pp. 95–99). Adelaide, AU: AAMT.

Rosenblatt, L. M. (1995). *Literature as exploration.* New York: The Modern Language Association of America.

Rudder, J. (1983). *Qualitative thinking: An examination of the classification systems, evaluate systems and cognitive structures of the Yolnu People of north-east Arnhem Land.* Unpublished M.A. thesis, Australian National University, Canberra, AU.

Thomas, R. (1996). Proto-mathematics and/or real mathematics. *For the Learning of Mathematics, 16*(2), 11–18.

Van Bendegem, J. P. (1993). Foundations of mathematics or mathematical practice: Is one forced to choose? In S. Restivo, J. P. Van Bendegem & R. Fischer (Eds.) *Math worlds.* (pp. 21–38). New York: State University of New York Press.

Vygotsky, L. S. (1962) *Language and thought.* E. Hanfmann & G. Vakar (Eds. and Trans.), Cambridge, MA: MIT Press.

Wells, G. (1999). *Dialogic inquiry: Towards a sociocultural practice and theory of education.* Cambridge, UK: Cambridge University Press.

Whorf, B. (1956). *Language, thought and reality: Selected writings.* Cambridge, MA: Technology Press of Massachusetts Institute of Technology.

Zevenbergen, R. (1996). Constructivism as a liberal bourgeois discourse. *Educational Studies in Mathematics, 31,* 95–113.

Zevenbergen, R. (2001). Mathematics, social class and linguistic capital: An analysis of mathematics classroom interactions. In B. Atweh, H. Forgasz & B. Nebres (Eds.) *Sociocultural research on mathematics education: An international perspective.* (pp. 201–216). Mahwah, NJ: Erlbaum.

THEME II

# COMMUNICATION: SOCIAL INTERACTIONS, SOCIAL SETTING, CLASSROOM ACTIVITY

Theme II contains three more chapters. Chapter 5 is written by Nadia Douek and is entitled: *Communication in the Mathematics Classroom: Argumentation and Development of Mathematical Knowledge.* Chapter 6 is by Inger Wistedt and Gudrun Brattström, and is called: *Understanding Mathematical Induction in a Co-operative Setting: Merits and Limitations of Classroom Communication among Peers.* Finally, Chapter 7 is by David Hewitt: *Conflicts and Harmonies among Different Aspects of Mathematical Activity.*

The focus in all three chapters is the study of communication in mathematics classrooms through a micro-analysis of social interactions (i.e., interactions between teachers and students or among students themselves). All three chapters have a serious interest in what constitutes productive communication in the mathematics classroom (i.e., communication that enables the production of mathematical knowledge and competencies) and are deeply concerned with pedagogic change. Nadia Douek, Inger Wistedt and Gudrun Brattström agree that productive communication is the product of deliberate efforts of interaction with more knowledgeable members of an educational setting, generally the teacher or the tutor. Hewitt takes an inclusive view on classroom activity, and highlights how social, emotional, and cognitive factors are present at

*Challenging Perspectives on Mathematics Classroom Communication*, pages 143–144

any time, creating conflicts and harmonies in students' *zone of opportunities* for learning.

The challenging message carried by all three chapters is that we cannot take communication for granted when we plan social interactions in certain classroom activities. Instead, one must realize that the social setting must be deliberatively organized to promote productive communication. However, the term "productive communication" is not free of values and politics. What is the meaning of "productive"? Who evaluates and what are the standards of evaluating the product of a communicative process? Do students and teachers share the same values and understandings concerning a productive communication? These questions are further explored in Theme III.

CHAPTER 5

# COMMUNICATION IN THE MATHEMATICS CLASSROOM

## Argumentation and Development of Mathematical Knowledge

**Nadia Douek**
***Institut Universitaire de Formation des Maîtres de Créteil, France***

### ABSTRACT

In this chapter, I consider different questions related to a productive development of argumentation in the mathematics classroom: what context of communication is suitable to develop argumentative activities, how to enhance conceptualization and the development of reasoning through argumentation and finally what are the problem solving competencies developed through argumentation. Teacher mediation is considered from this perspective. Communication is considered here as a condition for argumentative activity. I present a theoretical framework concerning conceptualization, in which argumentation plays an important role, and I support it with experiences and classroom observations.

*Challenging Perspectives on Mathematics Classroom Communication*, pages 145–172

## BACKGROUND OF THE STUDY

As an act of social belonging, engaging in communication is important for a student, from the perspective of classroom life, as well as class-shared knowledge construction. In particular, it can be seen as a way of developing linguistic representations of knowledge. These aspects have been widely considered in current literature over the last two decades (cf. Steinbring et al., 1998 for a representative set of orientations in the field of mathematics education). In this chapter, I consider communication as a condition of cultural development for the individual and the group of which one is part. Communication also reflects and influences the development of thought in relation to social interaction (see Vygotsky's comments about Piaget's internal language in Vygotsky, 1985, Chapter 2.).

The main aim of the chapter is to present a theoretical framework concerning conceptualization, argumentation and communication. In this framework, argumentation plays an important role for conceptualization, and communication is considered as a means of introducing students to, as well as a condition for, argumentative activity. The development of this theoretical framework is supported and clarified through the description of teaching practices. I refer to three classroom situations: first, the die example; second, the sun shadow example; and third, the wheat plants example. Aspects of my theoretical position and its hypotheses have been produced in a dialectical relation with observations, the planning of teaching experiments and their analysis (see Arzarello and Bartolini Bussi, 1998 for such research methodology). The theoretical construction is based on theoretical constructs derived from Gérard Vergnaud (1990) and Lev Vygotsky (1985). I shall justify the legitimacy of this combination in the context of this chapter, in the section entitled "Conceptualization."

Communication with its social dimension plays an important role in my theoretical construction as related to argumentation. In fact, communication situations within a "specialized" area of culture (e.g., school subject matters such as mathematics) are specific: when they are not limited to information, they must essentially be of an argumentative type, if we want them to have an impact on the development of knowledge. As a result, my hypothesis is that argumentation plays an important role in conceptualization, and I provide experimental evidence for this in the area of school mathematics. In particular, I analyze how various aspects of conceptualization evolved through argumentative activities (see the examples of sun shadow and wheat plants). From an epistemological point of view I do not separate mathematical objects of knowledge from the problem solving practices that are attached to them in mathematical activity. This position agrees with Vergnaud's definition of conceptualization that merges various aspects of a concept and the activities attached to it, as will be detailed in the section entitled conceptualization. I have argued elsewhere that argumentation is strongly related to proving (see Douek, 1999a, 1999c). I shall not develop this position here, but according to this perspective, students could be introduced to proving at very early stages. This point has already been developed by other authors who argue for the importance of developing argumentation in the early years of school (see Bartolini Bussi et al., 1999).

Through my long term experience working in primary school maths classrooms, I believe that students generally need to be introduced to productive communication by the teacher. On the one hand, they need to learn to express their ideas, and, on the other hand, they need to learn to listen and try to grasp, to situate themselves towards, and to react to the ideas of others. I analyze a situation where young students are introduced to argumentation (the wheat plants example). But before I do so, the die example will show a situation of "void" communication taken from a class where no special work concerning communication or argumentation was undertaken by the teacher. In line with Paolo Boero's theoretical construct of "fields of experience" (Boero et al., 1995), which structures the work done in the Genoa group classes, I consider how some specific everyday life contexts can be exploited to develop argumentative activities. In particular, I shall consider the role of the teacher in managing good communication situations related to those contexts as well as using them to develop conceptualization.

## CONCEPTUALIZATION

Vergnaud (1990, p. 145) has defined a concept as the system of three components:

(a) the set of its reference situations;
(b) the set of its operational invariants (like "theorems in action" for instance. Such "theorems" are detected by the observer because the subject behaves regularly as if he applied such a theorem, usually only implicitly, and for a given category of problems); and,
(c) the set of its verbal and non-verbal representations.

To illustrate this definition, let us consider the concept of measure of length. For the students involved in the teaching experiment of the wheat plants, the measuring of wheat plants and the ways in which the complicated problem was solved, can play the role of "reference situations" for this concept and some of its operational invariants. For instance, the solution involved the use of the additive property of the measure as an operational invariant. It was not known as a rule, but became the implicit means to solve the problem. The terminology introduced by the teacher, and the graphic representations used by students, enriched the set of its verbal and non-verbal representations. Vergnaud's definition can be very useful in school practices, because it allows teachers to follow the process of conceptualization in the classroom by monitoring students' development of the three components. From the research point of view, Vergnaud's definition suggests an analysis of the progressive constitution of the three components in relation to each other, and to other knowledge.

In the Vygotskian elaboration of "scientific" concepts in contrast to "everyday" concepts, awareness, intention, and systemic links between concepts are considered as crucial features (see Vygotsky, 1985: Chapter VI). Vygotsky's theory suggests specific questions: How can awareness about linguistic representations and operational invariants of a given concept be attained (e.g., as a condition for their appropriate and conscious use in problem solving)? How can systemic links between concepts be developed? The study of argumentation in relation to conceptualization can play an important role in tackling these questions. With reference to Vergnaud's and Vygotsky's elaboration on concepts, I consider *conceptualization* as the complex process of progressive construction of the components of concepts, construction of the systemic links between different concepts, and development of awareness about them. In this sense, the concept components themselves can be considered as systems.

The term "system" has been borrowed from two different theoretical frameworks (Vygotsky's and Vergnaud's). In a broad sense (and despite differences in details of use) the constructivist and the socio-cultural perspectives are compatible within the borders of this work. The description of concept that originates from Vergnaud does not impose any explanation on the origins of conceptualization. It can work if we focus on conceptualization as the individual's inner evolution adapted to socially constrained constructions, as well as if we emphasize its evolution through interaction with family at the beginning, and then within a wider social context. I am in favor of the second choice, so that I can integrate this description of concept development with Vygotsky's claims about the role of others' mediation in intellectual development, as well as with the Vygotskian social and epistemological view on systems of knowledge (i.e., "scientific concepts"). I claim that individual impulses move the subject towards communication and action. As soon as the impulse takes place, it is confronted, accompanied, and related to the movements of other individuals and entangled within social constructions. But I cannot view social construction as a whole detached from individuals who engage with it.

## THE IMPORTANCE OF "PRODUCTIVE COMMUNICATION" FOR CONCEPTUALIZATION

Let me illustrate the importance of productive communication (in the sense that it produces new knowledge) for developing mathematical knowledge, and the conditions of its existence, by giving an example of its absence in a situation where learning would have benefited from a "teacher-students communication situation."

### The Need for Communication: The Die Example

In a sequence (common in French schools) on classification of "solids" into polyhedra and non-polyhedra, in a first grade primary class (6–7 year olds), a teacher asked groups of 5 students to "put together the solids that are alike, and justify their choice." Students first took the objects in their hands and shared them, without communicating about the objects except for the sharing purpose.

After about five minutes, the teacher decided to clarify the task. Initially, he emphasized that the work was group work, and that the students had to work together, not to *share* the objects, but to *classify* all of them, in agreement with each other. He, then, gave the criterion of classification: some objects roll, others slide, and some can slide or roll depending on how they are put on the table. This generally is considered as a standard criterion in the beginning phase of work on this topic (it is found in some French mathematics education books as well): polyhedra can only slide, spheres always roll, cones and cylinders may roll or slide. Finally, he gave some examples. Showing a cylinder box, he said: "look at this box, now it rolls; but now, in this position, I can only make it slide."

The teacher then addressed a student and asked how the die should be classified. I could not detect the reason for the choice of the student, and, in general, the teacher did not interrogate many students. This student spoke easily, though with some shyness. The student said: "the die rolls." The teacher, disappointed, repeated the question and said: "are you sure," etc. And he finally explained: "no, the die slides..." pushing the die on the table to show that it slides. However, students' common knowledge is that the die is used for games and is always thrown to roll. Its corners and vertices are rounded. But the teacher was thinking that the die was a good example of a cube, and the cube is a polyhedron... so it should only slide!

Afterwards, the students returned to their "group work" with its ordinary difficulties. Some students watched, some did the work sharing and communicating more or less, depending on their personalities and their relations with each other. They classified the objects "correctly" in general, which shows that the application of the criterion was not difficult. Then, the teacher addressed the whole class, exhibiting some "good" examples, and took two big sheets of paper he had prepared: one with the title "solids that roll" and the other with "solids that slide." He stuck them each on a box and put various objects as examples in each box. Analyzing this episode, three issues seem to be of importance: first, is the epistemological position of the teacher; second, how he organizes communication; and, third, the role of communication in the process of conceptualization. All three are discussed below.

## The Epistemological Position of the Teacher

This situation raises an epistemological issue underpinning the educational position of the teacher. For the teacher, the root of knowledge seems to be the definition, and the educational choice is to present and teach the definition through examples. So, if the examples taken from reality are not mathematically good enough, the definitions will overcome the imperfections of reality and will be considered by the teacher as more convincing than the students' perceptions and interpretations. This issue is to be related to the modeling problem on the epistemological side, and to the teaching of modeling on the educational side. The mathematical solids (polyhedra, cones, etc.) can be seen as mathematical models of the objects offered to the students. The educational question concerns how to bring students to characterize the objects through criteria that can make way for a suitable definition of the mathematical objects. Maria Bartolini Bussi, Mara Boni, Franca Ferri, and Rossella Garuti (1999) provide an example of classroom modeling activity with 4th grade students (9–10 year olds), using combinations of gears. Those authors take a different epistemological position, as compared to the teacher in question here, on the relation between observable reality and mathematization. They provide a dialectical perspective for the evolution of this relation, and the activities have a strong communication component.

## The Organization of Communication

In the above episode, I noticed the absence of productive communication. The process and outcomes of the planned work were completely fixed by the teacher, who was not prepared to consider students' interventions beyond the context of a true or false test. In this situation the student could not act as a subject from the perspective of knowing and reflecting. I conjecture that the teacher expected that the proposed observation of the solids and their comparison were sufficient for students to grasp their characteristics as the teacher saw them. As a matter of fact, the teacher did not engage in making students express themselves. Students were not offered opportunities to share past experiences in such a way that these could constitute, for the group, common references to back reflections or to raise questions about the topic dealt with in the activity. It is likely that when someone is convinced that some characteristics are directly perceptible, the need for dialogue and communication about manifest objects is weak. In this case, the teacher might think that it is sufficient to teach the right vocabulary for what has been observed. In this way, the absence of communication is probably reinforced by the teacher's epistemological point of view.

This situation, once the definition was given the priority, offered no possibility for debate. It was extremely simplified, so that there was almost no need and no possibility for *argumentation*. Students had only to guess what was the best rule to solve the problem, but this guessing was difficult even though the desired classification was simple. To allow argumentation to take place it is necessary that some aspects of knowledge or of the application of knowledge on a particular situation can be challenged. Often complexity and/or doubt offer good challenges. In the situation that I am considering here, it was difficult for the students to bridge their observations, related to their experience and the observation of the solids in the light of the definition. It is generally difficult to provoke a reflection about the relation between examples and the definition, to provide justifications and explanations of how and why some elements of reality will be neglected and others emphasized, or to relate the knowledge of the definition to the elaboration of some new schemes or operational invariants during a problem solving activity, especially if there is no communication context. As a consequence, the conceptualization process will probably remain superficial in such situations.

## Communication and the Process of Conceptualization

The reciprocal relation between doubt and communication (through *argumentation*) can play the role of a dynamic support to encourage questioning, expression and evolution of conceptualizations, as I argue in this chapter with reference to different levels of schooling. The absence of communication, and particularly the absence of the combination of communication with interesting and relevant questioning, affects the role of experience in conceptualization. Specifically, it makes it weaker and denies students the possibility of relying on their own experiences and knowledge.

It prevents them from constituting common references and taking advantage from the possible shared experiences, and it does not encourage the students to use these references consciously to back their ongoing work. Problems like those related to the material construction of solids could play a role in the evolution of the geometrical conceptualization of solids, with the schemes (in the Piagetian sense of the term) that could be involved. But in the case of superficial communication, those problems may be experienced, by the students, from the cognitive point of view, as detached from the observation and classification work they did afterward. Representation of mathematical objects remains the main acquisition and the risk is that the students may not be able to connect the represented knowledge to other ideas or schemes. This way, learning may remain at the level of a collection of words related to some abstract images but not close to the process of conceptualization.

In the above analysis, I have related argumentation with communication, and both of them with conceptualization. In the following, I describe argumentation, and its relation to communication, as part of the theoretical framework provided in this chapter. Later, their relation to conceptualization is theorized.

## ARGUMENTATION AND COMMUNICATION

We cannot accept any discourse as argumentation (see Toulmin, 1958). Henceforth, the word argumentation will indicate two things. First, it can denote the process that produces a logically connected, but not necessarily deductive, (as it can happen when biologists discuss an experiment) discourse about a given subject. For example, the Webster Dictionary offers a definition: The act of forming reasons, making induction, drawing conclusions, and applying them to the case under discussion (1962, p. 100). Second, it can point to the text produced through that process, or as the Webster dictionary refers to it: Writing or speaking that argues. On each occasion, the linguistic context will allow the reader to select the appropriate meaning. In this chapter, the word "argument" is used as "A reason or reasons offered for or against a proposition, opinion or measure" (Webster, 1962, p. 100), including verbal statements, numerical data, drawings, etc. So, an "argumentation" consists of one or more logically connected arguments. Moreover, argumentation can be produced by people in individual tasks (oral or written) or in a group discussion. At this point, I want to distinguish between two interrelated perspectives, the cognitive and the epistemological, through which we can view argumentation.

*From the cognitive point of view,* I give priority to the subject's relation to knowledge (attained or in evolution). Often argumentative activities are dealt with by the subject through "transformational reasoning." Martin Simon explains:

> the physical or mental enactment of an operation or set of operations on an object or set of objects that allows one to envision the transformations that these objects undergo and the set of results of these operations. Central to transformational reasoning is the ability to consider, not a static state, but a dynamic process by which a new state or a continuum of states are generated. (1996, p. 201)

Argumentative reasoning can develop through the "semantic" complexity of the objects dealt with in the process of argumentation, especially when the multiplicity of meanings or interpretations of a situation or a word is involved. For example the expression "height of the sun" concerns the inclination of sunrays. When the knowledge regarding this subject is not clear enough, other meanings of "height" (e.g., vertical distance to Earth) can interfere. So, in the process of argumentation, these various interpretations of words and concepts, the activity itself, and the various possible incoherences are part of argumentation and dialectically put the meanings involved into question. This point is elaborated in subsequent sections.

*From an epistemological perspective*, I consider how pieces of knowledge are related, not specifically from the subject's point of view, in relation to his experience, but from what can be considered an objective point of view, in the sense that subject matter specialists acknowledge such relations according to established theories. Generally analogies and metaphors are accepted kinds of reasoning in an argumentation. For example, the word "zone" when used in the phrase "zone of proximal development," constitutes a space metaphor that plays an important role in reasoning about students' development of knowledge.

Argumentation uses references. I have introduced the expression reference knowledge to refer to "the knowledge which can be put into different forms (graphic, verbal, etc.), and agreed upon by the class, or used in argumentation in the subjective conviction that it is so" (Douek, 1999b, p. 92). Reference knowledge can include not only reference statements but also visual evidence, data, etc., assumed to be objective. Furthermore, argumentation is closely related to the complex cultural contexts in which it is situated. For example, the choice of suitable everyday life contexts can offer meaningful references for argumentation. Their complexity is important to provoke a true need for meaningful argumentation (see Douek, 1999b). A clearly formalized situation narrows the variety of arguments and generally requires a high level of mastery of technical knowledge, which is not always easily available to the majority of the students, especially in the lower grades. Often, a complex but familiar (or potentially familiar) context offers several occasions for argumentation at different levels, going from dealing with experimental evidence, to reasoning backed by formalized knowledge, through various uses of representations that may be drawings, linguistic descriptions, schemas, etc. In this manner, various aspects of knowledge cross and integrate. Students are induced to refine the problematic links between the components of the concepts at work as well as the systemic links between different concepts.

This is related to the theoretical construct of "field of experience," proposed by Boero Carlo Dapueto, Pierluigi Ferrari, Enrica Ferrero, Rossella Garuti, Enrica Lemut, Laura Parenti, and Ezio Scali (1995). This construct offers guidelines for following the long-term development of the "internal context" of the student (i.e., his or her conceptions, schemes, etc.) in relation to the "internal context" of the teacher (i.e., his or her conceptions, schemes, educational aims, etc.) and the "external context" (i.e., signs, concrete objects, physical constraints, etc.) made accessible in the instructional setting.

## Argumentation in the Classroom: Communication Roots of Argumentation and the Teacher's Mediation Role

It is clear that the development of argumentation is based on a personal need for clarification that a real "other" imposes, and thus implies a need for the experience of communication. Argumentation can be considered as a particular development of communication. It presupposes a contradictory listener, somehow an opponent (see Piaget, 1923, 1947), thus it is built by reasoning to prevent or to resolve contradictions. Therefore argumentation is naturally enhanced, and should be developed, in a communication with real "others" (the teacher, the other students or a text produced by someone else, see my discussion about the educational choices later in the text). This experience can then evolve towards an internal need for clarification, (i.e., talking to an abstract "other").

Here, I follow a parallel of Vygotsky's analysis of internal language. In his comments on Piaget's work, he shows how the young child internalizes language (Vygotsky, 1985). Vygotsky argues that the child "thinks out loud" in the way he talks with true others; then this will develop to become silent thinking or silent internal speech (pp. 105–106). One can imagine the individual reflecting on a problem by arguing with several voices, opposing or evaluating the different aspects of the problem under scrutiny. Therefore, it is important to undertake the work of constructing verbal argumentation and, at the same time, to pass from verbal argumentation with a true "other" to written individual argumentation by taking a position about the other's positions, given as texts. This, in turn, allows the production of a text that addresses an abstract "other," with its rules of coherency. In this manner, individual written argumentation is related to communication in a very deep sense, and necessarily requires long-term development.

The teacher's mediation is very important in order to introduce students into argumentative activity. The teacher needs to mediate actively (i.e., questioning, intervening, etc.) from the earliest stages of schooling (see Douek & Pichat, 2003). The teacher must face the child's difficulties in keeping to the subject evoked (when, for example, the child gets interested in marginal details, unimportant for the subject of work), in expressing his or her own ideas, in remaining coherent with his or her previous statements or known facts, etc. I cannot imagine approaching such goals if the teacher does not provide the child with a high quality communicative context, being really attentive to his or her utterances, hesitations, interests, or difficulties and challenging them through the choice of the problem solving situation or intermediate questionings, as will be seen below in the "wheat plants" example.

Coherent with the above-mentioned requirements, various components of the teaching and learning situation can play different roles in developing competencies in argumentation. For example, individual argumentation, classroom argumentation, and argumentation with the teacher can be seen as some of the educational choices. *Individual argumentation* in problem solving (considered as creative work) enables personal expression of ideas and intuitions to take place. It leads to the production of coherent (personal) texts. In this manner, it approaches one of the central elements of proving. *Classroom argumentation* gives a social dimension to personal ideas and expressions, where learners encounter others' ideas and expressions in a cooperative reasoning context, and where they develop argumentative skills through their necessary practice in a complex situation. This is, by the way, a natural context of communication, which enhances argumentation. *Argumentation with the teacher* (in individual interactions or in classroom discussions) has two important motivations. First, some important arguments can develop in response to a problem posed by another. The teacher can offer a multiplicity of "otherness" to push a student to progress in his or her zone of proximal development. Here I use the original definition of ZPD as a metaphorical space between autonomous performance and the performance attained with the help of the teacher or a more competent school fellow (Vygotsky, 1985). This happens through a specific mediation that the "natural" argumentation between students could not provide. Second, the teacher's practice of argumentation provides models of argumentation, showing its effectiveness in particular examples of its functioning, and implicitly showing its rules of coherency, of organization, etc.

I have already stressed the importance of good conditions of communication to favor the development of argumentation abilities. The contract that the teacher builds in the class is a determinant of that. Many students may be unable to make way for their personal elaboration when all their contributions have to occur in communication situations, while other students can be discouraged if all their performances must be individual. Therefore it can be helpful for most students to alternate communication situations with individual problem solving work. In the educational context, a communication situation is generally more interesting when it is challenging. Communication situations offer a good opportunity to nurture students' imagination with various elaborated ideas and intuitions, while individual creative work is needed to internalize new acquisitions and to develop personal ideas coming from one's own experience and imagination. It is important that students are aware of these two sources of enrichment.

## The Usual Phases of Teaching in the Genoa Group Project

Two examples are presented in the next section. They derive from classes whose teachers take part in the Genoa Group project for primary school (see Boero et al., 1995). Within this project, problems are frequently dealt with through sequences elaborated in accordance with the perspective given above, going generally through five phases:

1. *Individual production of written hypotheses on a given task.* The teacher can interact individually with the students while they are working, and can ask specific questions adapted to the student's needs, ask for more precision, etc. When the student struggles with producing written texts (as in the first grade or at the beginning of the second grade) the teacher interacts orally. At some stages of interaction, the teacher writes the conclusion of the discussion under the student's dictation and they copy it.
2. *Individual comparison of solutions.* Often the teacher chooses one or two texts produced in a precedent phase of individual work, and provokes further reflection about the problem by asking, in another phase of individual production, for some comparative work between the chosen productions and the students' own productions.
3. *Classroom comparison and discussion of student individual products, guided by the teacher.*
4. *Individual written reports about the discussion.*
5. *Classroom summary, usually constructed under the guidance of the teacher and written down by the students in their note books.* In most cases, classroom summaries represent the knowledge acquired by the students (with ambiguities and hidden mistakes). A final institutionalization phase (Brousseau, 1986) is only reached in few circumstances.

The organization of these phases is adapted to the situation. For example, a complex problem may need one more discussion phase, followed by individual reports. From a methodological point of view, this style of slowly evolving knowledge offers the opportunity to observe the transformation of students' knowledge in a favorable climate. Such transformations are normal, expected events, because no objective truth is presented by the teacher at the early stages. Moreover, argumentation fairly well reflects each student's level of mastery of knowledge, his or her level of use of references, and the common stable references for the class.

## ARGUMENTATION AND CONCEPTUALIZATION

Below, I analyze these relations between argumentation and conceptualization: a) Argumentation and the evolution of two of the three concept components, namely reference situations and representations, in relation to experience; b) Argumentation and the evolution of concept components concerning operational invariants; and c) Argumentation, discrimination, and the linking of concepts.

### Argumentation and the Evolution of Concept Components

How can we distinguish an *experience* from a *reference situation* when we (as observers) analyze students' behaviors? An experience can be considered a reference situation for a given concept when it is referred to as an argument given to explain, justify or contrast a proposition or a statement related to the concept (i.e., when it enters reference knowledge). This criterion applies both to basic concrete experiences related to elementary concept construction, and to high-level, formal, and abstract experiences (e.g., as in secondary or university mathematical work).

Thus, in order for an experience to become a reference situation for a given concept, it must be connected to symbolic representations of that concept in a conscious manner. In particular, a functional link that the student can master must be established between the constitution of reference situations for a given concept and its linguistic representations. The functional links involve operational invariants as semantic content of the knowledge involved. Argumentation may be the way by which this link is established. These points will be elaborated in the sun shadow and the wheat plants examples. (See also Douek, 1998, Douek & Scali, 2000.)

The student develops argumentation skills and constitutes reference situations for concepts through a dialectical process. Argumentation can be seen as a means to develop an experience into a reference situation for a given concept in two connected ways. First, by involving the subject's view and awareness about that concept in the experience. And second, by involving symbolic representations of that concept in the experience. These two ways create semantic roots for the representations relating the experience to the network of the subject's conscious knowledge. On the other hand, one clearly needs reference situations as arguments and backings in an argumentative process concerning a given concept. In this sense, a reference situation belongs to reference knowledge.

## Argumentation and Operational Invariants

Argumentation can foster operational invariants or schemes concerning a concept to be made explicit and ensure their conscious use. It can make inappropriate theorems in action explicit and help in correcting them. This function of argumentation strongly depends on teacher's mediation, and is fulfilled when students are asked to describe efficient procedures and the conditions of their appropriate use in problem solving. The sun shadow example shows the necessity of arguing to convince the "other" (in this particular situation it was not an abstract "other") of one's statement, and thus of one's implicit "theorem." This provoked the effort to make explicit the knowledge of what appeared to the observer to be a theorem in action (inappropriate, though). The inner nature of a concept as a "system" is enhanced through these argumentative activities: different schemes can be compared and connected to each other, and to appropriate symbolic representations. It is also an important way of developing awareness of the diversity of procedures as well as their common features.

## Argumentation, Discrimination and Linking of Concepts

Argumentation can ensure both the necessary discrimination of concepts and the systemic links between them. These two functions are dialectically connected. For example, argumentation allows the student to distinguish between closely related concepts, particularly through their operational invariants and reference situations (see Douek, 1998; Douek & Scali, 2000, and also the example concerning "height" in the "sun shadow" example). In the same manner, possible links between concepts can be established. As a provisional conclusion, I note that argumentation differs from other forms of communication and should have its specific place in mathematics education, because its function in conceptualization is crucial.

## METHODS OF ANALYSIS OF CLASSROOM SITUATIONS

Before moving to the analysis of the examples that illustrate the issues mentioned above, I would like to clarify the methods used for analyzing the classroom situations described in subsequent sections. In accordance with the theoretical framework for argumentation and conceptualizations, I have considered the following four steps: First, I listed, to the extent it was possible, the different forms of knowledge that appear, implicitly or explicitly, and play a role in students' work. This was a preliminary search necessary to provide elements for the subsequent analyses. Second, I composed a classification of the various statuses of knowledge which would be either reference knowledge, or knowledge in a construction phase. This was generally done in collaboration with the teachers, by studying past teaching situations, the students long term learning paths, and the specific goals of the actual teaching situation. Third, I noted the links that students make between various aspects of their knowledge, in particular by observing their use as operational invariants in problem solving strategies or their use (as reference knowledge) in argumentation, and the students' chosen representations for them. This called for "live observation" to detect the gestures that helped expression (see the sun shadow and the wheat plants examples) as well as material constraints that influenced students' understanding of the situations (such as in the wheat plants example). Fourth, I considered what I have called "lines of argumentation" and "jumps."

Lines of argumentation and jumps were identified in oral or written forms of argumentation. A "line" is a sequence of interventions or propositions related to each other by the fact that they remain within the same specific questioning and the same choice of representations. In some cases, this can be the whole argumentation work, as happens in the case of a written finished product (e.g., a mathematical proof). The line can evolve through the search of new references to back the reasoning, and this effort may change the representations at work. The "jumps" are the visible changes of lines. We can observe lines that are interrupted then re-established after elaboration on other lines. The crossing of lines of argumentation can lead to the development of a new line, or to the advance of one of the existing ones.

The main hypotheses of this chapter concern the potential of argumentation in developing conceptualization and the crucial role of the teacher in managing appropriate communication situations for it. In order to provide experimental evidence for these hypotheses, the criteria of analysis presented above are integrated in the presentation of classroom situations. Essential chronological descriptions of what happened in the classroom are provided in order to situate crucial elements surfacing in students' behaviors. Then, the same elements are reconsidered and put into relationship with our hypotheses. Below, two examples will be discussed which provide experimental evidence which serves to clarify and elaborate the issues discussed above.

## CONCEPTUALIZATION THROUGH ARGUMENTATION: THE SUN SHADOW EXAMPLE

The following task was set in a 4th grade class (ten year olds) engaged in a long-term activity of geometrical modeling of the sun shadow phenomenon. Here is the document, presenting the problem situation, as it was given to the students (see Figure 5.1)

At the beginning of the class project on sun shadows, Stefano (a VI grade student) thinks that shadows are longer when the sun is higher and stronger. Other students think the contrary. In order to explain his hypothesis, Stefano produces the following drawing:

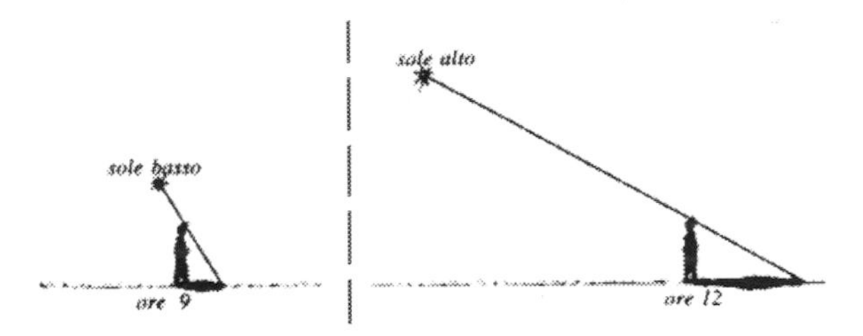


And writes: "As we can see in the drawing, the sun makes a longer shadow when it is higher, that is at noon, when it is also stronger."

We know very well that shadows are longer when the sun is lower (early in the morning and late in the afternoon). So, in Stefano's reasoning something does not work. What is wrong with Stefano's reasoning, and particularly with his drawing?

Try to explain yourself clearly, so that Stefano can understand.

Figure 5.1. Stefano's problem.

Stefano was a student in another class who really produced the drawing and the comment. The task of refuting Stefano's elaboration was not easy. The meanings attributed by Stefano to the words "high" and "low," were compatible with what students could see. It is true that at 9 a.m. the apparent height of the Sun from the ground (or the horizon) is always lower than at noon!

This task was given in the second half of a project spanning a two year period in the experience field of sun shadows. The students had made many observations and experiments concerning sun shadows and their movements during the day, during the year, etc. Due to these background activities, they knew that Stefano's statement was wrong, but the drawing could have been convincing. They were familiar with the "shadow schema" the drawing refers to, which is the triangle made of the sun ray, the obstacle and its shadow: they are gradually elaborating a model of the relation between sunray and shadow. And they had started working on the angles in plane situations.

There were a number of aims pursued with this task, agreed on with the teachers of the class. (There are usually two teachers who are responsible for a class of students through their primary school studies.) Our first aim was to get the students to question the statement "when the Sun is higher, the shadow is shorter," especially to reflect on the meaning of "high." In previous class work, "high" referred to the ordinary meaning of "height" (measured with the ruler or tape measure) of plants, children's bodies, etc., and also to a meaning not well defined operationally, when speaking about the "height" of the Sun in the sky. Moreover, the understanding of the movements of the Sun and Earth, and the relation to schematization of sun shadows, was just beginning to take place. A second aim was to approach and develop the concept of inclination of sun rays as a tool to explain and overcome Stefano's mistake. A third aim was to question some aspects of the conventionality of the "shadow schema" used by Stefano and shared by the class, in particular drawing a sun at the end of the segment representing the sunray.

The observation of student activities offered me the opportunity to analyze how argumentative activities, demanded by the task, involved some physical aspects of the sun shadows phenomenon. The students shared various experiences (many of them were reference situations), as well as gestural, graphic and verbal representations (Douek, 1999b, pp. 95–103) concerning the sun shadow phenomenon. I was also interested in investigating how those "situated" activities could support the process of conceptualization and refinement of the model represented by the "shadow schema" (pp. 104–105).

Following the usual organization of problem solving activities in the classes of the Genoa group, the first phase of the activity consisted of individual written problem solving concerning the task reported above. A classroom discussion followed, where the students presented their solutions and argued about their validity. This discussion offered many problems that were solved jointly through argumentation, and provoked evolution in students' understanding. A third phase was dedicated to individual written productions about "what is to be retained from the discussion." Finally, a second classroom discussion was organized. It was aimed at producing a collective summary of the shared knowledge newly attained. This was not an institutionalization phase; some fuzzy knowledge might remain in such a text. For example, up to the present problem solving situation, the ambiguity of the word "high" was inherited from past work. The text produced reflected the knowledge attained by the majority of the class. Some interesting results of some students appeared in the first discussion, but they disappeared from the collective summary, because they were not taken over by other students. I now concentrate on two argumentation situations that took place in the first discussion.

## Argumentation: Domenico's Episode

During the first discussion, Domenico said: "the length of the shadow depends on the position of the Sun." In formal terms, we might write: $L = L(P)$. Then, invited to explain at the blackboard (where Stefano's drawings were reproduced), he drew another position of the Sun (below the sunray, in the 9 a.m. drawing). He said: "*if I change the position of the Sun, the length of the shadow changes too.*" The content of this sentence, brought to a formal level, would be: if $P1 \neq P2$, then $L(P1) \neq L(P2)$. Two students (Aurelio and Luca) realized that it was not always true:

> Aurelio: I wanted to say another thing: if Domenico moves the Sun on the same ray, the shadow remains the same!

On the blackboard, they were able to show examples of different positions of the Sun, which did not change the shadow (being on the same sunray). This was not discovered at the formal level, but probably through the dynamic exploration suggested by Domenico's gesture accompanying his words, when he showed how changes of the position of the Sun changed the length of the shadow. But then Aurelio and Luca could imagine other positions compatible with the same length! The teacher encouraged the discussion about this discovery. Students spontaneously recalled this point twice, both during the first and the second discussions, as a *reference situation* for the operational invariant concerning the independence of the sunray inclination from the length of the segment representing the sunray.

There were two crucial moments of the argumentation work. First, when implicit knowledge was made explicit. Domenico's sentence reveals *a theorem in action* (in the sense of Vergnaud). Second, when an episode of transformational reasoning, which consisted in exploring, using imagery, the displacement of the Sun and its effects on the shadow, revealed the border of validity of this theorem in action. The communication context, organized in order to make the involved knowledge of several students explicit and conscious by the means of comparison, brought the whole class to awareness of these two points and to share the newly established knowledge. Argumentative activity rooted it to past knowledge, which contributes to sense making, and related experiences with geometrical representation and language representation.

## Argumentation: The Tall Man and Short Child Episode

Later on, in the first discussion, a student, Ambra, said that the height of the sun ("for us") could be seen from the "inclination" or the "height" of the arm pointing to the Sun. Steven (another student who had discovered, in his first text, that the prolongation of a "lower" sunray may overpass the "higher" sunray) proposed a correction: not "high arm," but "arm towards a high position". To support Steven's remarks, the teacher suggested observing what happened if he pointed to a low position of the Sun with his "high" arm, while Steven was pointing to a "higher" position of the Sun with his "low" arm. This observation (supported by many comments—especially concerning the idea of the prolongation of the two arms) created a new reference situation in the classroom, well represented in Figure 5.2 by the drawing from an individual report:

Figure 5.2. A child's drawing as a reference situation.

This reference situation and the opportunity for transformational reasoning it can offer, were shared through the communication situation.

## Evolution of Conceptualization in Relation to Argumentation

An important condition for argumentation is the *complexity* of the situation because of its importance in enhancing productive argumentational activity. An example is provided by the questioning of the clause "the sun is high" in the reference statement "when the sun is high the shadow is short". Both the different meanings the students gave to "high," and the modeling situation at the different levels at which the students grasped it, played important roles in the problem solving activity. When those meanings needed to be clarified, they offered several points that had to be opposed. This made clear to some students that there were good reasons to doubt their past knowledge, to reflect on the coherency between various statements, and to discuss. For example, an important contribution came from the second discussion, when students understood that Stefano intended "high" as an entity that could be measured through ordinary tools to measure lengths. During the discussion on this point, two lines of argumentation crossed; one referring to the distance from the "ground" (or from the "object" casting the shadow) in the shadow schema, the other referring to the Sun-Earth (or Sun-object) distance in physical reality.

At the beginning of the work, the word "inclination" had various meanings. It was a word evoking a physical experience (e.g., the arm pointing at the Sun) or, for some students, expressing intuition of the mistake in Stefano's schema whereby the slope of the sunray is steeper at 9 a.m. than at noon. Gradually, argumentation helped to make a connection between the two initial meanings of "height" and make explicit an important property (i.e., an operational invariant of the inclination concept): the position of the represented Sun along a given sunray does not influence the length of the represented shadow (see Domenico's episode). This property is related to the fact that the operational definition of "inclination" is connected to the concept of angle.

The argumentation made explicit another property: that the inclination of a sunray does not depend on the height of its graphical or physical representation. Steven realized it on paper through the schemas he drew in his first individual work, then transposed it into a physical representation, in the "tall man and short child" episode. Other students discovered the property during this episode, at the level of physical representation, then brought it to the level of graphic representation in their second individual texts. The discussion allowed the student to approach the connection with the angle concept, in the final part of the second discussion.

During the subsequent activities of planning and executing the measurement of the inclination in the three dimensional space situation, these properties allowed further clarification of the (not yet formalized) idea that the "width" (student's expression) of the angle does not depend on the length of the drawn sides. The students had already met this idea in a plane situation (two dimensional space) when they had to "reduce" a big shadow fan in their copybooks.

The progressive conceptualization of inclination occurred in connection with the differentiation of the meanings of "height," through the successive argumentation and discussion situations. Refinement of the model represented by the shadow schema occurred in strict relationship with progress in conceptualization and distinction between "inclination" and "height." For this reason one may speak of "reconceptualization" of the model. In particular, students grasped that the Sun-object distance in the shadow schema was a false variable for the model of the phenomenon, and that the relevant variable was the inclination of sunrays. Concerning the role of argumentation in conceptualization through the above examples, I want to stress the following points:

- argumentation transformed experiences (in particular, in the two reported episodes) into reference situations for two operational invariants of the inclination concept;
- it evoked past reference experiences and established knowledge deriving from past observations;
- it coherently related various representations (gestures, schemas, etc.) to each other and to the experiences;
- it revealed an operational invariant, the "theorem in action" mentioned in Domenico's episode, and adjusted it to its correct limits of validity; this was an essential tool in solving the problem;
- it was a step in linking knowledge into a system (in particular, by relating knowledge about inclination and about plane angles);
- it adapted the understanding of the geometrical model to the new knowledge.

## TEACHER'S MEDIATION: THE WHEAT PLANTS EXAMPLE

I shall now describe and analyze a classroom sequence consisting of five activities in a 2nd grade class concerning the same problem. Students had to measure wheat plants (which they grew in the classroom) in a pot, using rulers. They had already measured wheat plants taken in a field, now they had to follow the increase over time of the heights of plants in the classroom pot. (For further details, see Douek & Scali, 2000.)

The main difficulty was that rulers usually do not have the zero at the edge, and it was not allowed to push the ruler into the earth in order to avoid harming the roots. Children had to find a general solution (not concerning a specific plant). They produced two different procedures. The procedure of translating the numbers written on the rulers, using an *operational invariant* of the measure concept, which is the invariability of measure through translation. Let us call it the "translation solution." And the procedure of reading the number at the end of the plant, then adding to this result the measure of the length between the edge of the ruler and zero. In this second case, they use another operational invariant of the measure concept which is the additivity of measures. This will be called the "additive solution."

The *first activity* was an individual discussion with the teacher to find out how to measure the plants in the pot with a ruler that had (apparently) a one centimetre space between zero and its edge. Under the teacher's guidance (see later), students had to reach a conclusion on how to solve the problem and dictate it to the teacher, who wrote it down (because at this stage the students do not write easily), then gave it to the students to copy into their note books.

The main difficulties met by students were: a) *understanding that there was a problem.* Many students would have been satisfied by reading the number on the ruler corresponding to the top of a plant, if they had not been pushed by teacher's argumentation; b) *focusing on the problem.* There were many plants in the pot; some students concentrated on twisted plants; one worried (too much!) about not harming the roots and another insisted on introducing the use of a professional tape measure (with zero at the edge). The teacher interacted with each student, in argumentative style, in order to get the student to focus on the problem, as in the following excerpt. With the help of the teacher, the student had already discovered that the number read on the ruler corresponding to the edge of the plant is not the measure of the height of the plant.

Student: we could pull the plant out of the ground, as we did with the plants of the field
Teacher: that way, we could not follow the increase in height of our plants
Student: we could put the ruler into the ground, in order to bring zero to the level of the ground
Teacher: but if you put the ruler into the ground, you could harm the roots
Student: I could break the ruler, removing the piece under zero
Teacher: it is not easy to break the ruler exactly on the line of zero, and then the ruler is damaged

In the above excerpt, we have an example of teacher mediation. The teacher responds to students' difficulties; he asks questions that reveal the problem and challenges their thinking. Then, the students had to find some form of solution. It was difficult for some students either to imagine shifting the scale on the ruler, to bring zero to the edge of the ruler, or to imagine measuring the length between zero and the edge of the ruler. Some students switched during the discussion from the idea of pushing the ruler to imagining sliding the numbers on the rulers. Others from breaking the ruler to imagining putting this little bit of ruler on the top of the plant and reading the full measure. The familiar classroom practice on the "number line" had to be transferred to the new situation of measuring, and a transformation of the status of the ruler (from a measuring tool, to an object to be measured) was required. This was achieved, at least partially, through the argumentative reactions of the teacher during the discussion.

The final difficulty was to reconstruct the whole reasoning (built in interaction with the teacher) and dictate the resulting procedure to the teacher. During this phase of the activity the teacher discussed with the students the problem of representing their ideas about how to solve the problem. Reference was made (in many cases by the student, in some cases by the teacher) to preceding activities on the number line, and the "arrow representation" of addition and subtraction was negotiated as a possible, efficient representation.

At the end of the interaction, nine students out of twenty arrived at a complete solution (i.e., were able to dictate an appropriate procedure). Four were "translation solutions" (see at the beginning of this Section), four "additive solutions," while one solution contained an explicit indication of the two possibilities, "addition" and "translation." Moreover, four students moved towards a translation solution without making it explicit at the end of the interaction. The other seven students reached only awareness of the fact that the measure read on the ruler was the measure of one part of the plant, and that there was a "missing part," without being able to establish how to continue. Rita's solution is one of the four translation solutions:

> Rita: In order to measure the plant we could imagine that the numbers slide along the ruler, that is zero goes to the edge, one goes where zero was, two goes where one was, and so on. When I read the measure of the plant I must remember that the numbers have slid: if the ruler gives 20 cm, I must consider the number coming after 20, namely 21.

Alessia's solution is one of the four "additive solutions:"

> Alessia: We put the ruler where the plant is and read the number on the ruler, which corresponds to the height of the plant, and then add a small piece, that is, the piece between the edge of the ruler and zero. But first, we must measure that piece behind zero.

The *second activity* consisted of an individual written production. The teacher presented a photocopy of Rita's translation solution and Alessia's additive solution, asking the students to say who's solution was like theirs, and why. Rita's and Alessia's solutions were chosen by the teacher because they were clear enough and representative of the two main ways of solving the problem. This task was intended to provide all students (including, possibly, those who had not reached the solution) with an idea about the solutions produced in the classroom, and a possibility to situate their work in relation to someone else's work. With one exception, all the 13 students who had produced or approached a solution were able to recognize their solution or the kind of reasoning they had started. And 6 out of the other 7 students declared that their reasoning was different from that produced by Rita and Alessia.

The *third activity* was a classroom discussion. The teacher at the blackboard, and the students working in their note books (where they had drawn a pot with a plant in it), and using a paper ruler similar to that of the teacher, put into practice the two proposed solutions: first the "translation solution," then, the "additive solution." Meanwhile they discussed many problematic points that emerged. For example, the fact that the "translation" procedure was easy only in the case of a length (between zero and the edge of the ruler) of 1 cm (or eventually 2 cm); while the "additive" solution consisted in a method easy to use in any case. Another issue they discussed concerned the interpretation of the equivalence of the results provided by the two solutions ("why do we get the same results"?). Here is an excerpt from this whole class discussion.

Angelo: Rita's method is similar to Alessia's method

(many voices: "no, no…")

Ilaria: Rita makes the numbers slide, on the contrary Alessia ... she does not make it ... The two methods are not similar

Jessica: Rita says to make the numbers slide, but Alessia moves the piece of the plant

Teacher: Wait a moment, please. Jessica probably has caught an important point. She says: Rita makes the numbers slide... the measure of the plant for Rita is always the same, it is the numbers that slide … While, as Jessica says, Alessia has imagined taking a piece of the plant and bringing it to the edge of the plant, where we can measure it…

Angelo: I do not say that it is the same thing, I say that the two methods are a little bit similar

Giulia: It is like if Alessia would overturn the plant,… she would bring a piece over… a small piece went over and the plant seems to be hanging on… so we could measure it.

The *fourth activity* was an individual written production where students had to "explain why Rita's method works, and explain why Alessia's method works." With three exceptions (who remained rather far from a clear presentation, although they showed an "operational" mastery of the procedures) all the students were able to produce the explanations demanded by the task. One half of the students added comments about the comparison of the two methods; most of them explained in clear terms the limitation of the "translation" solution. Here is an example of a student's text:

> Rita's reasoning works, because it makes the ruler like a tape measure, because zero goes at the edge of the ruler. She imagined the same thing and made the numbers to slide. Where I see that the edge of the plant is at 8 cm, she says 'to slide' and sees that the ruler slid by one centimetre, and so she sees that 8 became 9. But this method works easily only if the ruler has a space of 1 cm between zero and the edge. Alessia's reasoning works because Alessia adds the piece of plant she carries to the height of the plant and adds 1 to 10 and she sees that it makes 11. One difference is that Rita leaves the piece of the plant where it is while Alessia carries it up to 10.

The *fifth activity* consisted in the classroom construction of a synthesis to be copied into students' notebooks. The good quality of many students' individual productions in the preceding activity allowed them to arrive at an exhaustive synthesis. In the above classroom activities, the "measure of the plants in the pot with the ruler" was recalled as a reference situation on some occasions, when students had to measure the length of objects not accessible to a direct reading of their length on the ruler.

## ARGUMENTATION, CONCEPTUALIZATION AND TEACHER MEDIATION: EDUCATIONAL IMPLICATIONS

The data discussed above show preliminary experimental evidence for some of the potentialities of argumentation in conceptualization at various levels, and the crucial role of the teacher in exploiting that potential through suitable communication contexts (i.e., situations, organized by the teacher, in which communication takes place). During the interactive resolution of the problem (first activity), argumentation with the teacher in a one-to-one communication situation brought students to grasp the nature of the problem. Gradually, it transformed some imagined practical solutions (e.g., as pushing the ruler into the earth or cutting it), into a *reference situation* for one of the operational invariants concerning the concept of measurement. It also extended the construction of appropriate verbal and symbolic representations. For instance, the "arrow" representations of addition and subtraction coming from preceding activities on the number line became (through the negotiation with the teacher) efficient tools in the new situation concerning measurement with the ruler. They were utilized by students to "represent" the idea of sliding the numbers along the ruler, and so were connected to a scheme for the measurement concept.

This communication situation favored the teaching of what are usually rather "personal" problem solving skills, like those involved in transformational reasoning. In order to see evidence of these skills, we can consider consecutive phases of a transformational reasoning process (guided by the teacher, in our example). Students' imagined solutions were inspired by possible actions without performing them. I regard this as a step towards constituting a scheme based on the inherent operational invariants. Subsequently, they made explicit the operational invariants by relying on past knowledge or reference situations. Then, they imagined some changes of the tool in its mathematical representation aspect (the graduation, in our example), thus constituting a scheme acting on a mathematical tool, and not only on a realistic situation. Eventually, some arrived at a schematic representation of the solution (as with the arrows). These phases are often smoothly integrated for a person more experienced in problem solving and generally constitute an individual cognitive activity. We must also recognize that such complex ability to transform various elements of a situation and go forth and through from a material situation to a mathematical representation and from a representation to a tool, plays an important role in building mathematical reasoning.

During the classroom discussion (third activity), argumentation fulfilled different functions. It allowed students to make explicit the two operational invariants ("translation invariance" and "additivity") of the measure of lengths, and the systemic links between them and other concepts (addition and subtraction) in the "conceptual field" of "additive structures" (Vergnaud, 1990). The careful description of the actions (e.g., sliding the numbers, or cutting the basis of the plant to bring it to its top where the numbers on the ruler are) was particularly important. It established within the class a shared link between the operational invariants, the transformational reasoning through which the students elaborated the schemes based on those operational invariants and the corresponding calculations.

During the last individual activity ("explain why...") personal argumentation helped the students to establish links between the two methods, revealing and enhancing the internalization of the acquisitions constructed (at the interpersonal level) during the preceding discussions. In this manner, students attained a first level of awareness about potentials and limitations of the schemes involved. It was important to favor a personal expression of the newly attained (or acquired) knowledge. Therefore, this situation can be considered as equilibrating the first activity where personal positions needed to be (more or less strongly) supported by the teacher.

Finally, we can suppose that the various argumentation situations proposed by the teacher all contribute to enhancing the students' argumentative potential. From various observations I noticed that argumentation is a difficult activity, especially at this young age, but at the same time I noticed that students working in this educational context succeed in performing astonishingly refined argumentation for their age level (e.g., Marco's text earlier).

The wheat plants example shows how performing argumentative activities needs an appropriate mediation. This concerns reasoning, transformational reasoning strategies (especially concerning moving from physical action to mental exploration of possibilities), and the developing of a quality of verbal expression. Dialectically, argumentation can be a means to develop and systematize knowledge (e.g., see how the conceptualization of measuring has evolved in the activities described above). The communication context plays a very important role in the "teaching" of argumentation (cf. Krummheuer, 1995, Yackel, 1998). The choice of suitable communication contexts is crucial to attaining the whole spectrum of the above-mentioned argumentative skills.

I have analyzed the role of argumentation in terms of conceptualization and in relating knowledge *into* a system. The consideration of argumentative behaviors in communication is relevant and important, because argumentation can be considered a particular development of communication. It presupposes a contradictory listener, thus it must naturally be developed in a communication context (and at an interpersonal level) and then interiorized. This is why it is important to provide various communication contexts where argumentation will develop. In this study, a number of classroom practices, such as individual oral argumentation with the teacher, individual written argumentation with punctual interventions of the teacher, collective classroom discussion managed by the teacher, and individual written argumentation in reaction to other students' texts can be considered as educational implications. As most intellectual skills, argumentation must also be developed into an individual activity, in order to become a way to produce coherent personal texts.

## REFERENCES

Arzarello, F. & Bartolini Bussi, M. G. (1998). Italian trends in research in mathematics education: A national case study from an international perspective. In A. Sierpinska & J. Kilpatrick (Eds.), *Mathematics education as a research domain: A search for identity* Vol. 2. (pp. 243–262). Dordrecht, The Netherlands: Kluwer Academic Publishers.

Bartolini Bussi, M.G., Boni, M., Ferri, F. & Garuti, R. (1999). Early approach to theoretical thinking: Gears in primary school, *Educational Studies in Mathematics, 39*, 67–87.

Boero, P., Dapueto, C., Ferrari, P., Ferrero, E., Garuti, R., Lemut, E., Parenti, L. & Scali, E. (1995). Aspects of the mathematics-culture relationship in mathematics teaching-learning in compulsory school. *Proceedings of PME-19*, Recife, Vol. 1. pp. 151–166.

Brousseau, G. (1986). Fondements et méthodes de la didactique des mathématiques. *Recherches en didactique des mathématiques, 7,* 33–115.

Douek, N. (1998). Analysis of a long term construction of the angle concept in the field of experience of sunshadows. *Proceedings of PME-22,* Vol. 2. (pp. 264–271). Stellenbosch, South Africa.

Douek, N. (1999a). Some remarks about argumentation and mathematical proof and their educational implications. In I. Schwank (Ed.), *Proceedings of the Conference European research in mathematics education (CERME-I)* Vol. 1. (pp. 128–142). Osnabrueck, Germany.

Douek, N. (1999b). Argumentation and conceptualisation in context: a case study on sun shadows in primary school. *Educational Studies in Mathematics, 39,* 89–110.

Douek, N. (1999c). Argumentative aspects of proving: Analysis of some undergraduate mathematics students' performances. *Proceedings of PME-23I,* Vol. 2, (pp. 273–280). Haifa, Israel.

Douek, N. & Scali, E. (2000). About argumentation and conceptualisation. *Proceedings of PME-24,* Vol. 2. (pp. 249–256). Hiroshima, Japan.

Douek, N. & Pichat M. (2003). From oral to written texts in grade I and the long term approach to mathematical argumentation. http://www.dm.unipi.it/~didattica/CERME3/.

Krummheuer, G. (1995). The ethnography of argumentation. In P. Cobb & H. Bauersfeld (Eds.), *The emergence of mathematical meaning.* (pp. 229–269). Hillsdale, NJ: L.E.A.

Piaget, J. (1947). *Le Jugement et le raisonnement chez l'enfant.* Neuchatel et Paris: Delachaux & Niestlé. (Original work published 1923)

Simon, M. (1996). Beyond inductive and deductive reasoning: The search for a sense of knowing. *Educational Studies in Mathematics 30,* 197–210.

Steinbring, H., Bartolini Bussi, M.G. & Sierpinska, A. (Eds.) (1998). *Language and Communication in the Mathematics Classroom.* Reston, VA: NCTM.

Toulmin, S. (1958). *The Uses of Argument.* Cambridge, UK: Cambridge University Press.

Vergnaud, G. (1990). La théorie des champs conceptuels. *Recherches en Didactique des Mathématiques 10,* 133–170.

Vergnaud, G. (1998). A comprehensive theory of representation for mathematics education. *Journal of Mathematical Behaviour,17* (2), 167–181.

Vygotsky, L. S. (1985). *Pensée et langage.* Paris: Editions Sociales.

Yackel, E. (1998). A Study of argumentation in a second-grade mathematics classroom, *Proceedings of PME-22,* Vol. 4. (pp. 209–216). Stellenbosch, South Africa.

CHAPTER 6

# UNDERSTANDING MATHEMATICAL INDUCTION IN A COOPERATIVE SETTING

## Merits and Limitations of Classroom Communication amongst Peers

**Inger Wistedt**
***Stockholm University, Sweden***

**Gudrun Brattström**
***Stockholm University, Sweden***

### ABSTRACT

In this study, undergraduate students solve a non-standard mathematics task confronting them with the concept of mathematical induction. The aim of the study is to investigate cases of reflective processes in which students understand and contextualise mathematical content in a co-operative setting. Six mixed-sex groups (3–5 students in each) from two Swedish universities were observed while solving the task. The tape-recorded group discussions were transcribed and analyzed from a constructivist perspective and by means of intentional

*Challenging Perspectives on Mathematics Classroom Communication*, pages 173–203

analysis, a method by which we ascribe meaning to the students' activities in terms of intent. In the analysis we focus on the students' ways of communicating, of relating to the subject matter, and of understanding induction as a method of proof. The results lead to a broadening of the view of classroom communication, as not only including communication of the subject matter, but meta-aspects of the subject as well. Despite the fact that the group-sessions were almost ideal from a communicative point of view, the students did not enhance their understanding of induction as a method of proof. We found that their difficulties rested on a meta-theoretical level. Since all of the students shared the same limited views of mathematical induction they could not manage to establish the prerequisites of inductive proofs by themselves. We point to the need for the students to interact with more experienced members of the mathematical community and stress its importance, not least if we view the results from an inclusive perspective.

## BACKGROUND, THEORETICAL PERSPECTIVES, AND AIM OF THE STUDY

Studying is, in part, a socialization into disciplinary cultures. Within specific "communities of practices" (Burton, 1995), for instance the practices of developing and furthering professional knowledge in mathematics, certain ways of approaching phenomena are cultivated, as are specialized means for communicating experiences (Säljö, 1999; Wertsch, 1995). Learning can be viewed as a process of becoming acquainted with such specialized practices—a process of developing professional expertise.

How do students acquire such expertise? In recent decades there has been a broad debate in Sweden, as well as in other countries, about the quality of learning in higher education. In Sweden, development projects have been initiated by the National Agency for Higher Education, combining two goals: a political goal of ensuring equal access to higher education, and a pedagogical goal of enhancing qualities in student learning by diminishing rote-learning and increasing the possibilities for students to develop a reflective attitude towards the subject matter (Wistedt, 1996). Collaborative work styles, problem- or project-based learning, are implemented within these projects, introducing the students to ways of learning formerly rarely practiced at the undergraduate levels of university education. The initiators hope that such work styles will be attractive to students with different aims and backgrounds, and make the best of their capabilities by presenting academic studies, from the very start, as a cooperative, meaningful, and socially rewarding activity.

The changing views of learning entail a redirection of focus from isolated learners, their traits, gifts or talents, to ways of learning mathematics which are relational with the learner, the situation, and the setting in which learning occurs (Halldén, 1994; Marton & Booth, 1996). In a learning environment which acknowledges differences in experiences and approaches to learning, student interpretations of the subject matter and of cultural expectations are regarded as crucial to the learning process (Marton, Hounsell & Entwistle, 1997). This study focuses on such interpretations, variations in ways of interpreting a mathematical task, and in ways of handling expectations which students at the undergraduate level of university education may associate with a mathematical culture. Communication plays a major role in highlighting such variations in ways of understanding a mathematical task and its academic setting. In this study we discuss the merits and limitations of classroom communication based on observations of students at the undergraduate level of university education as they try to come to grips with a concept known to be an obstacle to many undergraduates—mathematical induction.

## Analyzing Learning in Institutionalized Settings

The analysis of practices within institutionalized settings, such as communication of mathematical knowledge in a classroom environment, is often referred to as "discourse analysis" (see e.g., Bakhtin, 1981, p. 288; Bhatia, 1993; Brown & Yule, 1983), where the term "discourse" denotes the language use in a specific setting; e.g., a medical examination, a police interrogation, a legal proceeding, or a mathematical seminar (e.g., Drew & Heritage, 1992), and "language use" is understood as a social phenomenon: "social throughout its entire range and in each and every aspect of its factors, from the sound image to the furthest reaches of abstract meaning" (Bakhtin, 1981, p. 259). The somewhat narrower concept of "genre" is often used to denote and distinguish different kinds of discourses in terms of communicative intent. A genre can be described as:

> ... a recognizable communicative event characterized by a set of communicative purpose(s) identified and mutually understood by the members of the professional or academic community in which it regularly occurs. Most often it is highly structured and conventionalized with constraints on allowable contributions in terms of their intent, positioning, form, and functional value. These constraints, however, are often exploited by the expert members of the discourse community to achieve private intentions within the framework of socially recognized purpose(s). (Bhatia, 1993, p. 13)

Analyses of genres can be made from different points of departure, from a language philosophical perspective such as in the study of *speech acts* (Searle, 1969), from a sociological perspective in the study of *rules and conventions* that people use to produce and interpret communicative events in daily life (Garfinkel, 1967; Geertz, 1973; Goffman, 1959), or from an educational perspective in the study of cognitive *frames of reference* brought to the fore in institutionalized settings (Driver & Easley, 1978; Halldén, 1988a). In this study the perspective is educational and mathematical, which means that we focus on conceptions of mathematics and of mathematical concepts given topical interest in a situation where undergraduate students solve a mathematical task.

## Learning

Learning in institutionalized settings generally entails challenges to naïve and taken-for-granted notions (Säljö & Wyndhamn, 1988, 1990). Ideally such challenges should result in the appropriation of ways of understanding phenomena which are relevant to the subject studied. Within a constructivist perspective, such appropriations are often viewed as instances of *conceptual change* (e.g., Posner, Strike, Hewson, & Gertzog, 1982; Strike & Posner, 1992). Learning is considered to be a process of abandoning vague, crude, or inconsistent notions in favor of more precise and potent ones.

The constructivist approach has, however, been criticized for not paying attention to the situated aspect of learning, for separating what is learned from how it is learned and used (e.g., Brown, Collins, & Duguid, 1989). Within a socio-cultural perspective, researchers often refrain from introducing constructs such as "conceptions" or "cognitions." Instead, they describe changes in ways of acting in accordance with specific discursive rules, changes in students' ways of *knowing* a subject matter rather than in their *knowledge* about the matter at hand (Hawkesworth, 1996; Säljö, 1994).

We do not however, have to repudiate the constructivist claims in order to take into account the discursive nature of knowledge formation. It is true that constructivists attempt to understand and account for classroom communication from the perspectives of the individual participants and their meaning-making activities in a certain context. The researcher stands apart from the participants, observing what they do or say with the aim of identifying patterns that are outside the actors' awareness: "The implicit rules or social norms that the participants appear to be following can be formulated as a first step in explaining their mutual construction of the observed patterns" (Cobb, 1990, p. 207).

In a socio-cultural perspective, the term "context" above, refers to the situation in which learning occurs (Resnick, 1989, p. 1). In a constructivist perspective, however, the term "context" does not refer to the setting in which a certain task or concept is embedded, but to the individual student's interpretation of this setting. If we let the term "context" denote such personal constructions of concepts presented in a study situation (*cognitive context*), as well as constructions, or interpretations, of the setting in which learning occurs (*situational context*) and the discursive rules that regulate communicative activities (*cultural context*), we can talk about students' ways of appropriating new conceptions as a problem of "*contextualization*" (Halldén, 1994).

Intentions are, in such a perspective, a crucial aspect of contexts. Students within the same setting may, for different reasons, focus on quite different aspects of the social environment, a given task or concept. The setting could be the same but the contexts may differ depending on what the students interpret as being the purpose of the activities within it (cf. Cobb, 1990). When we are interested in variations in learning and understanding mathematics, not least culturally deviant ways of approaching the subject, such a perspective can prove fruitful.

## Communication

Learning is not only a constructive but also an interactive activity. It is our view that constructivists have paid too much attention to conceptual change as an individual, developmental process. Considerably less attention has been given to concept formation, that is, students' contextualisations of new concepts in the act of explaining something to someone else (Halldén, 1999). Much attention has been put on *cognitive conflict* as a driving force in the development of new concepts (see Pfundt & Duit, 1991 for an overview of the field) less on the role of *decentering*, that is, on how knowledge increases as a function of a refocusing of perspective (Inhelder & Piaget, 1958). Different interpretations of a concept or a task are applicable and relevant in different cultural settings. The problem of what constitutes "facts" that are to be explained, "phenomena" that are to be observed, or "focal events" that are to be understood can be viewed as a problem of finding interpretations that are applicable to a given situation, to find the relevant contextualization (Caravita & Halldén, 1994; Wistedt, 1994a, 1994b). The instructional problem is, in such a perspective, not regarded as a problem of finding ways of changing students' conceptions, but helping them pay attention to the relevance of different styles of thinking and acting within more embracing conceptual frameworks (cognitive context), in different situations (situational context), and within different speech genres (cultural context).

Communication clearly plays a major role in such a process. In communication with peers and tutors, students may modify their contextualizations. When a student is communicating his or her thoughts to others, those thoughts have to be expressed in a conventional form in order to be understood. Learning in a communicative setting is of necessity mediated by cultural tools, such as language and other social conventions, and tools in the ordinary sense of the word, such as books, graphs, and other artefacts, all permeated with cultural meaning. Communication may function as a means to appropriate a "mathematical attitude," and through communication we can ensure that the individually acquired knowledge will fit with the mathematics constructed by others.

### The Aim of the Study

The study presented here focuses on university students' contextualizations of the concept of mathematical induction. The aim of the study is to investigate cases of reflective processes in which students strive to understand mathematical content in a cooperative setting. Variations in ways of approaching and interpreting the concept, the setting and the discursive rules of mathematics are described and discussed in relation to classroom communication as a means to overcome limited and idiosyncratic ways of understanding the subject matter.

## METHOD

The study is based on empirical examples gathered within two Swedish development projects at the universities in Stockholm and Göteborg (see Wistedt, 1996, pp. 13–31). Six mixed-sex groups of 3–5 undergraduate students in each group were observed while solving a geometrical task confronting them with the concept of mathematical induction. Each group session lasted for about 45 minutes. The verbal interaction between the students was tape-recorded, later transcribed, and supplemented by sketches and drawings made by the students during the group work.

### Intentional Analysis

In the analysis of the data we try to uncover "the taken-for-granted interpretations, activities and practices that give rise to the observed patterns and norms" (Cobb, 1990, p. 208). We do this by means of intentional analysis (von Wright, 1971; Downes, 1984), a method by which we ascribe meaning to the students' actions in terms of intent.

Social life is vital to intentionality. In many cases the descriptions of acts are only intelligible in social or cultural terms. Many acts are, in fact, social institutions, e.g., problem-solving in a mathematics classroom, and proving or hypothesizing in specific mathematical contexts. The students may, however, understand such acts in ways that differ from cultural conventions. This means that we have to infer the acts. Faced with the problem that students may act in unconventional ways, we work backwards: What motivates an act that deviates from a conventional act in the observed ways? The inferences serve to connect the utterances to a context, and decide what the students are trying to convey. By virtue of what students say or imply in a situation, we gain a picture of what they believe, hold true or commit themselves to, a picture of their *contextualizations* of the content, the situation and the cultural demands (for a more detailed description of intentional analysis, see Downes, 1984, pp. 266–364). This means that the analysis is dependent on a mathematical perspective, but also on a perspective in which mathematical conventions are bracketed.

But do we not need to ask the students about their intentions in order to render a valid picture of their activities? The answer is no. First of all, the students do not respond to abstract phenomena such as "intentions" or "social norms" that the researcher projects into their world: "Instead, they act on the basis of the meaning that objects in their world have for them as they attempt to fit their activity to that of other people" (Cobb, 1990, p. 207).

Second, intentions are not ready-made, altogether rational and immediately accessible to the acting subject. Furthermore, intentions may change as a result of communication. We have all experienced this, for instance when we attend seminars or when our papers are scrutinized by peers. In such instances we may suddenly better understand ourselves and our intentions; we may experience a regeneration of previously held views. "Intention" is thus an analytical concept applied by the researcher in order to grasp the students' taken-for-granted interpretations, activities and practices as they are observed.

To assign intentional descriptions to actions is a normal, everyday way of interpreting and understanding others. We usually consider other people to be rational in their behavior, and when we observe people acting we "trace backwards" trying to connect overt action to inner states, like knowledge, beliefs, attitudes, and so on, the connections being made through practical inference (von Wright, 1971). However, as researchers using teleological explanations as a scientific method, we are faced with a lot of serious methodological problems: How do we come to grips with the multiplicity of intentions, the fact that people act for many reasons and that there are both primary and secondary intentions to any given act? How do we account for bias on the part of the interpreter, and for the problem of accessing information needed to make valid inferences? (For an elaborate discussion, see Downes, 1984). The first line of questions has to do with relevance, of narrowing the field of investigation to inferences that are relevant to the research problem; the other lines of questions have to do with accountability and the possibility of refuting proposed interpretations.

In this study the problems are handled as follows: Since the aim of the study is to describe how students construct learning goals and how such goals may be enriched and modified in communication with others, the descriptions are restricted to a task introduced by the researchers, and to the students' interpretations of that task. We describe how certain ways of contextualizing the setting are taken for granted and communicated to others, often indirectly. Six groups were observed while solving the task. The verbal interaction between the students was tape-recorded and transcribed in full. Below, excerpts from the transcriptions are given in the form of descriptive narratives, a somewhat unconventional way of presenting data. A "dramatized" form is chosen for the purpose of rendering a picture of the discourse as it unfolds and utterances are exchanged. In order to depict the chain of speech communication, direct quotations from the transcripts are linked together by interspersed comments. These comments are in most cases redundant (e.g., "said Marcus"), but in some cases they report observations of tones of voice, bodily movements, and the like (unconventionally stressed words are marked by italics). However, in the narrative part of the presentation interpretations are restricted to a minimum, thus enabling the reader to follow the successive analysis, and to some extent, put the proposed conclusions to the test.

## The Task

The students participated in the study voluntarily. Eleven students from Stockholm University and twelve from Göteborg University took part in the study, that is, about a third of the students in the respective programs. Six groups were formed on the basis of the students' own choice of peers. The task was distributed to the students at the beginning of each group-session (see Figure 6.1).

The task was of a somewhat unconventional nature. Prior to the group sessions the students had met inductive proofs in lectures, textbooks, and exercises where the examples used had been predominantly algebraic. This task did not involve algebra, but presented the students with a kind of puzzle, whose dimensions were dependant on an integer $n$. What was to be proven was that the puzzle was possible to solve regardless of the value of $n$. Our hopes were that the non-traditional form of the task would prevent the students from viewing it as a routine problem solvable by standardized methods. Rather we wanted them to view it as a genuine problem, promoting reflection upon the essentials of inductive proofs. The task was presented to the students in writing:

**Defective Chessboards and Induction**

Consider a "chessboard" with 2n × 2n squares where n ≥ 1. So for n = 1,2,3,... we have:

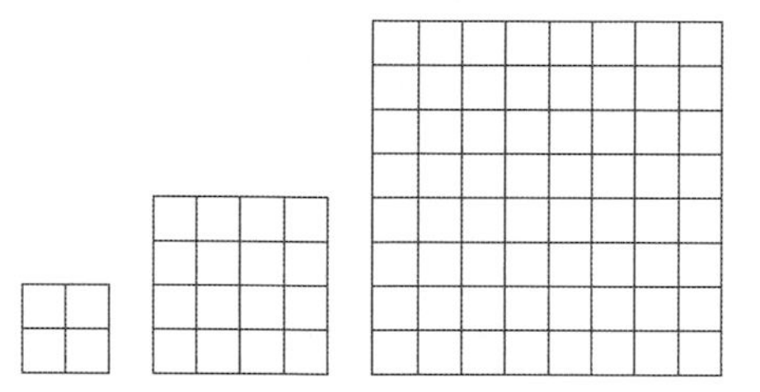

And so on. (Hence, for n = 3 we have a "genuine" chessboard.) Now let us remove one square somewhere on the chessboard. Like this for instance:

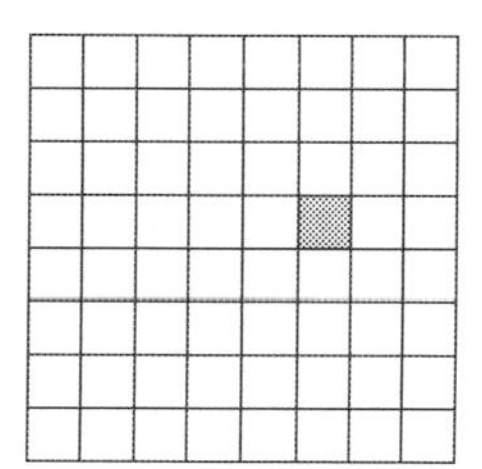

We call this a "defective" chessboard. We shall try to cover a defective chessboard by L-shaped pieces.

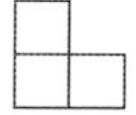

For example, below is a covering of a defective $2^2 \times 2^2$ chessboard:

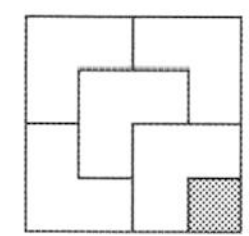

**Problem:** Using induction, show that it is always possible to cover a defective $2^n \times 2^n$ chessboard, for all n ≥ 1 and regardless of the position from which the square has been removed.

Figure 6.1. The task: Defective chessboard and induction.

## THE CASE STUDY

The non-algebraic nature of the task caused consternation in all six groups. None of the students seemed to view geometrical proofs as acceptable within a mathematical culture. In five of the groups the students managed to produce geometrical arguments, but although they seemed to be convinced by them they doubted their mathematical relevance: Was such a line of reasoning really an "inductive proof;" was it even "mathematical"?

There seemed to be a conflict between ways of solving the task which the students themselves found intuitively convincing, and ways of handling the task that they thought would convince expert members of a mathematical community. In the excerpts below we will focus on this conflict, since it highlights variations in ways of approaching the problem of acquiring conceptual knowledge in a social setting.

## Group A

In one of the groups at the University of Göteborg, three students were working together, Ellen, Linda and John (the names are fictitious). The students started out by reading the text (see Figure 6.1), and after a few minutes of silence Linda tried an interpretation:

> "If there are always three squares in an L, then you know how many Ls that will cover a certain chessboard ... we could prove *that.* And if you can always fit in the Ls, if there's never a piece left...*that* could be proved by induction."

Linda started writing but Ellen objected. You have to take "the shape of the piece" into account, she said. But Linda did not seem to hear the objection. She continued proving that the number of squares in a defective chessboard is divisible by three:

> "And then you assume that it works if this n equals p," Linda explained to the others. "Then we are to show that if... *if* this holds for p it also holds for p+1."
>
> "But since we assume that it holds for p, are we not supposed to *show* that it holds for p, in some way"?

asked Ellen, but Linda objected that this is an assumption you have to make in order to prove the statement. She wrote down the proof with the help of John and Ellen and when she had written down a formula that held generally, Ellen objected anew:

> "This may be irrelevant, but what we have shown is that you can cover a chessboard with *threes.* They don't look like that."

This time Linda listened to the objection and the three students designed a geometrical solution to the task.

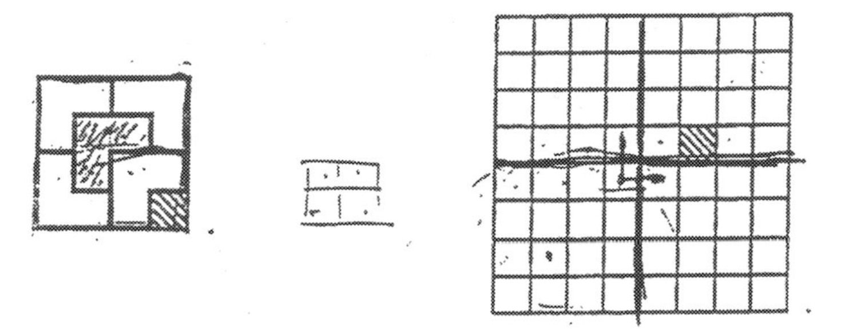

Figure 6.2. A sketch of the geometrical solution.

"The thing is," Linda said, "that in the four times four board you can put the defect wherever you like, since this bigger board is made up of four two times two boards. So if we want the defect placed here, we could always turn it around."

"Exactly," said John. "The shape means that you always have an even number of squares, or an even number of bigger squares, so you will always end up with three left where you can put in an L."

"This is what we are supposed to prove by mathematical induction," said Linda, and the others laughed and shuddered: "Ugh!"

Linda set out again on the algebraic track, where she tried to prove once more that the number of squares in a defective chessboard is divisible by three, this time considering the number of Ls and the number of defects in a given chessboard. John was helping her, but Ellen seemed to have lost interest:

"It would be nice if you participated in this," said Linda to Ellen.

"It would be nice if I understood induction," countered Ellen.

She said that she could understand the geometrical solution graphically, but the rest was lost to her:

"But that's only because you cannot accept that it is a *proof*, "said Linda.

"Exactly," said Ellen. "You cannot accept something…or *I* cannot accept something that I cannot see how it works, for in such a case I cannot logically see which steps that hold."

"And I have difficulty in explaining," said Linda, "since I think it's obvious." She started all over again: "If you think away numbers altogether."

"Hmm," said Ellen (making a sound which in the Swedish language may signify either a "yes" or "go on").

"Apart from this n. I can replace n with any integer."

"Hmm," Ellen said.

"And then you assume that it works," Linda went on. "And then you show that it holds for the successor, one step above. Agreed"?

"Hmm..."

"And *if* this p works, that we insert, if p works, p+1 works."

"But that could be a coincidence," Ellen objected. "I could just have happened to put it in a place that works."

"No, you couldn't," said Linda, "because if that p works, then you can rest assured that everything above p works as well, p+1 works."

"Yes...no, that is not clear to me," said Ellen.

"Yes, because we have proved it," asserted Linda.

"No," Ellen insisted, "that is precisely where I'm hung-up. I don't think that it is a proof. We assume that it holds for a certain p, and then we prove that it holds for p+1 and then we have, in addition, shown that it works for a couple of cases in the beginning. But it could be a coincidence that it just *happens* to hold for these cases. I don't understand why it is a proof, just because we insert a p. This is, like, an elementary thing that I have not understood."

"And I have not understood what you do not understand," concluded Linda.

## The Algebraic Interpretation of the Task

As we can see, Linda interpreted the task as a problem of proving that the number of squares in any defective chessboard is divisible by three, a weaker claim than the assertion that any defective chessboard can be covered by L-pieces. How are we to explain such an interpretation of the task?

In the beginning of the group-discussion, Linda stated that since an L-piece is composed of three squares you could prove that the number of squares in any defective chessboard is divisible by three: "we could prove *that*", and "*that* could be proved by induction" she said, two statements that indicate that Linda regarded such a problem as possible for her (and the group) to handle. Probably she already knew how to design an inductive proof of this kind. From her discussion with Ellen we can conclude that she had formed a sound knowledge of induction. Her understanding seemed, however, limited in scope and applicable only to algebraic examples. As we can see, she did not even hear Ellen's objection, that you have to take the shape of the L-piece into account. In Linda's interpretation of the task, shape was an irrelevant aspect. Later, when the students had completed a geometrical argument for the statement that any defective chessboard could be covered by L-pieces, Linda stated that "*this* is what we are supposed to prove by mathematical induction." Even though Linda had been taking an active part in the construction of the geometrical proof, she still suggested that if you are supposed to prove the statement using induction you would have to start from scratch. Her utterance tells us that she did not consider the geometrical argument as induction.

As mentioned, the task was unconventional in the sense that it did not involve an equality or inequality, examples that the students were familiar with from experiences of inductive proofs. However, Linda, and in fact the students in all six groups seemed to have formed the notion that inductive proofs should be algebraic. The tutors and textbook writers may well have had other intentions in using the examples. Few mathematicians would assert that algebra is a fundamental aspect of inductive proofs (tutors who had met the students in class were in fact surprised to find out that they had formed such a view). Algebraic manipulations are, no doubt, effective mathematical tools, but they are, nonetheless, tools and not essential aspects of proofs. But aspects that, for the professional mathematician, are form and background, may well for the student become content and pattern. If what is learned is integral to how knowledge is acquired and used, it is not remarkable, if teaching excludes other experiences, that students form notions of mathematics as the manipulation of formulae.

## Implicative Reasoning

The results thus far are not surprising. However, more surprising is that some of the students, nevertheless, interpreted the task as a geometrical problem. In all six groups this interpretation was suggested by a student and in five of the groups the students succeeded in formulating a convincing argument, convincing, that is, to most of the students in the group. Ellen, for instance, had a strong preference for the geometrical contextualization of the task. When it was suggested by John, Ellen's interest was instantly caught: "This is good," she said, emphasizing the words, "this is easy, because now you can explain in words what you see and you do not have to involve induction, which I'm not very good at."

Other students in the group tried their hand at constructing an inductive proof, but Ellen was unconvinced. If the statement holds for some $p$, then that might very well be the result of chance: "But it could be a coincidence that it just happens to hold for these cases."

In an almost Socratic dialogue with Ellen, Linda articulated all the elements of an inductive proof. However, it can be argued that what Linda produced is not a proof. According to Rotman (1988), what is needed for something to be a mathematical proof is a leading principle or narrative which transforms "an inert, formally correct string of implications" (Rotman, 1988, pp. 14–15) into a persuasive argument. Such a leading principle is clearly lacking in the string of logical statements that Linda makes: You convince yourself that a statement holds for a certain whole number (where $p = 1$, that is the $2^1 \times 2^1$ board). Then you assume that it holds for an arbitrary number ($2^p \times 2^p$) and then you prove that it holds for the following number ($2^{p+1} \times 2^{p+1}$). But can we just assume that the statement holds for an arbitrary number without having any grounds for our assertion: "...are we not supposed to *show* that it holds for p, in some way"? Ellen asked.

Her question is legitimate. What is the idea of an inductive proof? Inductive proofs are built on a step by step *hypothetical* reasoning: In this case we assume, without having any grounds for our assertion, as Ellen noted, that we can cover an arbitrary (but for the moment fixed) size of chessboard with L-pieces. Then we show, with the help of this assumption, that a chessboard of the next size is also possible to cover. If, following that, we can show that the smallest (2 × 2) board is possible to cover (which we easily can, since we just need one L to do it), everything falls into place. Now we know, thanks to our previous hypothetical reasoning, that the next board (4 × 4) also is possible to cover and the next, and the next, and the next... Up to this moment, however, the hypothesis is suspended. On the one hand we use the assumption *as if* it were true, on the other hand we must not think that it really *is* true, since this is what we have to prove. "It is just an assumption you have to make," as Linda said.

Ellen seemed to have difficulty accepting the hypothetical nature of the proof. One interpretation of her disbelief is that she demanded of a proof that it should provide some kind of lucidity or instant insight. This proof is not lucid in that sense. You can understand the proof and yet be unable to visualize a covering of, say, a 16 × 16-board. In a mathematical perspective this is a merit to inductive proofs, since you can handle complexities swiftly and easily without going into detail. To a student, however, the same characteristic may seem disqualifying, especially if the student strives to understand the fundamentals of the proof.

The advantages of inductive proofs do not present themselves readily to students. In our daily lives we are convinced when we have solid grounds for our assumptions, provided by direct observation or indirectly by authorities. In a common sense view, knowledge is absolute. If we assume that something holds for *p*, "are we not supposed to show that it holds for *p* in some way," said Ellen. In what way? Could the statement hold for *p* regardless of any assumptions? Such an absolute view of knowledge is common among students entering higher education (Perry, 1970), which makes it difficult for them to understand all the if–then assertions characteristic of academic ways of reasoning.

Ellen, however, insisted that her position was legitimate, and in a contructivist perspective it certainly was: "You cannot accept something...or *I* cannot accept something that I cannot see how it works," she said. The rephrasing of the utterance is interesting. Maybe some students can accept mathematical methods without reflecting upon their foundations. Ellen herself could not. She suspected that she has missed some "elementary thing," and until the logic became clear to her she was not willing to accept induction as a method of proof.

## Conflicting Views of Induction

One might argue that the geometrical interpretation of the task was suggested by some of the students because that was the required problem, that is, the problem that the authors of the text had in mind. Doesn't the text say so? (*Problem:* Using induction, show that it is always possible to cover a defective $2^n \times 2^n$ chessboard..., see Figure 6.1) Such an argument, however, builds on the assumption that the task "contains" a problem, ready-made for the students to solve. But the task is nothing but a text, and as a text it has to be interpreted, and as we have seen, this text, obvious as it may seem to a mathematically oriented reader, is not unambiguous: The word "induction" signals to the students an algebraic interpretation, while the shape of the figures involved signals geometry.

In the group-discussion above the conflict was resolved in the interpretative process. Linda interpreted the task in a way that fit her understanding of induction and, in a broader sense, her understanding of mathematics as a subject where you manipulate formulae, which means that she interpreted the task as a problem of proving divisibility by three. Ellen interpreted the task as a problem of proving that you can place L-shaped pieces in an arbitrary defective chessboard, which means that she gave up the prerequisite to use induction, since she did not feel comfortable with it. In other groups, however, the students found other ways to circumvent the conflict.

## Group B

In one of the groups in Göteborg, five students, Andy, Irene, Jacob, Louise and Robert cooperated in solving the task. They started out by updating what they knew about induction, and quickly moved on to the drawings made in the text:

> "The problem is to prove that it works, regardless of where you place the defect," said Jacob.

"You could always prove that the number of squares is divisible by three," said Louise.

"Yes, that's an easy thing," said Jacob.

"But look here," said Andy and pointed to the 2 × 2 board. "Solving this one is trivial," he added and giggled.

The others agreed cheerfully. That would be an easy task.

"So let's take this one and put it here," said Andy and pointed to the 4 × 4-board. "We can take this one..."

"No, wait a minute," said Louise. "We are not supposed to have more than one defect."

"No. You're right," agreed Andy. "They would be too many."

"That would give us four black squares," said Irene. "But on the other hand we would have three left."

"What? Three left"? Louise asked.

"Those three," said Irene and pointed to the drawing.

"And you can place those anywhere you like," said Andy.

"And then you can put an L in to cover them," concluded Irene.

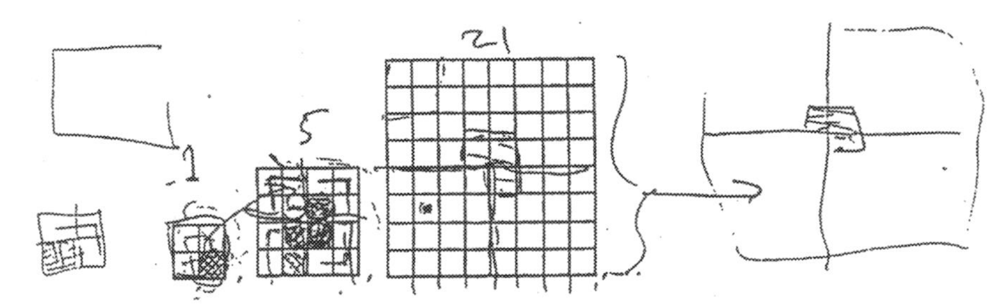


Figure 6.3. The geometrical sketches made by the students.

"And then we are supposed to describe this mathematically," said Louise and giggled.

"This is not a proof," said Irene, "this is just a way of showing how it will turn out."

"Yes, now we have done it logically," said Louise.

"Now we know that it holds," said Irene. "Now we believe in it."

The students tried to formulate a proof in algebraic terms. There are four possibilities to place the L on the smallest board. Could that be helpful?

"In order to cover this board ($2 \times 2$) you need one L, but there ($4 \times 4$) you need five, whatever that means," said Andy..

"But I do not understand how we are to describe where on the board the defect should be placed," said Irene.

"No," said Louise, "but if we do it like this..."

"Is that mathematical induction"? asked Andy.

"What the heck...induction," said Louise with a laugh.

" I think we have solved it now," said Andy, "By induction, almost."

"Yes, okay," Louise agreed.

"But the thing is," Irene objected, "that induction is to prove something..."

"...for k," Andy filled in.

"...and for the number of x," continued Irene. "We have just tested."

Jacob tried to formulate in words what the students had discovered:

"Let's say that one out of four squares is covered. On the other three squares we place the holes so that they are available to all three. It's as simple as that. One is covered by the defect and on the other three we can choose where to place the hole. And then we see to it that the holes are places so that they form..."

"...an L-shape," Andy filled in.

"Yes. But the problem is how to write it down mathematically. That is incomprehensible. It is almost impossible, "Jacob said.

"Yes, but as we said: this is, after all, induction. Really," Louise stated.

"It is induction, although not mathematical, said Andy and laughed.

"But it doesn't say mathematical in the text," said Louise. "It doesn't."

"No, you're right," giggled Andy.

"But it is induction, "Louise insisted. "We have shown that you can put it there."

"We have reasoned our way through the problem," finished Irene.

## The Preference for Algebraic Interpretations

To reason logically, to convince yourself that something holds generally, to argue in words for an assertion is not to prove it mathematically, seemed to be the conviction of all students in the group. To design a mathematical (that is algebraic) argument seemed, on the other hand, an impossible task. The conflict was resolved in a rather frivolous way. Louise found a loop-hole in the text. The author had neglected to write down the word "mathematical," which gave some latitude for the students to reinterpret the assignment.

From a mathematical perspective you could argue that the students had not understood induction. Mathematicians often talk about "understanding," referring to a generalized comprehension. What we understand are definitions, which work much like legal texts: everything that is not prohibited is allowed. This means, that if it is not explicitly stated that inductive proofs must be algebraic, such a restriction does not exist, even if *all* examples used in teaching are algebraic. Examples are supposed to aid the learner in developing a general understanding of a concept, they are not to be taken as constitutive.

The students, on the other hand, form their views of mathematics not only from what is said (and not said) in teaching, but also from how the subject matter is communicated and used. If tutors spend a lot of time calculating (since tutors often know that calculations can be stumbling blocks to students and hence call for a lot of practicing) the students may well come to view calculation as an essential characteristic of the subject, since so much time and effort is spent on manipulating formulae. If inductive proofs occur frequently (or solely) in algebraic situations, the students may easily form a view of induction as algebraic in essence.

The students in Group B, however, seemed to have understood the prescriptive nature of mathematics and they exploited the rules of the genre to pursue their own intentions. If it was not clearly stated in the text, that *mathematical* induction was to be used, you could feel free to chose any form of "induction," in this case "induction, almost."

Of course, you could always check later if your variant of "induction" would be acceptable to professionals. In one of the groups in Stockholm, three boys and a girl had worked out a geometrical proof which they did not only find convincing but also "darned neat," even "beautiful." But was it mathematical? When the observer arrived to collect papers and tape-recorders the students reported the results of their work:

## Group C

> "We have not found any workable formula," said one of the boys, "but we have a graphical proof."

The observer looked at the paper and the student explained the proof:

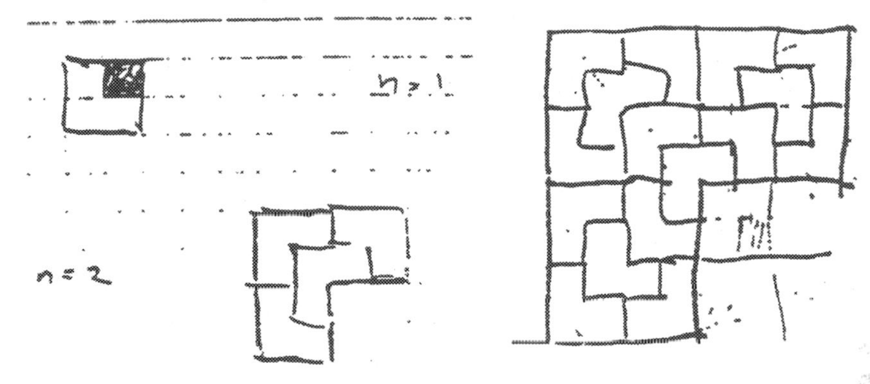


Figure 6.4. A sketch of the graphical proof.

> "...and the L that is left is similar to that L. You could always construct such Ls, until you reach the smallest square where you have only one L left."
>
> "Yes," said the observer, "but how do you know that you can always construct such Ls"?
>
> "Well," said the student, "we convinced ourselves graphically. Since you can construct one that is 4 × 4 and remove the L, and since you can do it on the 8 × 8-board too, which consists of 4 such figures, and on the 16 × 16-board, which consists of four 8 × 8-boards. And then it goes on."
>
> "Yes," said the observer. "May I keep this paper"?
>
> "Certainly," said the student.
>
> "Should we develop it a bit further in some way"? asked another student cautiously.
>
> "Well, no, you don't have to, since it seems to be a correct solution," said the observer.
>
> "Ah," said the student with a little laugh that sounded like a sigh of relief, "nice."

Arguably, the observer is taking liberties in accepting what the students have done as a proof by induction. What can be seen in the students' drawing, and in their explanations of it, is a way of viewing a defective chess board as consisting of four smaller ones, three of which have been made defective by the trick of removing an L-piece in the middle as indicated in the drawing. The trick is what the observer is looking for: if you know how to construct proofs by induction this is all you need in order to make it work in this particular problem. What the observer fails to consider is the very real possibility that the students do not know how to construct proofs by induction.

## UNDERSTANDING INDUCTION IN A COOPERATIVE SETTING

As mentioned in the introduction the assumption that cooperative work may enhance the students' understanding had formed the basis for the educational programs within which we collected our data, and the group sessions we have studied clearly had qualities of that kind. The students enquired into the specifics of mathematical induction, they discussed, debated and negotiated the meaning of the concept, and yet we found that the students did not always manage to enhance their understanding. In this section we will discuss these results in relation to the various ways in which the students interpreted the task.

### A Short Characteristic of the Group Discussions

In the narratives above, we have given accounts of ways in which rationality takes form in a social setting. In Group A we find an example of an almost classical argument, adversarial in style, yet not rival. Ellen and Linda both seek ways of countering each others" arguments by finding errors in thinking and reasoning, but the counter arguments are hedged in by reservations ("This may be irrelevant, but..." or "It is not clear to *me*") which emphasize the interpretative nature of the remarks. The arguments are refuted, but the knowing subject is respected. In Group B we find examples of a rationality that works collaboratively. An utterance is tentatively formed by one student, reformulated or filled in by another and completed by a third student.

To reformulate an utterance is, of course, indirectly to criticize the argument for being incomplete. Both the adversarial and the collaborative attitudes, often juxtaposed in discussions about fruitful ways of communicating among peers (see Coates, 1993), both entail a critical stance, an attitude which lies at the heart of mathematical problem solving and which is not necessarily linked to hostile attitudes.

Or is it? There is one side to mathematical problem solving that may encourage less considerate and thoughtful ways of socializing. The playful quality of the subject, the imaginary aspects, the "as if" character of the game, may get hold of the student. The play takes over. If you are focused on a problem, remarks that fall outside of your own contextualization of the task may never catch your attention. In the heat of the play you may easily trample upon playmates, for instance by ignoring their statements, as Linda did.

In a group discussion, however, such mishaps may be corrected. Miscommunication can in part be viewed as clashes of communicative styles. Different communicative styles have, however, different functions. In a setting requiring argumentation, a collaborative, supportive style may lead to submission, as it did in the group that never solved the assignment; the students were so preoccupied by agreeing with each other that they missed the chances of elaborating the meaning of the utterances put forward in the discussion (Wistedt, Brattström & Martinsson, 1996). An adversarial style may lead to negligence of fruitful arguments, as in the case of Linda in Group A above. Collaboration is as necessary to problem solving as challenging and refuting, and hence creating arguments calls for a broad communicative repertoire. Most of the students in our study make use of both styles of communication.

The communication described in the narratives above comes very close to the ideal of cooperative learning. This makes the examples ideal cases for investigating the relationship between the styles of work and learning outcomes, in terms of how the students understand the content of the task. Let us, therefore, turn to the students' contextualisations of the assignment and the concept of mathematical induction.

## The Cognitive Context

As mentioned above, mathematical induction relies on a step-wise hypothetical reasoning. In fact it relies on an axiom, the axiom of induction, which states the following:

> Suppose you have a set of natural numbers M with the following two properties:
>
> $$(1)\, 1 \in M \text{(1 belongs to the set M)}$$
>
> $$\forall n (n \in M \Rightarrow n + 1 \in M)$$
>
> (It holds for all $n$ that if a number $n$ belongs to the set M this implies that the successor $(n + 1)$ also belongs to M).
>
> Then M must be the set of all natural numbers N.

In our case, M is the subset of N, consisting of natural numbers $n$ such that the following statement P($n$) is true:

> P($n$): It is possible to cover any defective $2^n \times 2^n$ chessboards by L shaped pieces regardless of the position of the defect.

The set M consists precisely of those natural numbers $n$ for which P($n$) is true. What we want to do is to prove that P($n$) is true for all $n$, i.e. that M = N. In order to be able to use the axiom of induction to this end we need to prove that M satisfies conditions (1) and (2) above. Proving the first part of the conjunction ($1 \in M$) seemed to be an easy task for the students. The smallest board $2^1 \times 2^1$ (where $n = 1$) could easily be investigated and the students found that P(1) must be true (which means that 1 belongs to the set M, $1 \in M$). The second part of the conjunction, however, consists of an implication, and this seemed to be a major stumbling-block for the students. What does it take to prove an implication $A \Rightarrow B$? The statement tells you that *if* A is true *then* B is true, which means that we do not have to investigate the cases where A is false. The only case where the *implication* is false is the case where A is true and B false. In all other cases the implication is true.

In order to prove the second part of the conjunction ($\forall n(n \in M \Rightarrow n+1 \in M)$) we choose an arbitrary $n$. But how are we to choose such an $n$? This is what the students asked themselves in a variety of phrasings, and in posing such questions they told us that they were not trying to prove the *implication* but single statements about $n$. As mentioned above there is no direct method which proves $P(n)$ for a general $n$. Even if we tested our statement for a long chain of successive $n$'s we would still not have proven the assertion ("We have just tested it;" "This is not a proof. This is just a way of showing how it will turn out"). We may of course feel convinced ("Now we believe in it;" "We have reasoned our way through the problem"), in the same way that we are convinced that the sun will "rise" tomorrow since it has "risen" each day thus far. The sad thing is that we cannot be certain. Empirical induction is not a valid method of proof, and the students also formulated their doubts about such ways of reasoning ("...it could be a coincidence that it just *happens* to hold...," "How do you know that that holds for an infinite number of squares...?"). When using mathematical induction, however, we do not have to pose such questions. We just have to prove that *if* $P(n)$ holds *then* $P(n+1)$ holds. *If* an arbitrary defective $2^n \times 2^n$ chessboard is possible to cover, *then* the chessboard next in size, $2^{n+1} \times 2^{n+1}$, must also be possible to cover, since it consists of four defective $2^n \times 2^n$ chessboards (one defective and the other three made defective, where the virtual defects can be covered by an L). This is all we need.

The students did not pose their questions within a context where the axiom of induction was accepted as true. An "elementary thing" was missing as Ellen stated: the very foundation of inductive reasoning, the logic on which the proof rests. Deprived of such an understanding the students were at a loss when trying to grasp the point of the method of proof. Their understanding seemed to be built on examples introduced in teaching, and from such examples they had formed notions about the general characteristics of induction, i.e., that it had to do with algebra and to produce proofs according to certain prescribed conditions. However, it seems they had not been helped to make sense of what the axiom says outside the context furnished by the examples used in previous teaching sessions.

## The Situational Context

In most study situations, an overriding intention for the students is to fulfil obligations, to work through assignments, and to complete tasks in ways acceptable to their tutors. Such pragmatic ways of approaching the task can also be found within the groups in our study. When the students could not find ways to solve the problem they could at least find ways to solve the task.

Equally, tutors often take upon themselves to promote learning which is in accordance with locally approved knowledge, and to assess student activities in relation to institutionally accepted norms. Viewed in a constructivist perspective, however, knowledge formation is not a mere acquisition of habits and skills that are in accordance with culturally accepted rules. In such a perspective we have to consider the *continuity of experience* which means that we have to view learning in the broader context of knowledge and habits that the students have already formed and which constitute the basis for further development.

Not all experiences are, however, equally educative. In our study we found that the students all seemed to have formed ideas about mathematics that shut off their opportunities for mathematically relevant learning. The students were hooked on algebra, a way of reasoning which they seemed to identify with mathematical argument.

Algebraic reasoning is, no doubt, a forceful tool in mathematics. In such reasoning we have ways of signifying arbitrary numbers by introducing symbols which can be manipulated according to algebraic laws. Such concise means for signifying generalities make it possible to prove assertions about *all* numbers. Within the realm of geometry, however, we do not have such possibilities. We cannot denote an arbitrary $2^n \times 2^n$ board by a symbol and by some sort of algorithmic procedure assure ourselves that it is possible to cover. We have to picture a specific board, with specific dimensions and with the defect positioned in a specific place when we reason. How is it possible to prove that an arbitrary board can be covered from such a point of departure? This is the problem facing the students and the question raised above is a question the students asked themselves. How could it be done? How were they to know that they had considered every possible arrangement, every thinkable placement of the defect on every conceivable size of chessboard?

*Induction* was suggested as a method to handle the problem. The students, however, did not seem to find induction applicable. They knew how to formulate an inductive proof in algebraic terms, but they did not appreciate induction *as a method of proof*. A result, which stands out in the narratives presented above is the wide-spread sceptical attitude that the students hold against geometrical reasoning as a valid form of argument within a mathematical genre. In all of the groups the students found the geometrical proof convincing. But, when they wanted to assure themselves of the generality of the argument, they seemed to lack the necessary measures. This means that the students had no ways to negotiate the foundations of their beliefs; they could not, as Ellen said, assure themselves of the logic. Ellen was not willing to give approval to such knowledge. She needed to see how the logic worked before she could accept the argument as a proof.

Such a profound attitude is a hall-mark of mathematical reasoning, an attitude that most mathematics tutors would like to encourage in their students. The same attitude was exemplified by the students in group B, who refused to call an argument where you reason your way through the problem, a mathematical proof. The students posed relevant mathematical questions. They did not, however, find any answers to them within the group discussions.

## The Cultural Context

In order to be able to negotiate the meaning of induction, the students would have to broaden their conceptions of mathematics. One way of doing so is to pose their questions to more informed members of the mathematical community. On one occasion such a communication takes place. When the observer arrived to collect papers and tape-recorders the students took the opportunity to raise their questions. One of them cautiously asked: "should we develop it (the proof) a bit further in some way?"—an utterance that can be interpreted as an invitation to discuss the specifics of the solution.

The observer was, for the moment, not acting as a tutor, but we can use this instance to discuss what may happen when novices meet professionals and what such communication might mean to the students. The observer, who had a profound knowledge of induction, could easily judge how the method could be applied to the solution that the students presented. The sketch and the explanations that the students gave were interpreted from such an informed perspective and the answer "it seems to be a correct solution," summarized the outcome of such an interpretation. The perspective was, however, not made available to the students, who only had access to the verdict. The "tutor's" comment was not related to the students' interpretations of the task, and they had but to accept the conclusion at face-value.

Problems similar to the ones that our students encountered have been found in other studies (e.g., Halldén, 1988b) and with similar consequences for the process of learning. If the point of the instruction is lost for the students, they can only attend to single elements given in teaching:

examples, algorithms or rules that apply to local cases. In order to create contexts that link such elements to each other, and to a main line of reasoning, the students have to take more embracing conceptions into account. These include the beliefs that determine what kind of questions can be asked within the realm of the subject, what are acceptable means of answering such questions, and what constitutes the knowledge so constructed within the subject domain.

## THE MERITS AND LIMITATIONS OF CLASSROOM COMMUNICATION

The group work did not help the students to challenge their views of induction as algebraic in nature, and neither did it help them to establish the framework of the axiom of induction. Since all of the students shared the same limited notions of mathematics generally and inductive proofs specifically, there was no incitement in the groups to broaden their views.

The divisibility problem was, however, discarded in five of the six groups. In all of these groups the members objected that "We must consider the L-shape of the piece," or "This may be irrelevant, but what we have shown is that you can cover a chessboard with threes. They don't look like that." These are remarks which opened the interpretations of the task to critical reflection. In such cases the collaboration between the students served as a challenge to limited, taken-for-granted notions harbored by individual students.

The cases of classroom communication we have studied certainly created favorable learning conditions for the students: The atmosphere was cheerful and easy-going, the students willingly shared their views with each other and openly revealed their limited knowledge of the subject matter. They even took the liberty of approaching the mathematical discourse in a playful way, for instance when the word "trivial" popped up in the discussion in Group B, to the students' common amusement. The mathematically oriented reader will recognize this word as an expression often used, and sometimes misused, in mathematical conversation.

Cooperation between peers clearly has its merits. But limitations exist as well. An instructional problem arises when we introduce examples and concepts in teaching which are intended to be interpreted within a specific theoretical framework, while the students interpret them within personal frameworks based on their previous encounters with the content at hand. In order to know what aspects are of relevance in a particular context and which are not, the students must be acquainted with the theoretical context, in our case the idea of mathematical induction. In a situated perspective, such meaning-making is not primarily viewed as dependant on cognitive structure, but as something derived from the situation and the student's involvement in that situation, i.e., as a process of participating in cultural practices. However, the mere participation does not in itself guarantee that all students will understand the cultural practice in the same way, and certainly not that the students will understand it in the way their teacher intended. In a constructivist perspective, the meaningfulness of a situation might be different for different students, leading to different contextualization, and, hence, to different kinds of actions. In a constructivist perspective, we need to take such differences into account, to consider the various ways in which students may understand what constitutes knowledge within a particular disciplinary domain.

Above, we pointed out that such constitutive knowledge rests on a meta-theoretical level. Since this level determines the relevance of the information provided in the assignment, the students could not find their ways out of the dilemma facing them: either to prove a weak assertion (divisibility by three) by methods with which they were familiar and appreciated, or to prove a stronger assertion which they all eventually believed to be the interpretation intended by the constructor of the task (the possibility to cover a defective $2^n \times 2^n$ chessboard by L-pieces for all $n \geq 1$) by a method they did not regard as mathematical. When we scrutinized the students' formulations as they summarized their solutions to the task, we found that they had access to all the elements of an inductive proof: They knew that the $2^{n+1} \times 2^{n+1}$ board consisted of four boards $2^n \times 2^n$, all "defective" ("Let's say that one out of four squares is covered. On the other three squares we place the holes so that they are available to all three. It's as simple as that. One is covered by the defect and on the other three we can choose where to place the hole."). The consequence, however, was not linked to the antecedent—the if-then assertion on which the proof rests was missing in all the utterances stating the results of the students' investigations, and without this prerequisite, they were at a loss when trying to grasp the point of inductive proofs. Instead they contextualized the task within a different theoretical context—empirical induction. They investigated a number of defective chessboards. Even though, in several of the recorded discussions, this led them to the discovery of a general method to solve the problem for any given chessboard, they did not link their method to the abstract if-then structure of inductive proofs. Hence, they were unsure whether their arguments would pass for induction. They had to rely on authorities to check their ways of reasoning or they took the easy way out by reformulating the task as not requiring *mathematical* induction.

Meta-theoretical questions are rarely addressed in teaching. Often we regard them as too difficult for the students to grasp. By neglecting them, however, or by simplifying the issues, for instance by transferring the topics studied to a more concrete or applied level, the problem of contextualization will not be resolved. Since the meta-level is crucial for the understanding of what is communicated in a classroom, the students must have at least some access to the assumptions that render meaning to the information given. As we have seen, accessing meta-level information is hard for the students to manage by themselves. The students need to interact with more experienced members of the mathematical culture in order to overcome the limitations in their personal views of the subject.

This means that we have to broaden our view of "classroom communicatio" to include not only communication of the subject matter, but of metatheoretical issues as well. The students in our study were all newcomers to the mathematics departments. They had entered a new culture, and they brought with them experiences from previous learning. Some of their views of mathematics, for instance the view that mathematics has to do with manipulating formulae, were less productive in this new environment. Thus, even if we try to create favorable learning conditions, the students' previous experience may mean that these conditions are not fully utilized. Perhaps what is needed is to challenge the students' understanding of how knowledge is constructed and how meaning is created within a mathematical domain, and to establish the means by which varied understandings can be examined in order to make mathematical sense.

Such challenges are particularly important if we view our results from an inclusive perspective. As mentioned in the introduction, the developments initiated by the Swedish National Agency for Higher Education did not only have the goal to enhance student learning, but also to ensure equal access to educational programs. Female students are in the minority within most university courses in mathematics. This may make them particularly vulnerable to taken-for-granted notions, never spelled out assumptions, or "elementary things" that they suspect that they may have missed. Many of us share this experience of having learned in taken-for-granted contexts where the presuppositions for the reasoning were hidden from us (Wistedt, 1994a, 1994b). Students who are self-reliant may easily overlook such gaps in their prerequisite knowledge. They may feel comfortable anyway, trusting in the promises that all will eventually become clear. But students who belong to minority groups, or students who are less familiar with the cognitive practices of mathematics may feel less confident if they are left alone to figure out the fundamentals on their own. To refer to matters which "go without saying" or to state that a solution to a task is correct or incorrect without giving any ground for the verdict, may effectively exclude students who are unaware of the cultural norms, even unaware of the fact that such norms exist (Bergqvist & Säljö, 1994; Halldén, 1990; Wistedt, 1998). Research has shown that difficulties in discovering and utilizing taken-for-granted communicative tools co-varies with achievement level (Säljö & Wyndhamn, 1988, 1990). Students who are regarded as "low achievers," are often found to have problems deciphering information of a meta-theoretical kind. Cooperative work styles which stimulate communication between students and between students and tutors have one fundamental merit viewed from an inclusive perspective—such styles of work offer a variation in experiences and ways of viewing mathematics which may help the students to become aware of learning as a culturally related phenomenon.

## CONCLUSIONS

We have presented a case-study of how university students understand and utilize the concept of mathematical induction in a cooperative setting. We have taken into account the socio-cultural critique of the lack of attention to the discursive nature of knowledge acquisition. Socio-cultural theory with its focus on "situated learning" has, no doubt, provided insightful analyses of institutionalized learning, but socio-cultural researchers have yet to show how they can account for the continuity of learning, that is, how individual students enhance their personal understanding of the subject across settings and contents. Socio-cultural theory posits an inexplicable internalization process as the primary learning mechanism, which puts restrictions on the possibilities to view learners as active reorganizers of their personal mathematical experience. In order to take the discursive nature of knowledge formation into account, we have suggested "intentional analysis" as a method to uncover not only the students' understanding of the concepts brought to the fore in a learning session, but also their understanding of the setting itself, and of the cultural norms and conventions as they understand them. We have described how the students' previous encounters with mathematics have led them to believe that mathematical proofs are algebraic in essence. Our results show how such beliefs conflict with the nature of the task given to the students which calls for a geometrical application of inductive proofs. Although the communication between the students could be characterized as almost ideal, the students found no ways to contextualize the task under the conflicting premises of interpretation. We have argued that the solution to this problem rests on a meta-theoretical level, in our study exemplified by the students' problems to understand the implicative reasoning characteristic of mathematical arguments, and the problem of understanding the role of examples in mathematics instruction. We have also argued that this level is not readily accessible to the students as newcomers to mathematics departments. Since all of the students in our study shared the same limited views of mathematics generally, and inductive proofs specifically, they were at a loss when trying to resolve the conflict facing them. Thus, we have argued for a broadening of the view of classroom communication, which does not only include communication of the subject matter, but of matters concerning the nature of mathematical reasoning as well. To involve students and tutors in communication about the process of meaning-making in mathematics would benefit all students, particularly students who may feel less comfortable with mathematical ways of reasoning.

## REFERENCES

Bakhtin, M. M. (1981). *The dialogic imagination. Four essays by M .M. Bakhtin.* Austin, TX: University of Texas Press.

Bergqvist, K., & Säljö, R. (1994). Conceptually blindfolded in the optics laboratory: Dilemmas of inductive learning. *European Journal of Psychology of Education, 9,* 149–158.

Bhatia, V. (1993). *Analysing genre. Language use in professional settings.* London and New York: Longman.

Brown, J. S., Collins, A., & Duguid, P. (1989). Situated cognition and the culture of learning. *Educational Researcher,* 32–42.

Brown, G., & Yule, G. (1983). *Discourse analysis.* Cambridge, UK: Cambridge University Press.

Burton, L. (1995). Moving towards a feminist epistemology of mathematics. *Educational Studies in Mathematics, 28*(3) (Special Issue on Gender, edited by G. Leder), 275–291.

Caravita, S., & Halldén, O. (1994). Re-framing the problem of conceptual change. *Learning and Instruction, 4,* 89–111.

Coates, J. (1993). *Women, men and language.* (2nd ed.). London and New York: Longman.

Cobb, P. (1990). Multiple perspectives. In L. P. Steffe, & T. Wood (Eds.), *Transforming children's mathematics education.* (pp. 200–215). Hillsdale, NJ: Erlbaum.

Downes, W. (1984). *Language and society.* London: Fontana Linguistics.

Drew, P., & Heritage, J. (1992). *Talk at work. Interaction in institutional settings.* Cambridge, UK: Cambridge University Press.

Driver, R., & Easley, J. (1978). Pupils and paradigms. A review of literature related to concept development in adolescent science students. *Studies in Science Education, 5,* 61–84.

Garfinkel, H. (1967). *Studies in Ethnomethodology.* Englewood Cliffs, NJ: Prentice Hall.

Geertz, C. (1973). *The interpretation of cultures: Selected essays by Clifford Geertz.* New York: Basic Books.

Goffman, E. (1959). *The presentation of self in everyday life.* Garden City, NY: Doubleday.

Halldén, O. (1988a). Alternative frameworks and the concept of task. Cognitive constraints in pupils' interpretations of tutors' assignments. *Scandinavian Journal of Educational Research, 32,* 123–140.

Halldén, O. (1988b). The evolution of the species: pupils perspectives and school perspective. *International Journal of Science Education, 5*(10), 541–552.

Halldén, O. (1990). Questions asked in common sense contexts and in scientific contexts. In P. L. Lijnse, P. Licht, W. De Vos, & A. J. Waarlo (Eds.), *Relating macroscopic phenomena to microscopic particles.* (pp. 119–130). Utrecht: Cdβ Press.

Halldén, O. (1994). *Conceptual Change and Contextualization.* Invited address to the Symposium on Conceptual Change, Jena, Sept. 1–3, 1994.

Halldén, O. (1999). Conceptual change and contextualization. In W. Schnotz, S. Vosniadou, & M. Carretero (Eds.), *New perspectives on conceptual change.* (pp. 53–65). Amsterdam: Pergamon.

Hawkesworth, M. (1996). Knowers, knowing, known: Feminist theory and claims of truth. In B. Laslett, S. G. Kohlstedt, H. Longino, & E. Hammonds (Eds.), *Gender and scientific and authority.* (pp. 75–99). Chicago and London: University of Chicago Press.

Inhelder, B., & Piaget, J. (1958). *The growth of logical thinking from childhood to adolescence. An essay on the construction of formal operational structures.* New York: Basic Books.

Marton, F., & Booth, S. (1996). *Learning and awareness.* Hillsdale, NJ: Erlbaum.

Marton, F., Hounsell, D., & Entwistle, N. (1997). (Eds.), *The Experience of Learning,* (2nd ed.). Edinburgh: Scottish Academic Press.

Perry, W.G. (1970). *Forms of intellectual and ethical development in the college years. A scheme.* New York: Holt, Rinehart & Winston.

Pfundt, H., & Duit, R. (1991). *Bibliography. Students' alternative frameworks and science education,* (3rd ed.). Kiel: Institut für die Pädagogik der Naturwissenschaften an der Universität der Kiel.

Posner, G., Strike, K., Hewson, P, & Gertzog, W. (1982). Accommodation of a scientific conception: Towards a theory of conceptual change. *Science Education 66,* 211–227.

Resnick, L. (1989). Introduction. In L. Resnick (Ed.), *Knowing, learning, and instruction.* (pp. 1–24). Hillsdale, NJ: Erlbaum.

Searle, J. (1969). *Speech acts. An essay in the philosophy of language.* Cambridge, UK: Cambridge University Press.

Strike, K., & Posner, G. (1992). A revisionist theory of conceptual change. In R. A. Duschl, & R. J. Hamilton (Eds.), *Philosophy of science, cognitive psychology and educational theory and practice.* (pp. 147–176). Albany, NY: State University of New York.

Säljö, R. (1994). Minding action. Conceiving of the world versus participating in cultural practices. *Nordisk Pedagogik, 14,* 71–80.

Säljö, R. (1999). Learning as the use of tools. A socio-cultural perspective on the human-technology link. In K. Littleton & P. Light (Eds.), *Learning with computers. Analysing productive interactions.* (pp. 144–161). London, UK: Routledge.

Säljö, R., & Wyndhamn, J. (1988). Cognitive operations and educational framing of tasks. School as a context for arithmetic thought. *Scandinavian Journal of Educational Research, 32,* 61–71.

Säljö, R., & Wyndhamn, J. (1990). Problem solving, academic performance, and situated reasoning. A study of joint cognitive activity in the formal setting. *British Journal of Educational Psychology, 60,* 245–254.

Wertsch, J. (1995). Action in sociocultural research. In J. Wertsch, P. Del Río, & A. Alvarez (Eds.), *Sociocultural studies of mind.* (pp. 56–74). Cambridge, UK: Cambridge University Press.

Wistedt, I. (1994a). Everyday common sense and school mathematics. *European Journal of Psychology of Education, 1,* 139–147.

Wistedt, I. (1994b) Reflection, communication, and learning mathematics—A case study. *Learning and Instruction, 4,* 123–138.

Wistedt, I. (1996). *Gender-inclusive higher education in mathematics, physics and technology. Five Swedish development projects.* Stockholm: National Agency of Higher Education.

Wistedt, I. (1998). Assessing student learning in gender inclusive tertiary mathematics and physics education. *Evaluation and Program Planning, 21,* 143–153.

von Wright, G. H. (1971). *Explanation and understanding.* Ithaca, NY: Cornell University Press.

CHAPTER 7

# CONFLICTS AND HARMONIES AMONG DIFFERENT ASPECTS OF MATHEMATICAL ACTIVITY

**Dave Hewitt**
***University of Birmingham, UK***

## ABSTRACT

Communication is a complex phenomenon, with many social, emotional and conceptual dynamics. In this chapter, I develop the notion of a "Zone of Opportunities" (ZoO) within which communication takes place, and where the placement of attention of particular individuals can affect communication and the development of meaning. While acknowledging that cognitive, social and emotional aspects are involved at all times within mathematical activity, I consider each of these aspects separately from the perspective of a student's ZoO. I finish by analyzing times when students can experience a conflict between these aspects, and times when students can experience harmony between them, with each aspect supporting the students' learning of mathematics.

*Challenging Perspectives on Mathematics Classroom Communication*, pages 205–233

## COMMUNICATION

Many of my student teachers begin their course by feeling that what is important in teaching is that they "explain clearly." If they can do this, then all will be well and learners will understand. This is based upon an image of teaching where the teacher *gives* information ("clearly") and the student *receives* that information and so learns, represented perhaps by an arrow going from the teacher to a student. As a result such an image does not indicate the complexity that exists for a student where the teacher is not the only person in the classroom, and where there are various activities in which a student is involved which run parallel with any desire to learn from a teacher. These activities include learning about friendship and relationships with peers; the rules of the culture of the classroom, the school and society; learning about love and hate; learning about the cars which pass on the road, or the trees which grow in the field outside the window. It requires choice on the part of a student to attend to what a teacher is saying. This attention is not automatic, as anyone who has taught is fully aware. Inattention is not necessarily due to a student being deliberately awkward, but can be a consequence of there being many aspects of development in which a student is involved throughout their time in school. Their learning is not only concerned with the school's curriculum, but is also involved with, for example, the social challenges of learning about relationships with others. These are inevitably going to come into play when a student is singled out during a lesson to speak out loud, or told whether they are "right" or "wrong," etc. Helle Alrø and Ole Skovsmose (1998, p. 50) talked about different intentions which students have and identified what they call *underground intentions*:

> A student can have intentions about avoiding being noticed by the teacher, about sitting next to somebody, about joining the game in the next break, etc. Here we will talk about *underground intentions* which refer to the students' "zooming out" of the official classroom activity. [...] Apparently, the students are calculating studiously, but this is only when the teacher is in their close vicinity. When the students find that they can once again make themselves invisible in the abyss of the classroom, their activities change [their emphasis].

Even if a student is attending to what a teacher is saying, this does not mean that communication is successfully taking place. A listener has to be active when listening, since listening is not a passive activity. Communication, for example through speech, is not just about the receiving of sound waves, but is about the active participation by the listener to attend to, and work with, those sound waves to develop meaning. The same applies to written communication. Recall the number of times you have "read" a page of a book only to become aware of the fact that you do not know what was

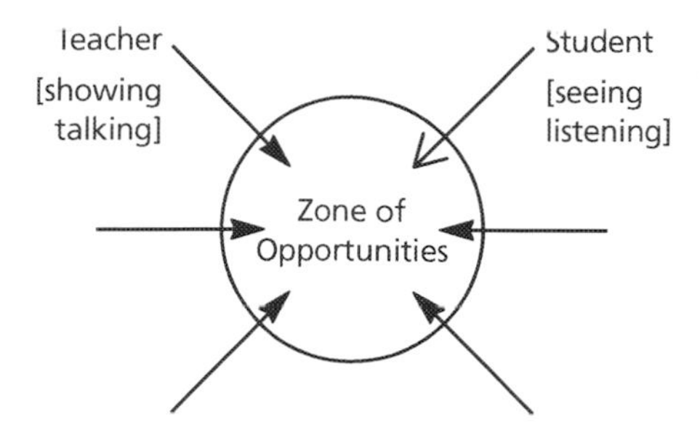


Figure 7.1. Dynamics involved in communication through the Zone of Opportunities when a teacher is making an input. Arrows indicate the giving of attention from all parties for communication to take place.

said on that page. Certain automatisms had been operating for your eyes to follow each line and you may even have mentally read each word on the page. However, if you did not attend to, and work with, those words, you would not have developed meaning for what was written on the page. You may even be doing this now!

Communication involves active participation on behalf of both the speaker and listener; writer and reader; etc. It is for this reason that the image I offer for communication has arrows coming from all participants in the communication process (see Figure 7.1). These arrows do not indicate physical entities such as sound waves traveling from one person to another, but the human attribute of attention which is required from all parties if communication is to take place. Thus, although sound waves may come from the voice of one person, for example a teacher, and go to the ears of another, say a student, the arrows do not reflect this, since they represent the student giving attention to those sound waves and giving meaning to them, as well as the teacher's active attention to speaking those words. Hence no arrow goes directly from one person to another. David Pimm (1995) talked about this idea of something coming *from* a person and giving meaning, in this case mathematical, to something which is perceived through a sense:

> Euclid, besides being a Greek geometer, also developed a theory of vision, according to which light rays emanated from the eye, striking the object, and that is how we were able to see them. Seeing involves projection [...] it is the same active sense that we *see* mathematics in the world, namely that we project mathematical forms onto it (p. 33, his emphasis).

Consider Figures 7.2(a) and 7.2(b) and I invite you to *see* them as I have indicated below each figure. It is the attention we give, and the meaning we project onto something, which enables us to *see* and make some sense of our world.

a.

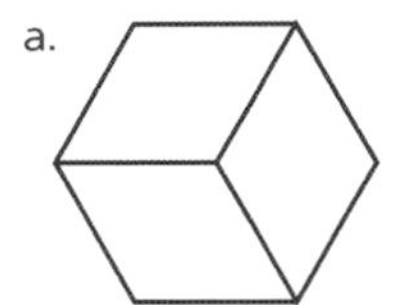

Figure 7.2a. A drawing of a cube.

b.

Figure 7.2b A drawing of a hexagon with lines going from the center to three vertices.

When a teacher speaks, or shows something, I describe this as an *offering*. It will be up to the student (either consciously or unconsciously) whether that offering is attended to and worked on. All a teacher can do is make offerings, only the students themselves can give their attention to such an offering and work with it in order to develop meaning. There are other arrows which appear as well as those from the teacher and the student. This is because there are other offerings which are involved when communication takes place. These may be from other people in the room, such as fellow students who may also be speaking at the same time as the teacher. Even the visual presence of a fellow student is an offering to which the student may decide to attend rather than to the words of a teacher. Offerings may be intentionally directed to the student or may be unintentional, such as the image of a passing car, and may involve any of the five senses. All offerings, whether intentional or unintentional, are *opportunities* for learning—whether or not that learning is related to any teaching objectives the teacher may have!

Figure 7.1 represents the fact that communication never takes place in isolation, and involves active participation by the listener/reader, etc. as well as the speaker/writer, etc. The difference in the type of arrow coming from the student compared with the other arrows indicates that this diagram represents these dynamics from the student's perspective. I have called this image the "zone of opportunities" (ZoO), since all offerings are potential opportunities for learning. If students focus attention onto a particular offering, then they turn that offering into *material* with which, by the act of placing attention, they begin to work. Thus, the zone of opportunities is a zone—determined by the student's attention—where offerings can be turned into material: *potential* opportunities for learning (offerings) can become *actual* opportunities for learning (material). A student would then need to work with that material for learning to take place.

The ZoO offers a perspective in line with, yet different from, that of Lev Vygotsky's "zone of proximal development" (ZPD), which considers "the distance between the actual development level as determined by independent problem solving and the level of potential development as determined through problem solving under adult guidance or in collaboration with more capable peers" (1978, p. 86). Vygotsky's ZPD, as it is commonly understood, places attention on the fact that a student can achieve more with expert guidance or support than if they were to carry out a task on their own. Thus it describes a difference in observed outcomes between working in isolation and working with a relative "expert." The ZoO, however, offers a metaphor for the opportunities available for a student in either of those situations (and indeed any other situation) rather than comparing the two. Whether working alone or with others, the ZoO recognizes the varying opportunities available (some which may be helpful to achieving a task, others which may be distractions in achieving the task), and that the student makes choices as to where attention is placed. These choices will have consequences as to what learning takes place. So a student, for example, may not be successful at a particular task due to the choice they made about where attention was placed. Another time, with similar opportunities, a different choice in the placement of attention can lead to greater success at the same task. The ZoO concerns a particular moment in time. It is a metaphor for the learning opportunities available for a student and emphasizes the role that attention plays in the nature of learning which takes place, no matter what the situation—whether in or out of a classroom, whether a teacher is present or not. Vygotsky's ZPD is a comparative zone across different scenarios—working with and without a relative "expert."

Each person has their own dynamically changing zone of opportunities since each person has different opportunities, due to such factors as: their different physical position which gives different available visual offerings; different sensitivities within each of the five senses which means that some sounds, for example, may be available to be heard by one person and not another; different access to language which can determine what is available for one person but is not available for another; different personal identities, such as being of a certain gender, race or class, and how those differences shape what is stressed or ignored; past personal experiences and awareness, which are internally available and are offerings as much as impacts from the five senses.

The ZoO describes dynamics at a personal level, the level of the individual in relation to the world and people around them. As such, it does not describe dynamics of communication as if viewed from outside, but describes such dynamics from the perspective of an individual. It offers an image to consider the impacts an individual receives through their senses at a particular moment in time (or short period of time) and the attention and meanings which may be given to each of these by that individual. This differentiates the notion from work which offers an overview of interaction in a classroom, such as Heinrich Bauersfeld's (1978) description of the *funnel pattern*, where increasingly narrow questions are asked by a teacher to lead students to a particular expected outcome, or Jörg Voigt's (1985) observations of patterns and routines in classroom interaction. The notion of the ZoO can, however, be used to consider a particular moment in such interaction from an individual's viewpoint. For example, Terry Wood (1998, pp. 170–171) gives an example of an interaction to illustrate the funnel pattern phenomenon. Here the teacher has just asked a student, Jim, to give an answer to 9 + 7:

Jim: 14.
Teacher: OK. 7 plus 7 equals 14. 8 plus 7 is just adding one more to 14, which makes__? (voice slightly rising).
Jim: 15.
Teacher: And 9 is one more than 8. So 15 plus one more is__?
Jim: 16.

Figure 7.3 represents a moment in time, from Jim's perspective, following the teacher's first response to Jim's answer of 14. Within the ZoO is Jim's own belief that 9 + 7 is 14, along with the question asked by the teacher. There are several opportunities here for Jim, and what he does will depend upon the attention and meaning he gives to these offerings. He might stress both his own belief and the teacher's comment and end up re-examining his own calculation. This can lead to him not answering the teacher's actual question but to making a comment which indicates that his first answer was wrong. Alternatively, he might stress only the question asked by the teacher and reply to that question, almost as if it were unrelated to 9 + 7. So, he may say 15 purely as a response to the new question of what is one more than 14. Yet another possibility is that he may stress the visual presence of a fellow student and be concerned with the social standing he has with his peer group. This may result in silence or behavior unrelated to the question being asked. The ZoO makes us realise that learning

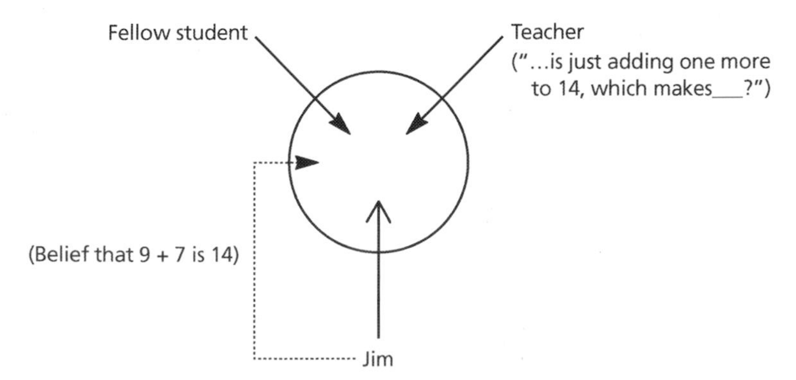


Figure 7.2. Jim's perspective of a point in time during an interaction with the teacher.

is dependant upon choices students make as to where they place their attention. If a funneling pattern of interaction is established, then this might shift a student's attention away from the underlying mathematics of the original task, and onto different, often more trivial, mathematical questions. The ZoO stresses the two key roles involved with teaching: that of adding to the quality of opportunities available for students; and that of trying to direct students' attention between those opportunities which are available.

The Zone of Opportunities concerns a dynamic localized in both time and perspective (freezing a moment in time and considering what opportunities exist in the moment, and acknowledging the different opportunities present in that moment for each person by considering the dynamic from one person's perspective) which can relate to notions describing communication on a more global level (considering a period of time, and commenting on phenomena occurring across a group of people). In this sense, the ZoO can be a companion to such notions, and as such does not discriminate between or challenge those notions. Its strength lies in giving a personal perspective to the dynamics involved in how individuals engage with the complex physical, social, and emotional world in which they live. It does not, however, comment upon personal or collective events which occur as outcomes of such engagements. Neither does the ZoO model teaching situations *per se,* since it does not require the conscious decision-making and actions of a "teacher." It can relate just as much to a baby in a crib engaging with the sounds and images from the surrounding world, as to a mathematics classroom. Consequently, the ZoO might be a companion to other notions which explicitly concern the teaching process, such as the *Didactical Contract* (Brousseau, 1997) or the *Teaching Triad* (Jaworski, 1991, Jaworski & Potari, 1998), while remaining distinct due to its localized nature.

Although the ZoO is general in its application, I consider next its relation to the teaching and learning of mathematics. While acknowledging that cognitive, social and emotional aspects are intertwined and present in all mathematical activity, I stress each of these three perspectives in turn before looking at times when they conflict, and times when they are in harmony with each other.

## COGNITIVE DYNAMICS THROUGH THE ZONE OF OPPORTUNITIES

Attention is a result of some things being stressed while others are ignored. This results in us being able to extract something from our otherwise highly complex environment. As a consequence, the things to which we attend are situated within a context. They have been extracted from surrounding sounds, movements, people, places, etc. Speech is said within a certain physical setting, and may have been said to a particular person or group of people. It is said with a certain auditory intensity and with certain physical gestures. Thus, as a listener, I can attend to the substance of what is said, the words and sentences, and the meaning I have for those. In addition, I can attend to the relationships between those words and other opportunities within the zone, such as gestures and surroundings. I can form, for example, associations between the content of what is said and the facial expressions and gestures made by the person who did the speaking. The simultaneity of the two events helps the establishment of such associations. Forming associations comes from stressing the simultaneity rather than stressing any particular one or other of the things which happened, such as the gesture or the content of the speech. Thus, the zone of opportunities contains within it much more than the sum of individual parts. There may be a certain number of sounds, movements, images, etc., available to me over a short period of time, but the opportunities available to me are so much more than this list since I can also attend to relationships between them.

Within a teaching context, students form associations assisted by things happening simultaneously. Sensitivity to this fact can lead to particular teaching strategies. For example, a training teacher was teaching a lesson about Pythagoras, drew a right-angled triangle, and wrote the following on the board:

$$H^2 = \ldots^2 + \ldots^2$$

He then said "The square on the hypotenuse is equal to the sum of the squares on the other two sides." However, this sentence was not just said. The teacher also pointed to certain aspects of the equation at the same time as saying the relevant part of the sentence. Thus, it went:

The square (point to the ' $^2$ ' next to the H)...
...on the hypotenuse (point to the H)...
... is equal to (point to =)...
... the sum (point to +)...
... of the squares (point to both "square" symbols on the right-hand side of the equation)...
... on the other two sides (point to the two sets of dots).

These actions, and their timing, meant that students following visually where the teacher was pointing, would look at the relevant aspects of the equation at the same time as hearing words within the sentence. This increases the possibility of associations being established between words such as "square" and symbols such as " $^2$ ", and so help students to "read" algebraic sentences when they are seen in symbolic form. I am not suggesting that this is a particularly good way of introducing Pythagoras, but I am highlighting the potential of acting in a way which is sensitive to the role that simultaneity plays in forming associations. Robert Schmidt said that "it is in the nature of signs to lead our attention away from themselves and towards the thing signified" (1986, p. 1). However, as Pimm commented in relation to Schmidt's work: "A sign names or points to something else, but bears no necessary relation to the thing named" (1995, p. 72). The symbol " $^{2}$ " is a sign which is indented to signify a process of "squaring." Yet in itself it bears no straightforward relation to such a process—it is not iconic—so its role of pointing towards this process is unlikely to be carried out without the help of someone who already knows this link. Since the new sign of " $^{2}$ " does not by itself point to the process of squaring, the teacher mediated the pointing by literally pointing to the sign " $^{2}$ " as the known word "squared" is said. The teacher takes on the role of pointing to the sign with the aim that in the future students will form an inverse "pointing" from the sign to the word "squared" and from there to the process of squaring.

Susan Pirie and Tom Kieren (1992) in their discussion of "constructivist teaching," nicely pointed out the difference between considering the knowledge and beliefs which guide teachers' behaviors (e.g., a teacher whose constructivist beliefs informs decisions being made), and classifying the behaviors themselves (e.g., certain approaches to teaching being described as constructivist). In the above example, I am more concerned with sensitivity to the dynamics within the ZoO which might inform an action of "pointing as one speaks," rather than suggesting that "pointing as one speaks" is always a good pedagogic behavior. There is a difference between a teacher demonstrating a sensitivity through their decision making, and carrying out particular behaviors *per se.*

There are so many impacts coming in to our senses at any moment in time. Some of these opportunities are available to several people as well as myself, such as voices (for those with a hearing impairment I could choose a different example), while others are private in that I might be the only person who can have direct experience of them, such as muscular aches. Past personal experiences are of this latter type and can be triggered by something which is happening in the present, and so "come to mind" and be available in the present as "memories." The triggering can happen so fast as to appear to be simultaneous, thus giving the opportunity for the new experience to be associated with past memories through simultaneity. Although one event may have happened long before a current event, attending to an aspect of the current event may trigger a past event which has a common feature. This brings the past event into the present as a "memory," and so both are available simultaneously within the zone of opportunities, and thus the link can be stressed. In this way, it is possible for someone to notice sameness over a period of time, something which is required in order to learn our first language. It is through noticing similarities in context when a particular word is said over a collection of experiences happening hours, days or even months apart that a meaning for that word is refined and becomes similar to the meaning others have for the same word (see Hewitt, 1988, for an example).

I had a conversation with a small group of year 10 students about the work they would be doing in the coming week. I mentioned something about trigonometry and one girl, Rachel, said she was not so keen on this because it involved angles and she's not much good at angles. I began rotating my hands and asked her to tell me when the angle between them looked like 60 degrees. She was able to do this successfully, and similar questions were posed by others in the group. She began considering angles larger than 180 degrees despite some difficulty in the physical rotation of the hands. Holding her hands flat together she said "But if I go to 360 degrees then it is exactly the same as when I started." I replied that this seemed to be true and so in the same way I might say that 720 degrees is also the same. "So this is like music," Rachel said, "in that a note can be A sharp or B flat." We continued talking about -60 degrees being the same as 300 degrees and she was sounding confident, appearing to have control of the angle discussion through the image she had developed. The attribute of equivalent ways of expressing angles seemed to be stressed, and this triggered a memory within music where a similar attribute was present.

Rudolf vom Hofe (1998) discussed the differences between *individual images* which a student may bring to a mathematical situation from their own personal experiences, and *basic ideas*—"Grundvorstellungen" (Oehl, 1962)—which are images or metaphors encapsulating the essence of a mathematical notion and which are purposefully used by a teacher to try to base mathematical concepts on familiar contexts and experiences. vom Hofe pointed out that

> the individual images of the students may differ from the "basic ideas" intended by the teacher [...]. Accordingly, it is desirable that the teacher not only try to teach basic ideas in a prescriptive way, but also that he or she develop a purposeful sensitivity for the individual images the students may in fact have (1998, p. 321).

The individual image Rachel brought from her past experiences with music assisted her in developing new meaning for this relationship between angles. Although there are differences between the equivalence, or otherwise, of A sharp and B flat and equivalence of angles in mathematics, this was a useful image for Rachel in her learning and it seemed important for me as a teacher to be sensitive to the usefulness of this image at that time. As vom Hofe commented, "basic ideas can be generated by students through modification of their individual images" (1998, p. 326) over time during interplay between their individual images and the teacher's basic ideas.

My recalling of this incident with Rachel is another example of how stressing an attribute of a current event can trigger a past event which has that same attribute. It was in writing about such triggering above, that I recalled the incident with Rachel (which shares this common attribute of triggering past events), which I have not thought about for many years. In fact I had to spend some time digging through old notes until I found my original record of this incident, written approximately nine years earlier.

## SOCIAL DYNAMICS THROUGH THE ZONE OF OPPORTUNITIES

There is much which has been discussed about the patterns of social interactions within a mathematics classroom (see, for example, Bauersfeld, 1988, Voigt, 1995) and generally this has been carried out through analysis of interaction between teacher and students, and between students themselves. This is analysis at the micro-social level where observation is on dynamics present within a collection of individuals. Such observations are not the focus for my discussions in this section, as initially I wish to focus at the individual level—to consider the potential *internal* dynamics for an individual within a social context—in order to consider social dynamics *through the ZoO.* Thus, I am placing my focus on the dynamics within an individual (in a social setting) rather than within a group of individuals (which constitute a social setting).

When I hear someone speak, I can attend to what they are saying, this being the content of their speech. At the same time, there are many other things that are available. For example, as well as *what* they spoke, there is *who* spoke, *how* they spoke, *when* they spoke and *where* they spoke. These are opportunities which are available in time within the ZoO. Mikhail Bakhtin said that

> Speech can exist in reality only in the form of concrete utterances of individual speaking people, speech subjects. Speech is always cast in the form of an utterance belonging to a particular speaking subject, and outside this form it cannot exist (1986, p. 71).

The meaning I develop from hearing what someone says to me can change according to whether it is said in hushed tones in the privacy of my home, or in a forthright manner as a guest speaker at a conference; or the contrast between someone appearing nervous as they speak versus someone appearing to brag while saying the same collection of sentences. There is much more with which I engage than the actual content of what someone says; I am also engaging with the person. So the original diagram I presented on the zone of opportunities (Figure 7.1) contains much more than I originally discussed. The student's attention to the content of what a teacher says also brings with it the way in which it is said, mannerisms, etc. The arrow from the student not only indicates that attention is paid to the content of what the teacher says, but also brings with it the knowledge and opinions the student has about the teacher, about normal teacher behavior, etc.

The social comes into play for a student when there is an awareness of others and the student attends to the relationships the student has with others. As Bakhtin's quote above indicates, speech comes from a speaking subject, and so the social is invariably present within any form of communication. Consider a situation where a student (A) has been working on the well-known problem of finding the number of handshakes when a certain number of people shake hands with each other. Having spent time working on this problem the student comes up to the board and writes the following:

$$\textit{number} \times \textit{number} - 1 \div 2$$

Then the student hears two people speak:

Person 1: That's not quite right, you need to use brackets.
Person 2: It doesn't matter, I know what you mean.

What might the student think? Before reading on, consider what you might think about these two comments. Does it change if the people are identified as:

Teacher: That's not quite right, you need to use brackets.
Student B: It doesn't matter, I know what you mean.

There could be several reasons why the teacher might want to stress the use of brackets, perhaps wanting to bring out the ambiguity of the statement without the use of brackets and possible misinterpretation of what the student meant. Again, there are several possible reasons for why student B replied in the way that she did, for example wanting to defend their fellow student, or simply to express that she understood what was meant even if it appeared to them that the teacher did not. What might student A think now? Alternatively, suppose the people were really as follows:

Student B: That's not quite right, you need to use brackets.
Teacher: It doesn't matter, I know what you mean.

Again, interpretations could be given to why the teacher and student B said what they did. How might this change the way student A thinks about the original written statement? Within the two different scenarios above, it is possible that student A might walk away with quite different learning experiences following his or her written contribution on the board. At one level, he or she may walk away feeling quite differently about whether has was "right" or not. It may also affect his or her self-image of themselves as a mathematician. It may affect his or her social standing within his or her peer group. There are all sorts of different ways in which the experience differs according to *who* said what, rather than just *what* was said.

The value a student places on what is said will depend to some extent upon who said it. In relation to cognitive dissonance, Leon Festinger said that "it is plausible to assume that the dissonance between one's own opinion and knowledge of a contrary opinion voiced by some other person is greater if the other person is important to one in some sense" (1957, pp. 180–181). A teacher is not the same as a fellow student in terms of the power relationships within a classroom. The authority of teachers is gained through a combination of how the teachers go about their role and also the respect, or otherwise, students give each teacher. If a teacher is an authority figure for a student, then this has to do with the student giving authority to that teacher, as well as the teacher holding status within a social structure. For many students, although certainly not all, their teacher is an authority figure in that an increased value is placed on what the teacher says about the subject matter of a lesson, compared with what is said by fellow students.

Recently, I used an activity (from *Whatever next? Ideas for use on A level Mathematics Courses*, Association of Teachers of Mathematics, undated) with student teachers where different trigonometric expressions on cards had to be sorted into equivalence groups. As students began collecting some of these statements together, I asked them how they knew they were equivalent. A common response I got was that they remembered and had been told these trigonometry identities by their teacher. I asked how they knew their teacher was correct. The students looked at me as if I was stupid and gave me replies such as: "well, the teacher isn't going to lie to me." This belief that the teacher is right, and learning is about accepting what a teacher says as correct (perhaps reinforced by the teacher's own behavior and beliefs), is one which is common inside many classrooms. This is, though, not always the case, and some students give a teacher authority due to the quality of questions and challenges the teacher offers. In this case, students listen carefully to what the teacher has to say because the students are in touch with their own sense of learning—"I listen to my teacher because I know I have been learning in this subject." This can be so, even if the teacher tells the students little but mainly uses techniques, such as questioning, to help students articulate their own thoughts about the mathematics. For example, Paul Cobb, Erna Yackel and Terry Wood described the authority of a non-traditional teacher:

> As in a traditional classroom, the teacher was very much an authority figure who attempted to realize an agenda. The difference resided in the way she expressed her authority in action (Bishop, 1985). When she and the children talked about talking mathematics, the teacher typically initiated and attempted to control the conversation. When they talked about mathematics, however, she limited her role to that of orchestrating the children's contributions [...] therefore, it makes sense to say that the teacher exerted her authority to enable the children to say what they really thought. (1989, p. 126)

If a teacher consistently proposes quality questions, then the student's stressing of what the teacher says can mean that the student's attention is reflected back onto their own thoughts and ideas.

Learning mathematics can be viewed "as the interactive construction of social practice" (Bauersfeld, 1995, p. 150) where "students arrive at what they know about mathematics mainly through participating in the social practice in the classroom" (p. 151). One implication of this is that students' attention is with other people and the practices which those others carry out, as much as with the content of what is said. For example, the "=" sign may be described by a teacher as meaning that what is on one side is equivalent to what is on the other side—a relation between one side and the other. However, the practice of the use of the "=" is often different, with it being used more as an operand by writing what is perceived to be "the answer" on the right-hand-side. This practice is strengthened through the use of calculators. Students often end up with the perceived practice of "=" as operand rather than relation and this can lead to difficulties in later algebra work.

This enculturation into a social practice goes further than developing meaning for certain symbols and conventions. The physical entry into a mathematics classroom, as opposed to a geography or science classroom, carries with it certain learned expectations of what takes place within such a classroom. As Resnick commented "learning is highly tuned to the *situation* in which it takes place" (1989, p. 1). Hendrik Radatz carried out research where students were asked to interpret a set of graphical representations (1986). The number of interpretations classified as mathematically acceptable dropped significantly when students were asked at the beginning of a religious lesson compared to when they were asked at the beginning of a mathematics lesson. The culture of the religious classroom sets different expectations and has different behaviours to that of the mathematics classroom. Thus, a student has within his/her ZoO, among other things, the teacher, the questions they are being asked to do, and the physical setting of the classroom. The meanings associated with each of these may be contradictory at times. It could even be that an art teacher is covering a geography lesson where there has been a room change with the lesson taking place in a science lab. A student will bring particular associated meanings to each arrow which comes into the ZoO. These are sometimes conflicting and the student will have to resolve such conflicts to decide how to act, or hold the tension of the conflicts (Breen, 1993) while still managing to act. Thus, it is not just the *what* and the *who*, but also the *where* which is significant in the learning of social practices within a given context. I learn about something new through noticing the way in which people engage with, or use, that something. For example, I have learned a lot of the language I use through the social context within which words are used. Kenneth Gergen commented that:

> Traditionally, the terms of our language have gained their meaning by their links to specific, real-world referents. However, for the constructivist, [...] the meaning of words largely derives from the relationships in which they play a part. Thus, the meaning of the term *aggressive* is not derived from a specific datum in the world, but from the linguistic context in which it is used by people who do things with each other (e.g., assign blame, prepare a reply). Thus, its meaning will change importantly depending on whether one is working with others to deploy troops, develop a business strategy, or combat cancer cells. (1995, p. 37)

The actual contextual situation within which some mathematics is taught is sometimes confused with the pseudo-context introduced as part of pedagogic decisions. For example, there is a belief by some teachers and text book writers that mathematics becomes more relevant when it concerns "real" life, such as reading gas bills or going to a fun fair. However, these are not real life for the students, in that they are not in the real contextual position of actually needing to deal with a gas bill within their life, or they are not actually at a fun fair. The real life is that the students are in a mathematics classroom reading from a textbook about such things. If a student does engage honestly in the real-life context this sometimes conflicts with the intended mathematical activity. For example, in a bottom set of 12–13 year olds, the text book being used asked students to write down the possible combinations of three main course and two desert course choices which were given. One of the desert choices was chocolate cake but one student said "I don't like chocolate cake so I wouldn't have it," and so did not include that in his list of possibilities. The teacher tried to explain to him that it is still a possibility, but he was determined that it was not a possibility for him and refused to include it. So students are really being asked to stay outside the real-life context and as such the context within which they are being asked to do some mathematics might remain quite mundane in the sense that they are just working through another page in the text book.

## EMOTIONAL DYNAMICS THROUGH THE ZONE OF OPPORTUNITIES

In a secondary school, the whole of Year 10 listened to two invited speakers from outside, each presenting their case: one for nuclear disarmament (a woman); and the other for the UK keeping a nuclear deterrent (a man). As I listened, I felt that the man presented a clear argument, whether I agreed with it or not. The woman, in my opinion, did not say anything particularly coherent, and offered few arguments. Later in the day the students (about 200 of them) were asked to vote for those with whom they agreed, having spent some time discussing the issues themselves in tutor groups. To my surprise, the vote came out overwhelmingly for disarmament. I asked several students their thoughts about what the two speakers had said and why they had voted as they had. Many of the responses I got indicated that they had not voted on the basis of argument, but on the fact that they felt the woman cared about her viewpoint, whereas the man had shown little emotion within his talk.

Students, particularly adolescents, are strongly affected by emotional aspects of personal relationships. Sadly, many people have a negative emotional experience of mathematics during their schooling. If mathematics is seen as a subject where there are right and wrong answers, then someone's experience of mathematics lessons can become associated with the experience of being right or wrong. Both of these can be negative experiences for students. If a student is asked to give an answer in front of the whole class and their answer is labeled as wrong by the teacher, then what can be felt is a sense of "I am wrong" rather than, "my answer is wrong." As a consequence, that student can begin to develop a negative self-image as a consequence, particularly if their incorrect response is a focus of derision or laughter from the student's peers. Frequent experiences of this kind can mean that a student can find the lessons unpleasant, or even frightening, and so associate mathematics with these emotions. In a study carried out by Manchester Metropolitan University (ESRC, 1999), of BEd students who were studying to become primary school teachers, 80% admitted they found mathematics intimidating when they were at school. Laurie Buxton's book *Do you panic about maths?* (1981) highlights many such cases.

Even students who are quite successful at mathematics can sometimes experience unpleasant emotional situations, where they are publicly congratulated for getting the "right" answer. Either way round—"right" or "wrong"—a student can become labeled as either "clever" or "thick"! This can be traumatic for a student who wants to "fit in" with their peers socially and does not want to be labelled as different. While mathematics is viewed as a subject of right and wrong answers, there is a danger of students having negative emotional experiences unless a teacher takes some positive action to reduce such possibilities. Cobb, Yackel and Wood (1989, p. 135) gave a transcript of a teacher who was working on this issue with her class:

> Teacher: (Softly.) Oh, okay. Is it okay to make a mistake?
> Andrew: Yes.
> Teacher: Is it okay to make a mistake, Jack?
> Jack: (Still facing the front of the class.) Yes.
> Teacher: You bet it is. As long as you're in my class it is okay to make a mistake. Because I make them all the time, and we learn from our mistakes, a lot. Jack already figured out, "Oops. I didn't have the right answer the first time" (Jack turns and looks at the teacher and smiles), but he kept working at it and he got it.

I recall a colleague, Laurinda Brown (personal communication), saying that she always tried to make her classroom a place where "it is OK to be wrong." This requires careful management of situations and a shift of emphasis away from answers (which might be labeled as either right or wrong) and onto methods and process.

Within all teaching situations there is also the personal relationship between teacher and student and, as with all relationships, emotions are involved. David Hawkins (1969, p. 25) talks of the important role of an "It" in the building of a relationship between an adult and child. He gives an example of how a relationship between a student and teacher can be initiated by the teacher bringing into the classroom an engaging "It" (the example he used was soap bubbles) which may captivate the student:

> The teacher has made possible this relation between the child and "It," even if this is just by having "It" in the room; and for the child even this brings the teacher as a person, a Thou, into the picture. For the child this is not merely something which is fun to play with, which is exciting and colourful and has associations with many other sorts of things in his [sic] experience: it's also a basis for communication with the teacher on a new level.

So, a relationship can begin to develop through the existence of an "It" which becomes a joint focus of attention and the basis of communication between teacher and student. Within a productive professional relationship between mathematics teacher and student, the It is the mathematical activity, which becomes the focus for mathematical discussion. The "It" with regard to mathematics is not an object as such, but the mathematical notions and awareness which form the basis of discussion. A physical object may exist, such as soap bubbles, but this is not mathematics *per se.* It is only by attending to particular aspects of soap bubbles, such as relationships within the geometry of the bubbles, that mathematics comes into play. In fact, there need not even be a physical object present; the "It" might be conceptual in nature, such as the commutativity of addition.

## CONFLICT AND HARMONY BETWEEN THE COGNITIVE, SOCIAL, AND EMOTIONAL DYNAMICS

The mathematics classroom is a complex environment where there are many opportunities for learning: sometimes learning is concerned with the subject matter a teacher might have as an objective for a lesson, at other times it may be concerned with, for example, learning about friendships within the peer group. Here I consider times when the cognitive, social and emotional perspectives conflict with each other and times when they operate in harmony.

### Conflict

Students' learning of mathematics can be viewed as the participation in particular social practices, "rather than through discovering external structures existing independent of the students. Across social practice, the participants produce and reproduce taken-as-shared negotiated regulations and norms for communicating about and acting on an accomplishment, which others may call *mathematical objects*" (Bauersfeld, 1995, p. 151, his emphasis). James Wertsch and Chikako Toma remark that for socio-culturalists "sociocultural processes are given analytic priority when understanding individual mental functioning, rather than the other way around" (1995, p. 160). So, the image here is that students are essentially involved in learning to participate in the social practices involved within the setting of a mathematics classroom. As such, the teacher's role is to set up an appropriate culture in the classroom, so that if students partake in the social practices established within this culture, then their behaviors are such that they are likely to be successful at the mathematical tasks which relate to those practices. However, conflict can sometimes occur when the culture of the classroom is one where mathematics is perceived purely as the carrying out of routines. This is acted out by: (a) teachers if they see their role as showing or telling students rules for carrying out certain behaviors without addressing the underlying mathematics behind those rules; and (b) students who learn to perceive mathematics as re-producing what a teacher shows or tells them. In this way, mathematics becomes a subject which, at best, is described rather than explained. This can create conflict between the cognitive and the social aspects of the learning situation.

Following the Third International Mathematics and Science Study (TIMSS) a video (U.S. Department of Education, undated) was brought out giving examples of lessons from Japanese, USA, and German mathematics classrooms. The first USA lesson on the video shows a classroom which appears to have a culture where the social practice is about students listening to the teacher describe a procedure, and then re-producing that procedure within exercises of a similar type. Thus an authority is given to the teacher as the person who provides information, and this authority can be transferred to the statements the teacher makes. Statements made by such an authority figure are taken as true, and the validity of the statement comes from who said it rather than the content of the statement itself. I describe this as "received wisdom" (Hewitt, 1999). So a statement about how to work out the angles inside a polygon (as is made by the teacher in this video) is received wisdom and accepted as true, not because of any cognitive analysis of the mathematics, but because of the authority given to the teacher. In this case, the social culture of the mathematics classroom, where practice is seen as re-producing what the teacher says or demonstrates, does not conflict with examples that students might carry out or know about. Any cognitive work they may do in a later lesson of trying this rule out for different polygons will not conflict with things they already know, such as the angles inside a triangle adding up to 180 degrees, or the angles inside a square adding up to 360 degrees. However, this way of working encourages students to view mathematics as received wisdom passed down from a teacher or text book, rather than to consider the underlying mathematics itself. This shift of attention away from cognitive considerations and onto the carrying out of a social practice can lead to conflict. A rule offered by a teacher in an attempt to help students in a particular situation may be incorrect for other situations which the teacher had not considered but which the students go on to explore. For example, I observed a teacher trying to help students who were struggling with using a 360 degree protractor. She asked them to remember that having positioned the protractor correctly "you use the inside scale if you are going down, and the outside scale if you are going up" from one line to the other. Having already shown the students to start by placing the lines and protractor in a particular orientation, she did not consider that some students might start with a different, although equally valid, orientation. With the start line hor-

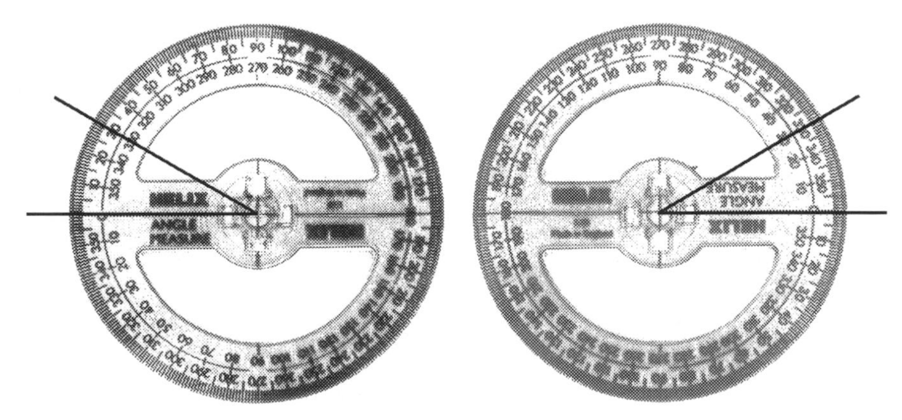

Figure 7.4. Measuring from the left-hand and right-hand side of the protractor, the rule of using the outside scale for "going up" gives different answers.

izontal, the rule is indeed true if you start at the left-hand side of a clock face, but becomes false if you start on the right-hand side. Some students who started on the right-hand side experienced a conflict between the rule accepted as part of the social culture of the classroom and the actual cognitive considerations of a particular situation (see Figure 7.4).

If the culture of the classroom is such that students perceive learning mathematics to be about following certain social practices carried out by a teacher, then these practices can sometimes conflict with the cognitive aspects of the mathematical activity.

On other occasions, conflict may not be present for a student at the time of working on an activity since their attention is only with the carrying out of a perceived social practice and they do not even consider cognitive aspects of the underlying mathematics. For example, I watched one group of 13–14 year olds working on area problems. I noticed that the students were having difficulties with the work, yet I also noticed that the students themselves did not appear to feel that they were having problems. One male student was finding the area of the shape in Figure 7.5.

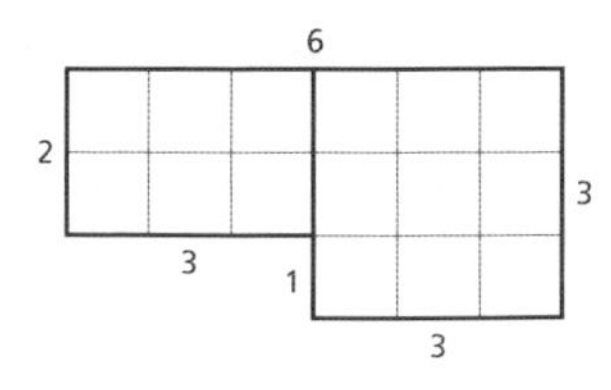


Figure 7.5. A problem of finding the area.

He appeared to be quite confident as he began to work out 6 × 3, 3 × 3, 3 × 1, 1 × 3, 3 × 2, and 2 × 6. I asked him to look at the squares, and he said there were nine squares in one part and six in the other. However, he then wanted to calculate 9 × 6. Along with several other students in that class, he appeared to want to multiply numbers together. It is as if he had abstracted the social practice of multiplying numbers together whenever the area needed to be worked out. Here, the adoption of a perceived social practice—*finding area means multiplying*—conflicts with the underlying mathematics of the situation. However, since some of these students were carrying out their perceived social practices without having cognate reasons for why those practices were so, they did not experience a conflict and would have happily completed a set of similar questions feeling that there was nothing wrong with what they were doing. Steinbring commented that

> The meaning of theoretical mathematical knowledge can be constituted in the course of interactive processes only if the meaning of mathematical knowledge is not confined to the actual interactive social practice in which students and teachers are involved and if it is accepted that there is some kind of epistemological meaning of mathematical knowledge "outside" the actual social practice (1998, p. 114).

Some of the students were not looking outside of the social practice. This means that they also did not really have an expectation as to what might be the size or nature of their answer and so they would not experience a conflict even if their final answer was wildly incorrect. George Mandler commented that:

> we must keep in mind that people frequently engage in actions that they believe to be correct (i.e., they proceed as intended, but the actions are in fact false, incorrect, illogical, etc.). In the case of unintended errors, the discrepancy arises because of a mismatch between what is intended and what occurs; in other cases, the mismatch is between an expected outcome ("I thought I did what would solve the problem") and the real-world response ("It didn't work"). (1989, p. 13)

Without some notion of an expected outcome, a conflict will not be experienced. The social practice appeared to have gained an element of automaticity, which, as Douglas McLeod stated, "has its disadvantages as well as its advantages in problem solving. The trick is to bring automatic processes back under conscious control when you need to apply them in non-routine situations" (1989, p. 26). The automaticity of multiplying dimensions when finding the area of a rectangle, for example, was being applied to the non-routine (in the sense that it is different to finding the area of one rectangle) problem presented in Figure 7.5. However, this automaticity was not brought to conscious attention and so was not examined cognitively.

There was, of course, a conflict later when the students had their work marked and found that most or all of it was wrong. Such conflict can lead to students having negative images of themselves as mathematicians since, having ignored cognitive considerations, they have no basis on which to understand why their work is wrong. A cyclical effect can occur with the student's lack of confidence, leading them to attend even more to social practices since trying to copy what the teacher does/says (or indeed what other students do) is at least an activity they do understand.

Colette Laborde (1994) studied interactions among small groups of students when there was conflict between their opinions about the mathematics of the task they were given. She commented that "conflicts are not always solved by rational arguments but also by authority arguments" (p. 151). This attention given to the social dynamics inside a classroom rather than the cognitive aspects of a mathematical activity can result in the social over-riding the cognitive for some students. Bauersfeld pointed out that often "The mathematical logic of an ideal teaching-learning process becomes replaced by the social logic of this type [covert social structure of classroom actions] of interactions" (1988, p. 38). Clive Kanes analyzed a lesson on complex numbers and found that

> it seems plausible that student responses were not primarily aimed at speaking the truth about the mathematical situation set before them. Instead, utterances were designed to resolve the awkwardness of a social interaction that was perceived, in some sense, to be going wrong. (1998, p. 134)

There are several offerings made within a student's ZoO, some of these may be concerned with their social development among their peers, while others are concerned with the mathematical features of an argument. As Alrø and Skovsmose (1996) pointed out, students have many "good reasons" which lie behind their contributions, only some of which are mathematical reasons. It may be that a student decides to place their attention with the social aspects and so come to an agreement which is based on the social dynamics with the other students concerned rather than any mathematical reasoning.

It is not always the stressing of the social which leads to conflict; ignoring social practices and stressing the cognitive also has consequences. One student, Shaun, became known by his teacher as someone who was not particularly good at mathematics and also did not put in much effort in order to improve. The basis of this judgment was that Shaun did not follow the expected social practices of someone doing mathematics in this particular teacher's classroom. In fact Shaun was an able mathematician who could carry out most of the work in his head and so never wrote anything on paper. He also considered the work so easy that he rarely bothered answering questions in class. His boredom led to him doodling in his book and becoming increasingly "difficult" in the classroom. Shaun stressed the cognitive aspects of the tasks and ignored many of the social practices which were part of the culture in that mathematics classroom. Conflict here is experienced between teacher expectations and the actual work that Shaun produced. Such students as Shaun often succeed in their examinations nonetheless. However, an extreme example of someone who does not adopt such social practices as showing their working, and using the agreed conventions and names accepted within the culture, may fail within the society's assessment procedures even if they are, in fact, very able mathematicians. Thus a balance between social practices and cognitive considerations is important.

## Harmony

The first Japanese lesson on the TIMMS video mentioned earlier is structured in the following way:

> Revision of a property developed within the previous lesson;
> Presenting a new problem which requires the application of the above property;
> Students asked to work individually on the problem for three minutes;
> Students discuss with friends, or collect a "hint card," or show solution to a teacher;
> Some students are selected to present their solutions to the whole class;
> The teacher reviews these solutions (all of which are still on the board);
> The next problem is presented and the above cycle continues.

The culture of the classroom is concerned with cognitive considerations of the mathematical problems. This is supported by the social practice of students being asked to work on problems themselves and then in groups, and for students to present solutions to the whole class on the board. The social practice supports cognitive considerations of the problems in three respects: first, the students were asked to work on the problems and were not just following procedures told to them; second, there appeared to be an emphasis on explanation and justification; and third, several solutions were offered and accepted which supports the notion that there is more than one way to think successfully about a problem.

Typically during the video, the teacher would ask students questions, such as "Is there a method from last time that uses the area of triangles?" or "Then how would you get the triangles with the same area?" These questions encourage cognitive considerations of the problem. When the teacher did

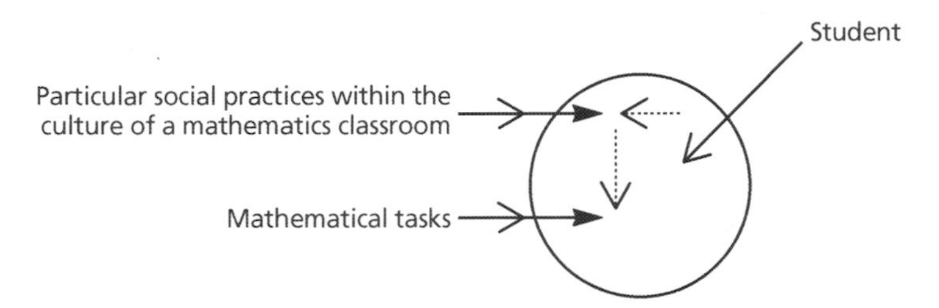


Figure 7.6. Student's attention to engagement in the social practice of a mathematics classroom can lead to success at mathematical tasks. Likewise, a similar figure would indicate that stressing the mathematical tasks would lead to successful participation within the culture of the classroom.

make statements rather than ask questions, such as during the revision of the previous lesson and reviewing solutions students had offered, attention was always given to reasons why certain things were so. The teacher's own practice within the lesson was an example of the way in which students were being asked to approach mathematics, through cognitive considerations of the mathematical properties inherent within the problems. The social practice and cognitive considerations of the mathematics ran in harmony with each other. Thus, even if a student stresses the social practices above the cognitive mathematical considerations, or vice versa, there is no conflict, since they run parallel and support each other (see Figure 7.6).

The emotional aspect of doing mathematics is sometimes a separate feature to the inherent mathematics within an activity. Some contexts are given to mathematical situations out of a desire to make the mathematics more interesting or relevant in the hope that "I like the context" is transferred to "I like the mathematics." These contexts are motivational matters and do not always inform a student about the inherent mathematics within an activity. However, activities can be devised which try to harmonize the emotional and mathematical aspects of the activity. There is a well known activity within the UK which is sometimes placed in different contextual settings, where there are horses, for example, numbered from 2 to 12 and a student has to throw two dice and add up their totals to get a number from 2 to 12. Whatever the total is, that numbered horse moves one square forward within a given race track represented by a grid. The horse which gets to the finishing line first is the winner. Students are asked to play this game several times and comment on which horses won, whether it was a fair race, etc. I have observed many classrooms where students have duly carried out the activity without showing any sense of surprise that certain horses kept winning, or of considering aspects of the underlying mathematics of the situation. I tried to bring an affective dimension into this activity by having twelve people be a particular "horse" number from 1 to

12. I made the throwing of the dice a fun activity through using large foam dice which could really be thrown some distance in the room. Thus, each of the twelve students enjoyed the prospect of throwing the dice. However, I set up a rule whereby the person who threw the dice was the person whose number had just come up on the previous throw. As the game proceeded, gradually some people began to complain:

> Horse 1: Sir, I can't win.
> Student A: Sir, he can't win.
> Student B: No, you can 'cause someone else is going to throw your number.
> Student C: No, not number one.
> Student A: Not with two dice you can't.
> Student B: It could be a one and a one couldn't it?
> Student A: That makes two.

A few other students representing horses complained with comments such as "I ain't getting anywhere." The student who was horse number 12 gave a big reaction when finally his number came up and he got a chance to throw the dice. This was followed by a big groan from the class as seven came up yet again. More complaints followed such as: "I recon' the dices are weighted down" and "The dice are fixed."

By the end of the lesson, there were a variety of emotions within the room, with the kind of emotions being related to the mathematics of the situation. There was frustration from students who were horse numbers 2, 3, 11, and 12. There was embarrassment from the student who was horse number 7 because she kept on throwing the dice. The student who was horse number 1 had become uninterested in the game as he realized that he was never going to get a throw of the dice. Students who did not represent horses still got involved commenting on how unfair that game was, and several had begun accounting for why this was so.

These emotions had links to the mathematics of the activity. According to the social dynamics between individuals, someone who was horse 7 may have been bragging at the end of the game, or, as happened in this particular lesson, have been embarrassed by the continued attention she got. Obviously these emotions have to be handled carefully. However, the emotions brought to the fore and magnified the mathematical "data" from carrying out that number of throws. With the differences in results so clear, there was considerable motivation within the room to explain these differences.

This revised Horse Race game acknowledges that adolescent students are naturally exploring and developing emotional aspects, and makes use of the understandable attention they give to these aspects of their lives. The parallel nature of the likely emotions produced through being

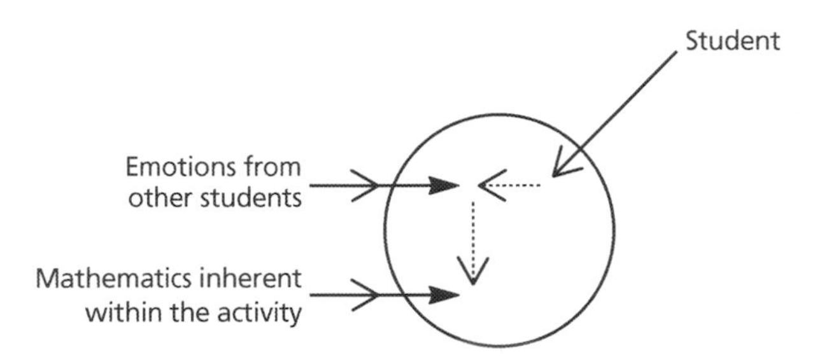


Figure 7.7. The revised Horse Race game: Attention to the emotions of fellow students also offers insights into the mathematics inherent within the activity.

involved in the game, and the inherent mathematics within the activity, means that when a student naturally pays attention to the emotions of fellow students, this also offers insights into the mathematics within the activity (see Figure 7.7).

## SUMMARY

To represent some of the complexities involved with communication, I have developed a notion, the zone of opportunities, which reflects the personal attention required for both speaker and listener, writer and reader, etc., for communication to take place. The arrows within this image represent the attention each person needs to give in order to listen as well as to speak. The fact that communication does not take place in isolation is represented by several arrows appearing within someone's ZoO, and this highlights the fact that a student has a choice of where they place their attention. All offerings, whether deliberate or otherwise, are opportunities for learning, whether that learning be concerned with the social development of that student among their peers, or the awareness of different sounds that car engines make as they travel along a road, or the subject matter to which a teacher might be referring. At times, a teacher may incorrectly assume that a student's response to a question is primarily based upon consideration of the curriculum subject matter under discussion, when the student's response may have more to do, for example, with the social dynamics within the class.

The social is stressed when attention is shifted onto *who* is speaking as well as, or instead of, *what* they are speaking. Different emphasis can be given to one person rather than another, and in particular, if authority is given to a teacher, then the authority for the person can be transferred to statements made by that person. This can result in mathematical properties being accepted as "received wisdom" from statements made by an authority figure. In this way, properties are taken as correct on the basis of who said them rather than cognitive consideration of the inherent mathematics. When a student sees learning as taking on board the perceived social practices within the mathematics classroom, rather than cognitive considerations of the inherent mathematics, this can lead to conflict and a lack of success at mathematical tasks. Indeed the ignoring of social practices can also lead to a lack of success within a cultural assessment structure which has certain expectations as to how people carry out mathematical activity. I have given examples of when this can lead both to success with the mathematics tasks and also lead to the student experiencing conflict between the social practices and cognitive considerations of the mathematics. A third possibility can occur where a student perceives a social practice which is mathematically incorrect, however no conflict is experienced since the student's attention is with the carrying out of the processes involved with this perceived practice and not with any cognitive considerations of expected outcomes from those procedures.

Students are involved with affective issues within their own development as a social person, and these play a significant role in the mathematics classroom. Mathematics can be viewed by both teachers and students as a subject where answers are right or wrong. If a student is told that their answer is wrong, then "my answer is wrong" can be taken as "I am wrong," and lead to negative emotions being associated with mathematics and themselves as mathematicians. A teacher placing emphasis on methods and questions as much as answers, can help establish a classroom where it is "OK to be wrong" and help avoid the negative emotions associated with "being wrong." I have also given examples where the social practices and emotional issues run parallel with cognitive considerations of the mathematics. In such situations each supports the other, and a sense of harmony can be produced for students working within the mathematics classroom.

## REFERENCES

Alrø, H. & Skovsmose, O. (1996). Students' good reasons. *For the Learning of Mathematics, 16*(3), 31–38.

Alrø, H. & Skovsmose, O. (1998). That was not the intention! Communication in mathematics education. *For the Learning of Mathematics, 18*(2), 42–51.

Association of Teachers of Mathematics (Undated). *Whatever next? Ideas for use on A level mathematics courses.* Derby: Author.

Bakhtin, M. M. (1986). *Speech genres and other late essays.* (C. Emerson, Ed. & Trans.), Minneapolis, MN: University of Minnesota Press.

Bauersfeld, H. (1978). *Kommunikationsmuster im mathematikunterricht—eine analyse am beispiel der handlungsverengung durch antworterwartung.* In H. Bauersfeld, et al

(Eds.), *Fallstudien und analysen zum mathematikunterricht* (pp. 158–170). Hannover: Schrödel.

Bauersfeld, H. (1988). Interaction, construction, and knowledge: Alternative perspectives for mathematics education. In D. A. Grouws, T. J. Cooney & D. Jones (Eds.), *Perspectives on research on effective mathematics teaching*, Vol. 1. (pp. 27–46). National Council of Teachers of Mathematics, Reston: Hillsdale, NJ: Erlbaum.

Bauersfeld, H. (1995). The structuring of the structures: Development and function of mathematizing as a social practice. In L. P. Steffe & J. Gale (Eds.), *Constructivism in education.* (pp. 137–158). Hillsdale, NJ: Erlbaum.

Bishop, A. (1985). The social construction of meaning: A significant development for mathematics education? *For the Learning of Mathematics, 5*(1), 24–28.

Brousseau, G. (1997). *Theory of didactical situations in mathematics,* (Ed. & trans. N. Balacheff, M. Cooper, R. Sutherland & V. Warfield). Dordrecht, The Netherlands: Kluwer Academic Publishers.

Breen, C. (1993). Holding the tension of the opposites. *For the Learning of Mathematics, 13*(1), 6–10.

Buxton, L. (1981). *Do you panic about maths?*, London: Heinemann.

Cobb, P., Yackel, E. & Wood, T. (1989). Young children's emotional acts while engaged in mathematical problem solving. In D. B. McLeod & V. M. Adams (Eds.), *Affect and mathematical problem solving. A new perspective.* (pp. 117–148). New York: Springer-Verlag.

ESRC (1999). *Primary student teachers' understanding of mathematics and its teaching.* UK: Author.

Festinger, L. (1957). *A theory of cognitive dissonance.* Evanston, IL: Row, Peterson and Co.

Gattegno C. (1971). *What we owe children. The subordination of teaching to learning.* London, UK: Routledge.

Gergen, K. J. (1995). Social construction and the educational process, In L. P. Steffe & J. Gale (Eds.), *Constructivism in education.* (pp. 17–39). Hillsdale, NJ: Erlbaum.

Hawkins, D. (1969). I, Thou, it. *Mathematics Teaching, 46,* 22–28.

Hewitt, D. (1988). Fickle. *Mathematics Teaching, 125,* 14–15.

Hewitt, D. (1999). Arbitrary and necessary: Part 1 a way of viewing the mathematics curriculum. *For the Learning of Mathematics, 19*(3), 2–9.

Jaworski, B. (1991). *Interpretations of a constructivist philosophy in mathematics teaching.* Unpublished doctoral dissertation, Open University, Milton Keynes, UK.

Jaworski, B. & Potari, D. (1998). Characterising mathematics teaching using the teaching triad. In A. Olivier & K. Newstead (Eds.), *Proceedings of the 22nd Conference of the International Group for the Psychology of Mathematics Education 3.* (pp. 88–95). University of Stellenbosch, South Africa, pp. 88–95.

Kanes, C. (1998). Examining the linguistic mediation of pedagogic interactions in mathematics. In H. Steinbring, M. G. Bartolini Bussi & A. Sierpinska (Eds.), *Language and communication in the mathematics classroom.* (pp. 120–139). Reston, VA: National Council of Teachers of Mathematics.

Laborde, C. (1994). Working in small groups: A learning situation? In R. Biehler, R. W. Scholz, R. Strässer & B. Winkelmann (Eds.), *Didactics of mathematics as a scientific discipline.* (pp. 147–158). Dordrecht, The Netherlands: Kluwer Academic Publishers.

Mandler, G. (1989). Affect and learning: Causes and consequences of emotional interactions. In D. B. McLeod & V. M. Adams (Eds.), *Affect and mathematical problem solving. A new perspective.* (pp. 3–19). New York: Springer-Verlag.

McLeod, D. B. (1989). The role of affect in mathematical problem solving. In D. B. McLeod & V. M. Adams (Eds.), *Affect and mathematical problem solving. A new perspective.* (pp. 20–36). New York: Springer-Verlag.

Oehl, W. (1962). *Der rechenunterricht in der grundschule.* Hannover. Schroedel.

Pimm, D. (1995). *Symbols and meaning in school mathematics.* London, UK: Routledge.

Pirie S. & Kieren T. (1992). Creating constructivist environments and constructing creative mathematics. *Educational Studies in Mathematics, 23,* 505–528.

Radatz, H. (1986). Anschauung und Sachverstehen im Mathematikunterricht der Grundschule. *Beiträge zum mathematikunterricht 1986.* (pp. 239–242). Hildesheim: Franzbecker.

Resnick L. B. (Ed.) (1989). *Knowing, learning and instruction. Essays in honor of Robert Glaser.* Hillsdale, NJ: Erlbaum.

Schmidt, R. (1986). On the signification of mathematical symbols. Preface to Bonasoni, P. (trans. Schmidt) *Algebra geometrica.* (pp. 1–12). Annapolis, MD: Golden Hind Press.

Steinbring, H. (1998). From "Stoffdidaktik" to social interactionism: An evolution of approaches to the study of language and communication in German mathematics education research. In H. Steinbring, M. G. Bartolini Bussi & A. Sierpinska (Eds.), *Language and communication in the mathematics classroom.* (pp. 102–119). Reston, VA: National Council of Teachers of Mathematics.

U.S. Department of Education (undated). *Attaining excellence: TIMSS as a starting point to examine teaching eighth-grade mathematics lessons: United States, Japan, and Germany* [video tape] U.S. Department of Education.

Voigt, J. (1985). Patterns and routines in classroom interaction. *Recherches en didactique des mathematiques, 6,* 69–118.

Voigt, J. (1995). Thematic patterns of interaction and sociomathematical norms. In P. Cobb & H. Bauersfeld (Eds.), *The emergence of mathematical meaning: Interaction in classroom cultures.* (pp. 163–201). Hillsdale, NJ: Erlbaum.

vom Hofe, R. (1998) On the generation of basic ideas and individual images: Normative, descriptive and constructive aspects. In A. Sierpinska & J. Kilpatrick (Eds.), *Mathematics education as a research domain: A search for identity.* Book 2. (pp. 317–331). Dordrecht, The Netherlands: Kluwer Academic Publishers.

Vygotsky, L. S. (1978). *Mind in society. The development of higher psychological processes.* Cambridge, MA: Harvard University Press.

Wertsch, J. V. & Toma, C. (1995). Discourse and learning in the classroom: A sociocultural approach. In L. P. Steffe & J. Gale (Eds.), *Constructivism in education.* (pp. 159–174). Hillsdale, NJ: Erlbaum.

Wood, T. (1998). Alternative patterns of communication in mathematics classes: Funneling or focusing? In H. Steinbring, M. G. Bartolini Bussi & A. Sierpinska (Eds.), *Language and communication in the mathematics classroom.* (pp. 167–178). Reston: National Council of Teachers of Mathematics.

THEME III

# COMMUNICATION: PRACTICE, COMMUNITY, IDENTITY, POLITICS

Three chapters are included in this group. Chapter 8 by Wenda Bauchspies is entitled, *Sharing Shoes and Counting Years: Mathematics, Colonialization and Communication*. In Chapter 9, Dalene Swanson writes about *School Mathematics: Discourse and the Politics of Context*. And finally, Chapter 10 by Tony Cotton, discusses *Critical Communication in and through Mathematics Classrooms*.

The chapters in this theme discuss communication and students' access to knowledge from sociological and political perspectives. Bauchspies in her ethnographic study discusses how students in a mathematics classroom in Togo, West Africa, resist mathematical learning and develop individual patterns of communication. Swanson provides a detailed examination of the politics played by the school discourse in constructing students with disadvantaged identities, and creating lack of communication concerning main goals, values and requirements of mathematical activity. It is almost impossible for such patterns to enable any access to a productive mathematical communication. Finally, Cotton explores an alternative model of classroom communication that takes into account issues of social justice.

As mentioned already, what one means by productive mathematical communication is not a neutral statement. It entails value-laden issues such as: Who decides what is a valid mathematical product? What is a valid process of construction? Who sets the criteria for evaluation? Who controls

*Challenging Perspectives on Mathematics Classroom Communication*, pages 235–236

which aspects of the mathematics classroom? Chapters in this theme provide enough food for thought to start considering these issues in our discussions of communication in mathematical classrooms. The challenging dimension in all three chapters is that they highlight success in classroom communication as being a matter of students' (and teachers') awareness of and access to the values and politics of the mathematical knowledge that is required in particular school contexts.

CHAPTER 8

# SHARING SHOES AND COUNTING YEARS

## Mathematics, Colonialization, and Communication

**Wenda K. Bauchspies**
***The Pennsylvania State University, USA***

### ABSTRACT

I begin with a discussion of the ethnographic voice first to help set the scene for my discussion of communication, colonization and mathematics in Togo, West Africa, and second to highlight issues of power, culture and social relations in communicating information. In exploring issues of voice—for the researcher, teachers, and readers—I identify issues of communication, meaning, and culture that also appear in mathematics classrooms and mathematics research. In the next section, I establish the role of words and numbers as windows or doors into social relations and worldviews. I then describe an "average" middle school mathematics classroom in Togo to see what fences and pathways are being created by the community to elicit certain responses from its members. I end with a discussion of repeat students who manipulate number to reconstruct their reality. My discussion of classroom communica-

*Challenging Perspectives on Mathematics Classroom Communication*, pages 237–259

tion shows how it affects all individuals. Communication unites students with its movements and symbols. These shared movements and symbols create the joint activity of mathematical knowledge. I use the lens of colonization to discuss aspects of education that we may be blind to in our own culture. I challenge educators to pay attention to how numbers and words are used, what roles they are given, and what roles they are not given.

## THE ETHNOGRAPHIC VOICE

I begin with the premise that former colonies have adopted the social institutions of colonialism. This premise guides and shapes the telling of my stories. In addition, I filter the stories through a western theoretical understanding of word and number (I will not capitalize the word "western" when it is an adjective. I will capitalize "West" when it is used as a noun). I use a western theoretical lens because, unfortunately or fortunately, I am of western culture, and those are the theories most accessible to me. However, the textile that I weave reflects my boundary walking. I use my observations of Togolese classrooms because they help me to see what I am blind to in my own culture's classroom. As a stranger, I am able to see social relations, classroom communications and actions differently than in classrooms where I am an insider. I am able to hear words and watch their usage and impact from the stranger's stance. It is from this place of difference that I perceive, interpret, and do my research (I first went to Togo in 1991–1993 as an educational consultant and learned French during this time. I later returned in 1996 to do my dissertation research). I want to highlight this "place of difference" or stranger stance because it identifies a borderland (Anzaldúa, 1987). My stranger's stance and the boundaries I experience and observe are a reflection of me: a wanderer, a child of western culture, a social theorist, and a friend.

It has been suggested by other social scientists that I include in my research commentaries from Togolese students/teachers, or that I co-author publications with Togolese teachers. When my project started I did not have the support for multiple authors. If I had had this opportunity, I would have welcomed it, and it would have clearly changed the nature of my project and my publications. In addition to financial obstacles there were also language and cultural barriers to co-authorship. My language of scholarship was English and theirs was French. Thus my project was embedded in social, cultural, and economic relations that have shaped what I present here.

I cannot "give voice" to any of the individuals I spoke with or observed. One teacher specifically stressed that I tell her story so the western world would know that her life and the lives of her students were difficult. Even as

I write this paper and you read it, it is not her voice. It is my understanding of her voice, communicated via words, language, and social practices. Her voice exists within a social context that "I" cannot capture. For example, in reading the reader assigns meaning to what he or she reads. My voice and my words exist within social relations and change as the reader brings his or her social relations to them to create understanding.

Similar issues are at work when studying communication and mathematical meaning within the classroom (see for example Lins, 1996). Paul Cobb reminds us that a "teacher can never know with absolute certainty what is going on inside each student's head" (1988, p. 92). I would like to extend that to the mathematics researcher as well. He or she can never know with even partial certainty what is going on inside the mathematics classroom. However, by observing, listening, inferring, and asking questions, the researcher or the teacher can create a model of what is going on "inside" while remaining flexible to alter, shift, and augment that model as more information is gathered. Thus, as researchers and teachers we use empirical data, teacher's/student's actions and teacher's/student's words to constrain and expand the construction of new knowledge structures (Cobb, 1988).

I ask my readers not to read my stories as "their" stories but to read them as "my" stories or perhaps "our" stories. They are stories in a socio-cultural context. I do not wish to remove them from that context by claiming authority, objectivity, validity or Truth. The authority, objectivity, validity and Truth does not exist from me the writer, but from the reader when the words resonate with a story from the reader's world, and provide the reader with a clearer understanding of his or her own socio-cultural context. I am attempting to write sociology/anthropology in the tradition of Dorothy Smith (1987) that develops consciousness of society and social relations. My research voice is neither one of standing outside society and social relations and talking about it objectively, nor is my voice that of an insider writing about social relations and society from within as informant. My voice is that of an outsider who is within a society and has social relations that are reflected in my understanding and production of experiences and knowledge.

I write in the shadow of traditional anthropology and "new ethnographies" (Rabinow, 1977, Taussig, 1987, Behar, 1993, Martin, 1994, Rapp, 1999). Social scientists are trying to move beyond studying "the Other," as Malinowski (1954/1948) did, and to produce reflective, reflexive, democratic, and feminist research. However, it is still "research" that carries with it western culture and western institutions. I am aware that "I" write about another culture and other people and that officially their voices are not present by co-authorship. Neither option is the perfect solution, as there is the risk of "colonizing the other" when "their" voices are included or excluded. This is what makes research potentially so empowering and disempowering. Thus, I am arguing that we need to be just as careful of "colonization" when we include the Others' words, commentaries and stories directly or indirectly. The current methodology for doing this is through reflective and reflexive research where the researcher is clearly present. This is to admit that I, as researcher, am a cultural construction of my culture and the institutions I work within, or in other words, that I have been colonized by my culture and its social institutions. By clearly positioning myself in this paper, I (as researcher, communicator and teacher) am negotiating this colonization that can serve as a venue for speaking and for silence.

Teachers would often ask me to "critique" their lessons, as their superior and a western expert. This is a role that I would always refuse because I did not want to be a western authority. I wanted to acknowledge the teacher's expertise and knowledge. Instead I embraced the role of ethnographer and apprentice as defined by James Spradley and David McCurdly:

> [i]nstead of studying people, the ethnographer learns from them.... When "subjects" become teachers who are experts in understanding their own culture, the relationship between investigator and informant becomes quite different. The investigator will ask those he/she studies to become her/his teachers and to instruct him/her in the ways of life they find meaningful. (Spradley & McCurdy, 1972, p. 12)

In asking Togolese female math educators if I could be their apprentice and in becoming one, I altered the traditional power relationship between "yovo" and Togolese. As a "yovo," a Westerner or foreigner, my historic role in Togolese culture would have been one of "teacher," "source of material wealth," "protector," and "colonizer." Rarely in Togolese memory have foreigners taken the posture of apprentice; typically their role is expert, colonizer, and authority. The apprentice system is alive and well in Togo, and is used to educate mechanics, hairdressers, seamstresses, and tailors. Its strong existence provided a cultural metaphor that helped to facilitate my role as ethnographer. Therefore, I seemed to be accepted by many women as a person to talk *with,* not as an expert to defer to. Women who teach mathematics and science are respected leaders in their communities and often counsel community members on a variety of issues. As they counseled and educated others, they educated me about their lives, their students, their schools, and their culture. In exchange I shared with them my life, my experiences with students, teaching, and schools in my culture.

## INTRODUCTION: NUMBER AND WORD

> Let us introduce the refinement and rigor of mathematics into all sciences as far as this is at all possible, not in the faith that this will lead us to know things but in order to determine our human relation to things. Mathematics is merely the means for general and ultimate knowledge of [humans]. (Nietzsche, 1974, p. 246)

"All is number" is a concept that my students recently encountered through the assigned reading outlining how the history of science interweaves with culture and religion (Wertheim, 1995). In the class discussion some students were sceptical. Therefore, we listed things that did not appear to be "number," like, English, the language we were using to discuss the very idea of all is number. I remember thinking, "okay, are we going to disprove 'all is number' with our first example"? Instead of sharing my doubts, I asked the students to think of a way in which language relates to number. After several moments of silence, one student suggested singular and plural or the idea that language communicates quantities. This example appeared to make believers of the sceptics and the students quickly recognized the remaining examples as relating back to number.

The students found the "English" to be persuasive because they are used to thinking of english and mathematics as separate disciplines. Students learn early in western mythology that one is either good in mathematics and sciences or good in English and humanities but not both. Since English passed the "all is number" test, the students were satisfied with Wertheim's claim, and our classroom discussion moved on to other things. The acceptance of "all is number" occurred because the students found the evidence compelling. It was compelling precisely because it was based upon a popular dichotomy and brought it into question.

In that particular discussion my students and I avoided, side stepped or had no need of a discussion on the similarity of words and number or language and mathematics. Linguists show us how language and metaphor are cultural constructions and reflect/shape how we think, act, and feel (Lakoff & Johnson, 1980). I suggest that number and words share the same role. Therefore, in this paper I want to address word and number in order to think about mathematics, communication and the classroom in Togolese secondary schools and to highlight what is occurring in a postcolonial classroom that has adopted and adapted the social institution of the colonializer.

As a sociologist my discussion is not unique to mathematics classrooms because it focuses on understanding and highlighting social relations. As a sociologist of mathematics, this discussion is of particular importance to mathematics educators and researchers because of the popular belief of mathematics as universal, abstract and pure.

## MATHEMATICAL WINDOWS

If we take Friedrich Nietzsche (1974) at his word, then it is through the use of mathematics that we can understand our relationship to things. Mathematics can teach us about ourselves. As a social theorist I understand mathematics as a special case of a human construction that leads us back to ourselves even as it is a tool of knowledge production and communication. It has the ability to instruct us precisely because it is our construction. In western culture mathematics has always had a very unique stature in the knowledge tool kit. It is so omni-present in western thought that statements like "all is number" are regarded sceptically and then accepted by the young initiates while the idea of "all is poetry" is rejected and declared inconceivable by those very same youth. Mathematics has traditionally been viewed by scholars and lay persons alike as "neutral" or "culture-free" (value-free) knowledge. This view has been challenged by those who understand mathematical knowledge as a social construction (Bloor, 1976, Ernest, 1991, Restivo, 1992). Our increasing appreciation that "World" mathematics is "Western" mathematics, implicates mathematics and number in the imperialist and colonizing adventures carried out by the European powers in the 19th and early 20th centuries (see Bishop, 1990, cf. Restivo & Bauchspies, 2000). As well as being a cultural tool for expansion, perhaps mathematics can also be a tool for introspection. Or, in other words, mathematics is a window or a door into human relations or the worldview of humans. For the western body, mathematics is a well-trimmed sail that provides insights into our relations to the world. Our mathematics reflects our culture, communication, and selves if we allow ourselves to view it from this perspective.

> Mathematical number contains in its very essence the notion of a mechanical demarcation, number being in that respect akin to word, which, in the very fact of its comprising and denoting, fences off world-impressions. ... It is by means of names and numbers that the human understanding obtains power over the world. (Spengler, 1962, p. 43)

In determining human–world relations, mathematics, sciences, and inquiry bring order to our world specifically through the use of words and numbers that "fence off" the world into ordered knowledge that we build upon through communication. Nietzsche reminds us that we do not communicate thoughts, only movements and signs whose pathways create thoughts (1967). These fences and pathways are structured by the social content in which the words and numbers are repeatedly used, thus turning paths into well traveled highways with well established fences. It is the job of the social theorist to point out these fences and pathways so that their "permanence" and "foundations" are observed or understood from new perspectives. This may strengthen, weaken, or highlight some aspect of the fences and pathways previously unnoticed or hidden. Names and numbers give human comprehension "power over the world," and simultaneously these symbols are cultural triggers that require specific responses in community members. I would like to look at these specific responses and how they work within the culture of the classroom and larger community.

> They are not bare words, but words that do answer to certain responses; and when we combine a certain set of symbols, we inevitably combine a certain set of responses. (Mead, 1962, pp. 268–269)

Thus I propose to pay attention to words and numbers in Togolese classrooms to see what fences and pathways are being created by the community to elicit certain responses from members, and to learn about the relations between ourselves, things, and other cultures. However, I would like to remind myself and the reader of the power of symbols, be they words or numbers, precisely because they are boundary makers and markers.

> Compared with music all communication by words is shameless; words dilute and brutalise; words depersonalise; words make the uncommon common. (Nietzsche, 1967, p. 810)

## Setting

In the early 1990s I was an educational consultant to middle school Collège d'Enseignement Général (CEG) teachers in Togo. Later I returned as an ethnographer to do research on women, science, and education in West Africa. My research and this paper are a reflection of my experiences living in Togo and working in the educational system. During the course of my study, I interviewed and visited twenty female science and mathematics educators at their homes and schools. I had contact with an additional twenty-five women through questionnaires. Before I would meet a woman for the first time, they had the possibility of self-selecting themselves out of meeting me, and many did. Some women taught in public schools and others in private schools. Rather than using a pseudonym for each woman and her story, I have created a composite who I call Mme Fazao or Mme Kousé. I use the names interchangeably through out the paper.

Formal schools in Togo have a social and cultural history that reflects two different colonizing nations, several missionary traditions, and its own traditional education system. Attending school in Togo is a rite of passage to community power, to financial stability, and into the "educated" class in contemporary Togo. When Western style formal education was first introduced by the missionaries and colonizing nations it was not immediately embraced by all Togolese (Bauchspies, 1998a). A hundred years after western style schools were first introduced, the processes of acculturation have been at work. Now, school maturation, particularly in the disciplines of science and mathematics carries high status, as it does in western nations. Schooling in Togo has come to represent the opportunities found in western nations, and a means to the power that the West has symbolized in West Africa since it first colonized it with guns and "superior" technology (Adas, 1989). There is a fetishization of the West in Togo that has implications for the disciplines and discipline of the classroom (Bauchspies, 1998b). The individual who successfully completes studies of mathematics and science in Togo can then move into the prestigious culture of the West and the powerful realm of the Togolese functionary. Formal education in the last eighty years has been an economic ticket that has entitled the holder to well paying work for the government or a Western organization. In the last decade, Togo has experienced political and economic troubles, and not all educated individuals have jobs or jobs in their field. In the twenty-first century the post-colonial relationships between financial stability, educational attendance, and educational achievement are shifting.

## Classroom and Lesson

In 1996 I visited the classrooms of female CEG teachers teaching physics, biology, agriculture, physical science, chemistry, and mathematics throughout Togo. Below I describe an "average" mathematics class in an "average" sized community in order to present a sense of what I observed and experienced. This class is a sixième class, or the first level of middle school. There were approximately ten rows of benches, three columns wide, with two students at each bench (sixty students in total.) This is a medium sized class for Togo, where classes can be as large as ninety or a hundred students. The female-male ratio is approximately forty-sixty. The teacher, Mme Fazao started the class with a review of multiplication:

example 1: $6 + 6 + 6 + 6 + 6 = 5 \times 6$
example 2: $3 + 3 + 3 = 3 \times 3$
example 3: $x + x + x + \dots |x| = x \times \mathrm{b}$

When Mme Fazao asked the students questions about examples 1 and 2, numerous hands were raised. Only a few hands volunteered to attempt to answer questions on example 3. The first student suggested the answer "x × 9," to which the teacher responded "Why the number nine? We do not know that number." The second student's (also a male) answer was accepted as correct. Mme Fazao ended the discussion by reciting a definition that students immediately began to copy into their notebooks. She provided additional help by writing some of the phrases of the definition on the blackboard. The teacher is a master at working with numbers, and teaches her apprentices the art of number working through examples, question/answer, and lecture. She is both master number worker and leader of the class. Her power resides in her ability to organize numbers and students.

Mme Fazao's next example was: $4.18 \times 3 = 12.54$. She asked the students to name the factors. The second student who responded identified the factors correctly. Mme Fazao then proposed if $b = 1$ then $a \times b = ?$ and if $b = 0$ then $a \times b = ?$ Students indicated their desire to answer the questions by raising their hands. After the correct answers were identified, Mme Fazao asked them, "how did you get your answer"? In this way they had a brief discussion of the properties of multiplication. During this discussion Mme Fazao gave the following example: $3 \times a \times y \times 22 \times 0 = ?$. Students responded with 66, 22 and 0. I was watching the girl who said 22 when another student said 0. Upon hearing the third answer, the girl's body posture changed and appeared to reflect her new understanding that 22 was not the right answer while 0 was.

The primary actor in the classroom is the teacher. The students become secondary actors when they raise their hands, are called upon, and respond to questions. These secondary actors are learning the practice of numbers from the teacher. However, many students are silent. Some are silent by choice, others raise their hands and remain silent. The size and structure of the class makes it difficult for all students to be actively involved. Many students become silent observers of formal number working while a few become observable participants.

The discussion was interrupted at one point by Mme Fazao intercepting a note being passed between students. She read the first line of the note out loud and then reprimanded the student for writing notes and not paying attention. The guilty student hung his head in shame during the reprimand. Not only does the teacher guide, direct and instruct students in the specifics of working with numbers, but she also guides, directs and instructs in proper behavior.

A female student volunteered to go to the board to calculate $1.25 \times 12$. She did not have her shoes on and did not have the time to put them on, so she headed to the board shoeless. Wordlessly, another student quickly put her sandals in the aisle and the shoeless student stepped into them on route. This happened in a matter of seconds with no apparent communication.

The student wrote:

```
  1.25
  x 12
 -----
   250
  125
 -----
 15.00
```

The student's work prompted Mme Fazao to ask, "how do you know where to put the decimal point"? This provided a chance for a student to state the rule for placing the decimal. Mme Fazao stressed the importance of counting to find the place for the decimal. The lecture/discussion progressed to the commutative property of multiplication. After writing out the definition she gave an example. The students carefully wrote the definition and example in one notebook, and did the calculations for the example in a second notebook. This was done in order to keep the first notebook clean and orderly. It is an orchestrated dance that all students do almost simultaneously, cued by the words and movements of the teacher. Near the end of the hour, Mme Fazao assigned a homework problem: calculate two ways of finding the product for $3 \times (1 + 0.4) = ?$

Before dismissing the class, she passed back a homework assignment graded in the French style, with the highest grade in this class being 16/20 and the lowest zero. Most of the grades were between 6/20 and 10/20. Several were between 0/20 and 2/20. Mme Fazao announced the grade of each student as she returned his/her paper. At one point she asked the students "how many of you are returning to elementary school?" This appeared to both chastise students and to encourage them to excel because some students became silent while others sat taller on their benches (Bauchspies, 2000a).

## MOVEMENTS OF NUMBERS AND STUDENTS

In this particular sixiéme class we find numbers being manipulated by definitions, theorems and properties that are determined by an external expert number worker, a mathematician, or a text. The manipulation of the numbers by these definitions is enforced by the local number worker expert, the teacher. The apprentices or students are being trained to apply these properties to groups of numbers in particular formations. The students are expected to recognize and identify the groups of numbers and their formations in order to perform the called for response at the desired moment. All over Togo, we can find a similar pattern in mathematics classrooms. Perhaps all over the world, we can find a similar pattern in mathematics classrooms based on world mathematics. This is a common pattern for transferring knowledge within a hierarchical social structure and has been well researched and criticized elsewhere (Apple, 1982, Freire, 1971). It provides order and a pattern for human relations. It creates the well-traveled pathways of communication, through imitation of signs, symbols, and movements for students, teachers, and mathematicians. Remember that the process of traveling these pathways is one of practice, imitation, and shifting understanding.

The manipulation in the classroom is not limited to numbers. The classroom communicates both mathematical content and proper decorum. Content and behavior are often thought of separately. However, they work together to create and to reinforce the pathways and fences of the larger community. The use of numbers by teacher and student at the blackboard and on paper mirrors the shuffling of students into places and numbers. This is exemplified when the teacher reads the students' scores out loud. The communication of and about numbers in a mathematical classroom is not a one-way movement, for the teacher is shaped by the authority she carries, the knowledge she communicates, and the responses of the apprentices and experts. The classroom communication affects all individuals and unites them with its movements and symbols.

## MATHEMATICS AND LANGUAGE

> One never communicates thoughts: one communicates movements, mimic signs, which we then trace back to thoughts. (Nietzsche, 1967, p. 809)

Neela Sukthankar points out the difficulties of learning mathematics, the "universal language," in a language that is not a student's native tongue (1995). Students in Papua New Guinea, like Togolese students, are learning mathematics in the national language (English or French) that is often students' second, third, or fourth language. Sukthankar concludes that this has a "substantial impact on Papua New Guinean students' development of many higher-order mathematical concepts and procedures" (1995, p. 139). In addition teachers may depend upon rote learning because students do not have a "satisfactory comprehension of the prerequisite ideas" (Sukthankar, 1995, p. 139). I observed Togolese teachers using rote learning to help students learn the discipline of science and mathematics in a second, third or fourth language because of the scarcity of textbooks and mathematical literature.

The teacher's careful review of the placing of the decimal, and the definition and properties of multiplication serve to illustrate the necessity of teaching the established mathematical language and way of thinking to perpetuate the culture of mathematics. The numbers and symbols appear to be universal, however, they only become so through their "shared, common, understood meaning, which is not up to [the student or teacher] to choose ... The social meaning of 'two,' 'three,' and 'five' forces us to agree that 2 + 3 = 5" (Hersh, 1994, p. 18).

The mathematics classroom is an "evolving social form of life" where communication goes beyond words and the direct dialogue between student and teacher (Ernest, 1998). Communication is occurring in the forms of students passing notes, sharing shoes, and raising or not raising their hands. This matches Paul Ernest's description of school mathematics' discourse as composed of living conversations, personal interactions, written interactions, and semiotic artefacts (1998). "Conversation can be understood on three levels," according to Ernest (1994, p. 36). The first level is interpersonal communication that occurs between individual and individual or individual and text, as seen by students passing notes or teacher-student dialogue. The second level is cultural, and is the summation of interpersonal conversations, such as the class moving in unison to take notes. The third level is internalized private conversations, as illustrated by the student whose facial expression changed as she compared her answer to other potential answers. This is not "simply communication"; it is part of mathematics communication, knowledge, and culture. "Mathematics is inescapably conversational and dialogical in an immediate and overall way, for by its very nature it addresses a reader" (Ernest, 1994, p. 38). It can be difficult to see sharing shoes, passing notes, and raising or not raising ones hands as mathematics. However, these are all part of the joint activity of thinking collectively or of creating mathematical knowledge. Neil Mercer argues that "language is a tool for carrying out joint intellectual activity, a distinctive human inheritance designed to serve the practical and social needs of individuals and communities and which each child has to learn to use effectively" (2000, p. 1). Classroom language and communication of all sorts is not simply a vehicle of information transfer but a means to think collectively. The sharing of shoes symbolizes this collective thinking.

Reuben Hersh reminds us that "mathematical objects are a certain variety of social-cultural-historical objects. They're distinctive. We can tell mathematics from literature or religion. Nevertheless, mathematical objects are shared ideas, part of the culture of a significant subculture" (1994, p. 16). For most students, mathematical objects are not restricted to numbers, equations, graphs and symbols. They also include the test scores, the passed note, the raised hand, and the shared shoes.

## MOVEMENT, SYMBOLS, AND MIMICING

Anna Sfard's (1994) work on communication and the classroom establishes that there are different epistemologies, ontologies, metaphors, and practices being used by students and teachers. She suggests that, in addition to teachers increasing their awareness of these differences, students be provided with as many experiences to practice mathematics as possible. The resources of teachers and students directly influence the experiences available to "practice mathematics." In Togo, where books and paper are an expensive commodity and not available to all, Mme Kousé would often instruct her students to practice mathematics at home on the household blackboard. Every Togolese home with school age children has a blackboard and chalk because these are affordable educational resources. Mme Kousé instructed her students to change the numbers in the problems she gives them in class or they find in the text, and then to solve the problem. This was her strategy to provide additional mathematical practice and experience with the goal of increasing student success on exams.

David Bloor describes "the ostensive learning of number concepts [as not consisting] solely of being shown samples of two things or three things or four things. It also consists in training in the technique of counting and adding." (1994, p. 24). Sfard argues that students experiencing and practicing mathematics is more important than simply hearing about mathematics (Sfard, 1994). Jill Adler furthers Sfard's discussion of explicit mathematics language teaching by highlighting how a teacher grapples with "transparency" in communicating mathematical knowledge to her class (1999). Adler suggests that teachers need to help students to balance the visibility of mathematical language and the invisibility of mathematical language in mathematical practice.

Numbers, words, and discipline are visible aspects of mathematical language in a Togolese classroom. Teachers stress the careful placement of mathematical symbols verbally and by their attentive stance to student communication. At one point I wondered about this balance of the visibility and invisibility of mathematics in the lesson described above. It is rare for a student to have access to the text because books are expensive and not many people/institutions own them. However, even if a textbook were available to a student, she/he would find few word problems applied to each concept taught. These word problems may or may not be visible applications of mathematical practice to students. Institutional structures like large class size, limited resources, heavy teaching loads, and end of year exams also work against the teachers making math visible. (I have heard teachers on at least three continents identify and complain about these four issues.)

In Togo, both teachers and students appeared comfortable with the practice of mathematics based on numbers, equations, formula, theorems, and properties. Mathematics was something to be practiced in the classroom, as a sacred activity that is performed at the shrine with sacred items: blackboard, chalk, pen and paper by an elite group. By definition and practice, mathematics appeared to be very separate from everyday activities. In the West, a popular concern of educators is how to make connections between classroom knowledge and everyday activities. As a western educator I asked teachers if there was harmony or conflict for secondary school students negotiating between school and home. One teacher replied that there is no conflict, only harmony for the good student. I did not understand this at the time, and thought that she had misunderstood me. I later realized that it was me who had not understood, because of my western sight. The "good" student of Togolese or Western culture, is the one who is able to negotiate the well traveled paths and see the fences for what they are, cultural boundaries that contribute to a community's communication, knowledge, and consciousness.

When I first entered a Togolese classroom in 1991, I had the feeling that I was observing only half of what I thought education to be. I asked myself, the educator: where is the higher level thinking, the critical thinking, and the innovation? Since my first visit to Togo, I have trained as an anthropologist, sociologist and social theorist. I saw, experienced, and observed what other postcolonial scholars had named; the adoption or imposition of colonial social institutions on the colonized. However, what I comprehend now that I did not understand then, was that calling colonized education banking education or rote memorization is another act of colonialism. As a Western stranger in a former colony, what I saw most clearly was the reflection of "my educational system" in theirs. This allowed me to see aspects of my education that I was blind to in my culture: banking education, the discipline of the body, and power through the use of words and numbers. Thus, Western schooling in a western culture provides cultural blinders that the "good" Togolese student does not have.

In western schools in western cultures, power is buried deeper or differently behind different language, cultural triggers and responses. However, banking education, the disciplining of the body, and colonialization of the mind are occurring (Bauchspies, 2000b). Thoughts are being communicated in "our" schools and "theirs." As with thoughts in any place, we can theorize about what is actually occurring, based upon our observations, conversations we hear and the contexts in which we observe. What we see, the symbols and movements, are put into context with relations based upon our well-established pathways. The trick is to learn to see the "others'" well-traveled pathways while experiencing one's own. This is not an easy task. We need to recognize that in the mathematics classroom many sorts of relations are being dictated through symbols and movements. It is the lump sum of them that amount to communication, thoughts, and consciousness. It is precisely because we do fence off our pathways that we can communicate, and it is also the communication that broadens, narrows, blocks, and opens up new routes of consciousness. I have learned that as carefully as I might try to avoid the pathway of judging the "other" as different and inferior, it is a very deep one. I can now name that which has haunted me: my own colonialism. Through naming it, I risk diluting and depersonalizing it. However, I also can use it to gain control of it.

## COMMUNICATION, CLASSROOM AND COLONIALISM

Typically colonization is used to identify an exclusive and exploitative relationship between two nations in the economic arena. During the process of colonization one group or nation may become acculturated or assimilated when that nation becomes indistinguishable from the other. When talking about individuals, students, and schools, we generally refer to it as socialization or perhaps enculturation. Socialization is defined as a lifelong process that prepares individuals to participate in social systems, to reproduce and maintain the system, and to create a social self (Johnson, A., 1995). In other words, schools socialize students to be members of society. It can also be said that schools enculturate students and children by teaching them skills and behaviors that they will need in order to survive within the society.

When I talked to individuals that knew the French system of education, they knew precisely what I was talking about when I described the Togolese classroom with its strict attention to language, notebooks, exams, memorization, discipline and hierarchy. I was told Tunisian classrooms, French classrooms or Senegalese classrooms shared these characteristics that are attributed to the socio-cultural history of being French colonies (Manning, 1988). A major instrument of colonization was the educational system, language of education and disciplines of education.

Education was disciplining the "other" in the ways of the "self." The colonized nation was colonized by a foreign educational system. Students of the colonized nation were colonized by the teachers and administrators of this foreign educational system. To put it another way, education colonizes students. I am purposively using the term colonization over socialization or enculturation, because I want to emphasize issues of power and culture. Renuka Vithal and Ole Skovsmose discuss the need to include culture and power in discussions of ethnomathematics/mathematics (1997). One way to do this is to describe and discuss education as colonization in order to highlight what we normally do not see. One reason why a discussion of education as colonization uncovers the hidden is because colonization has a negative value, while education has a positive value in western culture. It is easier to critique something that has negative connotations than it is to critique something we value highly.

The classroom I described previously happened to have been a Togolese one. However, the pattern of the teacher presenting examples, asking students questions, defining a property/characteristic, and giving practice problems is a common pattern found in mathematics classrooms around the world. Another common pattern is the community of learners working together to learn the discipline, in a common language that gives them the tools to learn the new language of numbers. In order to learn this language of numbers, the learners and the teacher must be colonized. The young girl said twenty-two and was forced to reconsider when zero was pronounced correct. I "saw" it as "the light bulb going on." Could I just as easily have described it as the colonization of her mind/body?

What does this mean for mathematics educators and researchers? In my introduction, I talked about the problems of giving voice to the teachers I worked with and about validity and objectivity. Mathematics is the one western discipline that is held to be neutral, abstract, objective, and pure. Western culture values describing the world by the use of numbers, for example in the behavior of fluids, atoms and disease. An equation is considered pure, unquestionable and trustworthy when telling us about the best gasoline, compound or medicine for a particular purpose. We believe in numbers. They provide answers. We are colonized by numbers. Western culture readily acknowledges "all is number." We recognize the colonization of Africa. Do we recognize our colonization by numbers? It occurs in the classroom, in the market place, in the home, and in casual conversation. We are not immune to the power of numbers. Every time we as educators give a definitive answer based upon a number, we are reinforcing this colonization. It does not mean that we stop teaching. It does require that we pay attention to our mathematics communication in and outside the classroom, that we pay attention to how we use numbers and words, what roles we give them in our everyday lives, and what roles we do not give them.

## ROLES OF NUMBERS: REPETITION AND RESISTANCE

Students in the Togolese secondary classroom may range in age from 10–30 years old. The older students are generally "repeat" students who have been at that particular grade level for three or four years (or possibly longer). Officially, students can repeat a year twice at the same school, but it is not unusual for a student to repeat a class four or five times by switching schools. Technically, there is an upper age limit on school attendance, but that is easily bypassed by forging identity documents. One school administrator explained that the beginning of the year was very difficult because everyone seeks places in already overcrowded classes. In her role as *grand frère* she feels obliged to help.

The older students are often viewed as troublemakers by the teachers. Repeat students disrupt the class by asking questions designed to distract, annoy or entertain. They generally sit at the back of the class because they are the tallest students. They are the furthest removed from the teacher and sit next to similarly fated peers. Should a teacher want to walk to the back of the room to supervise them, it may be impossible to do so because of the large number of students in a limited space. The students in the back of the room tend to talk and make noise that may prevent other students from hearing dictation. The repeat students already have notebooks for this class and for this subject matter. Therefore, they are less inclined to take careful notes.

In one class, I watched an older boy hassle a smaller boy sitting directly in front of him. The older student was sloppily dressed, with his shirt unbuttoned. (Generally, school authorities criticize students for sloppy dress and it is unusual to see a student with an unbuttoned shirt in the classroom.) He was clowning with his neighbor across the aisle. He tapped and poked the younger boy in front of him with a pen when the boy leaned back. This went on for a few minutes until the younger boy fought back by grabbing the pen and throwing it on the ground. The older boy just laughed at the younger one.

Sometimes students interrupted the lesson by asking questions that caused the class to laugh or created a tangent in the class discussion. In a biology class, Mme Fazao was discussing the structure of the skin and pigmentation. A student, well aware of my white skinned presence, asked her why white-skinned people lie in the sun to tan themselves and black skinned people do not. Students giggled nervously when he asked the question. She attributed this question to a repeat student who was trying to create a diversion and hinder the progress of the class. Another time a student answered a question with an incorrect French word that was a cognate to the correct word. Students laughed and a few minutes of dictation were lost. After class the teacher identified the clowning student as a repeat student.

Repeat students remain in school to avoid working in the fields, becoming apprentices, or doing other labor reserved for uneducated people. Repeat students are refusing the life outside of school. However, they are also resisting school by not being "serious" students. Teachers often have little sympathy for repeat students. Teachers claim they are lazy, and the repeat students claim teachers are lazy. Repeat students attempt to disrupt or hinder the class, which ultimately hurts their fellow students if the curriculum is not finished by the end of the year. One day, Mme Fazao walked out of class because the students in the back of the room were being too noisy for her to talk. Even if they are silent, their bodies contribute to the problems associated with overcrowding in the classroom. Their presence may contribute to other students not passing and becoming "repeat" students. It is like a logjam; once one log gets caught, it prevents other logs from advancing. As repeat students resist the life outside of the classroom, they also resist the culture of the school.

Schools are starting to group students by age in the same grade to separate first time students from repeat students. Mme Fazao taught three troisième classes in which students acted and behaved differently. She explained that the class with smaller, more immature students was composed of first time students. Mme Fazao said she spoke gently and slowly with this class, because it was their first time in troisième. She described these students as polite and respectful, and they paid attention when she rebuked them. Mme Fazao's other troisième classes were described as older, more rowdy and disrespectful. She felt that their attitude said "I already know this, I don't need to listen to you." One school administrator was surprised to find "older" students in the "younger" students' classroom. The repeat students had realized that the school administration separated younger and older students by class in the same grade level. They resisted this by lying about their age in order to be placed in a "younger" class. Students in the "younger" classes were more likely to pass the final exam than students in the "older" classes. First time students who happen to be the younger students are more likely to pass exams than repeat students. As a result, repeat students believed that their presence in a "younger" class increased their chances of success.

Mme Fazao explained that her school was strict about repeat students. Often they would be asked to leave, and if transfer students did not have a certain grade-point average they would not be accepted. To insure her own children's success in an overcrowded CEG, she enrolled them in a private primary school with small classes to give them a solid foundation. When her children are ready they will go to a public CEG. Mme Fazao explained that public secondary schools offer better-educated teachers, and if her children are strong students they will learn more, even if the class is overcrowded.

The manipulation of numbers by repeat students is a direct reaction to the classification of individuals by the "learning machine" which supervises and groups its objects (Foucault, 1979, p. 147). The categorization by the school authorities resulted in the repeat students' resistance. Repeat students resisted being made objects of power while creating their own hierarchies, that are in turn parasitic on the existing system. These parasitic hierarchies are fluid, as the students may eventually be pushed out of the system, and simultaneously new repeat students are being created by the learning machine. As parasites on the system, repeat students have become disciplined and are part of reproducing the privilege of successful students (cf. Willis, 1977). Repeat students exist for teachers and administrators in the hedgerows between the accepted roles and pathways of teacher and students. They disrupt the well-traveled pathways and accepted fences while simultaneously reinforcing them.

Number practices have defined theorems and properties that dictate how to work with the numbers. When working with a time line, numbers progress in a uniform manner through time or space. The concept of age is in the same family as timelines that measure equal quantities of time by augmenting the unit by one. However, the more practiced number worker can design and utilize specialized forms of timelines that appear to defy the average number worker's concept of a timeline. These repeat students are master timeline workers and understand the numbers as defining relations within the community. The forging of documents that indicate a student's age is a manipulation of number as a linear concept. There are students who do not know their actual age or date of birth because documents are so frequently forged that they cannot be relied on. One student told me that he had two ages and explained how to calculate them. My linearly minded mathematics never mastered his system, while it worked well for him and others.

Gary Urton pointed out that: "every system of arithmetic and mathematics is, in the end, an 'art of rectification'." (1997, p. 218) Number manipulation within a culture is done to create "equilibrium, balance, and harmony in relations between and among number sets, groupings, and collections" (Urton, 1997, p. 218). Urton came to this conclusion because in Quechua society the goal of arithmetic is to establish balance in the distribution of resources. Thus the students' number practice of determining their age was a way of accessing resources in their communities and a way of colonizing numbers to serve them. During the course of a repeat student's mathematical education, he or she has encountered mathematics not as a unity, but as a "multiplicity of different and distinct although sometimes overlapping social contexts [that are] in and out of school, and in and out of the institutions of professional mathematics" (Ernest, 1998, p. 234). The manipulation of their ages illustrates the control they have over number and their ability to work with it as a rectifying element.

Renuka Vithal and Ole Skovsmose discuss four strands of ethnomathematics (1997). I want to focus on the third and fourth strands with regard to understanding the repeat students. The third strand is described as exploring "the mathematics of different groups in everyday settings showing that mathematical knowledge is generated in a wide variety of contexts by both adults and children" (Vithal & Skovsmose, 1997, p. 134). Repeat students re-create their age, in order to generate a new relationship between the number or symbol and themselves. No longer does age symbolize the exact "number of years" that the student has lived. It now has a new meaning that co-exists along with the "number of years" meaning. A subgroup of Togolese students has generated numerical knowledge that works for them in their everyday world. Vithal and Skovsmose's fourth strand outlines connections between mathematics in the everyday/night world and the formal classroom. The repeat students' shifting of the meaning of "number of years" creates new relationships in the formal classroom, in the school and in the community between first time students and repeat students. "Mathematics not only creates ways of describing and interpreting reality but becomes a means for reconstructing reality" (Vithal & Skovsmose, 1997, p. 143).

## CONCLUSION

The sixième mathematics classroom seems quite typical of mathematics classrooms. A property of number is communicated by the teacher through examples, definitions, questions, and practice. Information about numbers is communicated in a process we normally call education and that I am calling colonization to highlight that ideas are being imposed upon students and teachers in an orderly fashion to provide them with the mainstream or dominant worldview in which to interpret, predict and order their experiences. I told the story of the "repeat" students who are using number and redefining it to work within their experience, and to help them both resist and reproduce the educational system. These students have been colonized by number, and through this colonialization have developed means to resist it by altering their social relations. In other words, the repeat students have colonialized number to re-create social relations. Thus in the classroom, even as we colonize our students there is the possibility that the students will colonize the knowledge and re-arrange its social usage. This reminds us that knowledge is embedded in a social context with social relations, even when we do not see it. We can see it by looking for where it is being reformed, often in the margins of our classroom, perhaps by the trouble maker, the "light bulb" student, the first hand to be raised, or the one who offers her shoes to a shoeless peer.

## END NOTE

I returned to Togo in the summer of 2001 and asked about the repeat students. Every teacher I asked provided a different answer: it is no longer a problem because the schools are not permitting students to repeat more than once; it is still an issue for the schools and communities. The current economic situation is very difficult and may be limiting the ability of students to repeat repeatedly and/or may be forcing students to keep trying school because the options outside of school are severely limited. The story that I have told here about repeat students manipulating number to create new social relations does not appear to be a form of mathematical knowledge that will have a long life and become institutionalized into the field of mathematics. However, it reminds us that mathematics is a means to "determine our human relations to things" (Nietzsche, 1974, p. 246) and ultimately to knowledge of self.

## REFERENCES


Adas, M. (1989). *Machines as the measure of men: Science, technology, and ideologies of Western dominance.* Ithaca, NY: Cornell University Press.

Adler, J. (1999). The dilemma of transparency: Seeing and seeing through talk in the mathematics classroom. *Journal for Research in Mathematics Education, 30*(1), 47–64.

Anzaldúa, G. (1987). *Borderlands La Frontera: The new Mestiza.* San Francisco, CA: Spinsters/aunt lute.

Ascher, M. (1991) *Ethnomathematics: A multicultural view of mathematical ideas.* Pacific Grove, CA: Brooks/Cole.

Apple, M. W. (1982) *Education and power.* Boston, MA: Routledge & Kegan Paul.

Bauchspies, W. (1998a). *Togolese female science educators: Instruments, bridges or innovators?* Rennselaer, NY: Rennselaer Polytechnic Institute.

Bauchspies, W. (1998b). Science as stranger and the worship of the word. *Knowledge and Society, 11,* 189–211.

Bauchspies, W. (2000a). Images of mathematics in Togo, West Africa. *Social Epistemology, 14*(1), 43–53.

Bauchspies, W. (2000b). Cultural re-constructions of an adoptive child: Science. *Cultural Dynamics, 12*(2), 237–260.

Behar, R. (1993). *Translated woman: Crossing the border with Esperanza's story.* Boston, MA: Beacon Press.

Bishop, A. (1990). Western mathematics: The secret weapon of cultural imperialism, *Race and Class, 32*(2), 51–65.

Bloor, D. (1994). What can the sociologist of knowledge say about 2 + 2 = 4?. In P. Ernest (Ed.), *Mathematics, education and philosophy: An international perspective.* (pp. 21–32). London, UK: Falmer.

Cobb, P. (1988). The tension between theories of learning and instruction in mathematics education. *Educational Psychologist, 23*(2), 87–103.

Ernest, P. (1994). The dialogical nature of mathematics. In P. Ernest (Ed.), *Mathematics, education and philosophy: An international perspective.* (pp. 33–48). London, UK: Falmer.

Ernest, P. (1998). *Social constructivism as a philosophy of mathematics.* Albany, NY: SUNY Press.

Ernest, P. (1991). *The Philosophy of mathematics education.* London, UK: Falmer.

Foucault, M. (1979). *Discipline and punish.* New York: Vintage Books.

Freire, P. (1971). *Pedagogy of the oppressed.* New York: Herder and Herder.

Hersh, R. (1994). Fresh breezes in the philosophy of mathematics. In P. Ernest (Ed.), *Mathematics, education and philosophy: An international perspective.* (pp. 11–20). London, UK: Falmer.

Johnson, A. (1995). *The Blackwell dictionary of Sociology: A user's guide to sociological language.* Oxford, UK: Blackwell Reference.

Johnson, B. (1995). Mathematics: An abstracted discourse. In P. Rogers & G. Kaiser (Eds.), *Equity in mathematics education: Influences of feminism and culture.* (pp. 226–234). London, UK: Falmer.

Lakoff, G., & Johnson, M. (1980). *Metaphors we live by.* Chicago, IL: University of Chicago Press.

Lins, R. (1996). *Struggling for survival: The production of meaning.* Sheffield, UK. BSRLM Day Meeting.

Malinowski, B. (1954). *Magic, science and religion and other essays,* New York: Doubleday Ancho Books. (Original work published 1948)

Manning, P. (1988). *Francophone sub-saharan Africa 1880–1985.* Cambridge, UK: Cambridge University Press.

Martin, E. (1994). *Flexible bodies: Tracking immunity in American culture from the days of polio to the age of aids.* Boston, MA: Beacon Press.

Mead, G. H. (1934). *Mind, self, and society: From the standpoint of a social behaviorist.* Chicago, IL: University of Chicago Press.

Mercer, N. (2000). *Words and minds: How we use language to think together.* New York: Routledge.

Nietzsche, F. (1967). *The will to power.* New York: Vintage Books.

Nietzsche, F. (1974). *The gay science: With a prelude in rhymes and an appendix of songs.* New York: Vantage.

Rabinow, P. (1977). *Reflections on fieldwork in Morocco.* Berkeley, CA: University of California Press.

Rapp, R. (1999). *Testing women, testing the fetus: The social impact of amniocentesis in America.* New York: Routledge.

Restivo, S. (1992). *Mathematics in society and history: Sociological inquiries,* Dordrecht, The Netherlands: Kluwer Academic Publishers.

Restivo, S. & Bauchspies, W. (2000) (Eds.). *Cultural Dynamics, 12*(2), 131–260.

Sfard, A. (1994). Mathematical practices, anomalies and classroom communication problems. In P. Ernest (Ed.), *Constructing mathematical knowledge: Epistemology and mathematics education.* (pp. 248–273). London, UK: Falmer.

Smith, D. (1987). *The everyday world as problematic: A feminist sociology.* Boston, MA: Northeastern University Press.

Spengler, O. (1962). *The decline of the west.* New York: Modern Library.

Spradley, J. & McCurdy, D. (1972). *The cultural experience: Ethnography in a complex society.* Prospect Heights , NY: Waveland Press.

Sukthankar, N. (1995). Gender and mathematics education in Papua New Guinea. In P. Rogers & G. Kaiser (Eds.), *Equity in mathematics education: Influences of feminism and culture.* (pp. 135–140). London, UK: Falmer.

Taussig, M. (1987). *Shamanism, colonialism, and the wild man.* Chicago, IL: University of Chicago Press.

Urton, G. (1997). *The social life of numbers: A Quechua ontology of numbers and philosophy of arithmetic.* Austin, TX: University of Texas Press.

Vithal, R. & Skovsmose, O. (1997). The end of innocence: A critique of "ethnomathematics." *Educational Studies in Mathematics. 34,* 131–157.
Wertheim, M. (1995). *Pythagoras' trousers: God, physics, and the gender wars.* New York: W.W. Norton & Co.
Willis, P. (1977). *Learning to labour: How working class kids get working class jobs.* New York: Columbia University Press.

CHAPTER 9

# SCHOOL MATHEMATICS

## Discourse and the Politics of Context

**Dalene M. Swanson**
***University of British Columbia, Canada***

*Western culture has made, through language, a provisional analysis of reality and, without correctives, holds resolutely to that analysis as final.*

—Benjamin Whorf, 1956, p. 244

## ABSTRACT

The aim of this chapter is to provide a sociological framework for the issues that influence mathematics classroom discourse and practice across varying contexts. The discussion locates these issues within the broader social domain at various levels of discourse and does not limit the discussion to classroom configurations and interactions only. My arguments address concerns about the way in which students, teachers, and members of the pedagogic communities are socially constructed in relation to mathematics discourse and practice and how these instantiate themselves within classroom communication, practices, and the differentiated forms of mathematics knowledge which are made available to them.

*Challenging Perspectives on Mathematics Classroom Communication*, pages 261–294

## CONTEXT, POWER, AND CONTROL IN SCHOOL MATHEMATICS DISCOURSE

School mathematics is not "neutral"! It is historically, culturally and socially constructed. These constructions are realized differently within diverse classroom settings and communities of practice, especially within different geographical and/or socio-political contexts (see Swanson, 2000). The specific mathematics classroom, within the context of the school and broader social domain, produces its own realizations of mathematics discourse and practice in relation to a variety of other mathematics classroom settings located within different communities. Consequently, classroom mathematics communication is not merely about "transmission" of mathematical ideas within a neutral context, or a conduit of "mathematics language" outside of school mathematics. Rather, it is a set of activities, interactions, or practices which are socio-culturally and politically situated, and serve to produce and reproduce, or contest, certain relations of power and control from within the broader social domain.

Power and control translate into principles of communication, which differentially regulate forms of consciousness (Bernstein, 1993). Power relations "create boundaries, reproduce boundaries... between different categories of groups, gender, class, race,... discourse,... agents" (pp. 116–117). These power relations are realised in contextually specific ways in classroom practices and discourse within diverse mathematics classroom locations, establishing boundaries, and reproducing and "recontextualizing" forms of social difference in the classroom and school. Classroom practices in different socio-political contexts articulate the ideologically produced priorities and selectivities of discourses within the realm of the social domain outside of the mathematics classroom. These broader discourses (re)shape the discursive and somatic practices within the classroom. In other terms, ideological resources are recruited from dominant social discourses, but recontextualized into school mathematics within specific pedagogic contexts, producing selectivities and silences. By *recontextualizing*, I mean, "the process whereby knowledge is delocated from its context of production and is relocated, transformed and elaborated in another" (Ensor, 1996, p. 2). It mediates the structural and material conditions of the broader society in contextually specific ways. This practice is less available or effective in reverse, in that the discursive power invested in mathematics is self-regulating and serves to police its boundaries vigilantly. Examples of recontextualizing practices are the use of domestic shopping practices to exemplify the usefulness of mathematics in daily living, or the mathematizing of cultural artefacts for the purposes of showing mathematics' "relevance" or "hidden nature" to "other cultures" (see Gerdes, 1986). Another common example of the recontextualization of practices outside of mathematics into its own practices lies in the conventional wisdom that mathematics contributes to modern industrialization, and is essential to economic and technological "advancement": mathematics as the passage to socio-economic "success," purportedly to serve both individual and national interests.

The agency of school mathematics and the role which *context* plays in establishing subjectivity (identity or personae) within the classroom is pivotal to this discussion. Power relations are not only communicated *through* the discourse of school mathematics, but are constituted *within* school mathematics discourse and practice in contextually different ways, albeit resonant instances of practice and discourse are produced across contexts. Forms of agency, which inhere in discourses within the broader social domain, are realized in the distributions of message and constructions of meaning which produce subjectivity within mathematics classroom contexts (see examples in Dowling, 1993, 1998; Ensor, 1991; Swanson, 1998; Walkerdine, 1988, 1989).

I now outline some theoretical perspectives of sociological approaches to pedagogic texts or activities. These perspectives have influenced my own approach to sociological issues in mathematics classrooms. In particular, I include some key concepts of the work of Basil Bernstein in his description of pedagogic contexts and discourse. I also focus on concepts of discourse, context, and subjectivity through the work of Paul Dowling in producing "a language of description" for the empirical analysis of pedagogic texts. Particularly, this language facilitates an analysis of school mathematics in the contexts of its production. This, along with some key concepts in Bernstein's model, enables a reading of an empirical study, which I undertook in a South African school. The focus of this small-scale study was to examine the construction of disadvantage in relation to the discourse of school mathematics within the context of the research school, with particular emphasis on the role of the categories of "race," social class and culture in assisting with the formulations and maintenance (production and reproduction) of such constructions. In this study, I showed how a group of students referenced in terms of race, culture, class and language difference, amongst other constructions, were positioned in terms of "deficit" and "difference" with respect to mathematics. In this way, socially constructed difference translated into discourse and practice within the school, which alienated these students from school mathematics.

## A SOCIOLOGICAL ANALYSIS OF SCHOOL MATHEMATICS IN CONTEXT

In order to engage with the sociological concerns I wish to address in this chapter, it is necessary for me to provide a few theoretical pointers to the conceptual issues underpinning them. This serves two functions. It provides some theorizing of the issues that will inform my discussion (and facilitate a critique of an empirical work in a later section), as well as provide some key concepts in the theoretical framework of my empirical study discussed in the next section. Thus, this section is organized in the following way: First, consideration is given to sociological descriptions in defining the role of cultural context or *ethos* of the school and the role of school mathematics in context; and, second, a discussion is provided focusing on discourse, context and subjectivity, particularly through Dowling's work.

### Sociological Descriptions: Some Theoretical Antecedents

Bernstein's model (1993, 2000) provides a strong theoretical perspective on how social relations are (re)produced and recontextualized within pedagogic discourses. (Some of Bernstein's terms have been adapted or refined over the years. For the most part, I am using his 1993 work here, discounting any nuanced shifts in meaning over time, as I am concerned with the core concepts to facilitate my discussion.) This model presents a description of the culture of school context and provides an interpretation of criteria by which certain positions of "strength" or "weakness" are held by students in relation to the culture of the school. Bernstein refers to how *instructional discourse* is embedded in *regulative discourse.* This way, school mathematics is an instructional discourse within the *discursive order* of the school and is embedded within the regulative discourse of the school's *social order.* Bernstein explains that the rules of the "social order" refer to expectations about conduct, character and manner, which index codes of class, gender, ethnicity and other forms of social difference (1993). This embeddedness refers to how discourses within the social domain are refigured in the classroom. Consequently, social domain discourses, such as those that index gender, social class, language difference, culture, ethnicity and intellectual ability, articulate with each other according to principles of power and control, mediating and transforming them within and through school mathematics and establishing identities for students and teachers with respect to mathematics. The construction of these identities in the mathematics classroom, works in conjunction with the particular cultural *ethos* of the school (see Bernstein, 1975), and the broader socio-political context, as well as conceptions of the "nature" of mathematical knowledge within specific communities of practice (Ernest, 1991). Particularly, certain prevalent constructions of "good practice" in mathematics teaching and learning are referenced within the social domain and recontextualized within school mathematics, serving to inform classroom practice. Constructions of "good" and "poor" practice work in consonance with constructions of "success" and "deficiency" in the classroom, indexing gender, ethnicity, class and "ability," thereby producing subjectivity and establishing identities for students and teachers in terms of these constructions and practices. These relationships are consonant with Bernstein's (1993) description of positions of "weakness" and "strength" held by students and teachers in relation to the culture of the school and, more specifically, school mathematics discourse. Thus, mathematics' position in relation to other instructional discourses within the context of the school assists in establishing these identities.

School mathematics, in general, holds a position of dominance within a hierarchy of instructional discourses. It maintains its dominance through the principle of a complex "social division of labor of discourses" (Bernstein, 1993, p. 118). Mathematics discourse possesses an "elaborated coding orientation" (Bernstein, 1993), or can be described as a case of "high discursive saturation" (Dowling, 1998, p. 103) as it displays a high concentration of symbolic content, level of abstraction and produces generalized utterances. Hence, mathematics, as a high discursive saturation activity, generalizes its field of reference to other activities, thereby mythologizing them (Dowling, 2001, p. 25). Mathematics castes a gaze over other ideologies, social practices and discourses, and subordinates them to its "regulating principles" (Dowling, 1998, pp. 121–136). As Dowling remarks: "Mathematics is a mythologizing activity to a degree that is probably unparalleled on the school curriculum" (1998, p. 2). This is based on the all-pervasive "universal truth" that mathematics is "useful," "necessary," or "relevant" to most, if not all, other disciplines, and possesses the power to describe the salient attributes of the "real" world. Consequently, school mathematics is often elaborated and recontextualized through discourses in the social domain that reify rationalism, science and technology, and lobby for its utilitarianism, economically and/or socio-politically. This can be extrapolated to include political and cultural practices, especially as they refer to political imperatives that view school mathematics as texts for producing social change. An example of this is the discourse on *People's Mathematics* that was most prevalent during the period of South Africa's anti-Apartheid struggle. Historically, school mathematics, more often than other subjects, has lent itself to social differentiation, through the power of its recontextualizing/mythologizing practices. Consequently, the positional discourse unproblematically connecting school mathematics to the aims and ideals of social justice and democracy can be utopian. (Refer to Dowling, 1998, for discussion on "the myth of emancipation" [pp. 11–17] and related contradictions in certain educational and social justice policies. Also, see Skovsmose & Valero, 2001, on debate about mathematics education's "intrinsic resonance," "intrinsic dissonance" or "critical relationship" with democracy and social justice agendas.)

Positions of "success" or "failure," established in accordance with constructions of social difference, are particularly divisive with respect to school mathematics, which most often possesses a "special voice" or "strong classification" (Bernstein, 1993, pp. 117–118) within schools. The *strength of classification* or *specialization* of a discourse is one of power and lies in its insulation from other discourses. This is particularly visible in schooling contexts where the cultural ethos of the school is more *stratified* (having a more pronounced hierarchy of discourses) than *differentiated* (with greater horizontal, and less vertical, separations) (Bernstein, 1993; 2000).

## Discourse, Context, and Subjectivity through the Work of Paul Dowling

While Bernstein's discourse is useful in providing a sociological description of the schooling context and the texts (activities) it produces, Dowling's work in particular, allows for an analysis of how subjectivity is realized in pedagogic texts. He considers how mathematics discourse is constituted and elaborated within different contexts, and what activities, practices and articulations of meaning apply. His work brings in an extension and refinement of Bernstein's concept of classification. It is in varying the strength of classification of a discourse/practice such as school mathematics (the delimitation or extension of its boundaries), with respect to *content* and *mode (expression)* that Dowling's "domains of practice" are arrived at (see Dowling, 1998, pp. 132–137).

Dowling's (1993) study is concerned with "the production of a language for the systematic sociological description of pedagogic texts" (p. 2). In particular, this can be applied to mathematical texts. In this study, Dowling defines the esoteric and public domains of a practice, such as school mathematics. The *esoteric domain* is that domain which achieves high specialization in both expression and content, and is highly categorized/bounded with respect to other practices. Within this domain are sub-domains, which we could refer to as *topics*. In school mathematics, these "topics," such as algebra, Euclidean geometry, transformation geometry, trigonometry, and so on, are evident. While these topics are interconnected, each maintains a degree of positivity and separateness. The esoteric domain regulates the practices of an activity (Dowling, 1998, p. 135). The *public domain*, on the other hand, refers to practices that exhibit greater context-specificity and the public domain appears as a non-specialized practice. It remains subject to the principles of the esoteric domain that regulate it. Lacking generalizability, these principles are not well expressed in the public domain. Nevertheless, it is through this domain that apprentices enter the activity before specialization (Dowling, 1998, p. 136). (Over time, there have been some refinements to Dowling's "domains of practice" and to the elaboration of his language of description. My concern here is with the main sociological principles.)

Dowling (1993) refers to the positioning of subjects (positioning strategies) and the distribution of messages (distributing strategies) to these positions as *textual strategies.* They have analytical separation but work concomitantly with each other in practice. *Apprenticeship* and *alienation* are textual strategies achieved by the "relative presence or absence, respectively, of apprenticing text" (Dowling, 1992, p. 6). These textual strategies are achieved by the *extension* or *delimitation* of the esoteric domain. Pedagogic texts are designed so that *dominant* voices or subject positions are *apprenticed* into the esoteric domain of an activity, such as mathematics. *Subordinate (or subaltern)* voices are *alienated* from the esoteric domain so that context-specificity is maximized for these voices, while context specificity is minimized for relatively dominant voices (Dowling, 1992).

In describing how meaning is articulated, Dowling introduces his theoretical concepts of discourse and procedure. These concepts enable description of the distributions within the esoteric domain of mathematics discourse, the differing modes of expression, and the nature of the discourse made available to dominant and subordinate voices. First, discursive elaboration of mathematics discourse includes making available and explicit the "regulating principles" of this discourse, while procedural elaboration hides meaning and obscures the principles of the discourse. Discursive elaboration makes principled connections between and within topics thus providing access to the discourse. It refers to the specialization and generalization of the discourse. Procedural elaboration, by contrast, hides the relationships and teaches algorithms, facts and procedures. It often provides the student with mere rules. Procedurally elaborated discourse may embed the student voice in the mundane or within the confines of their situated context, disallowing access to the regulating principles of the discourse. Often, for these alienated voices, mathematics is transmitted as an "abbreviated" text so that little or no continuity of meaning exists and the connection between concepts is not made. Strategies that alienate and construct voices in a subordinate position are referred to as *localizing strategies*, while strategies that apprentice voices into the esoteric domain of mathematics as a relatively dominant voice are referred to as *generalizing strategies* (Dowling, 1993, p. 105).

Briefly, to exemplify an application of this descriptive language: At a mathematics education conference held in Vancouver, Canada (June, 2000), the focus of which was "Visualization in Mathematics Education," a speaker described how she taught simplifying with square roots to students with difficulties in mathematics as a "jail break," (an unfortunate metaphor in itself). In this way, prime factors became "convicts" which, when grouped together in twos, (prime factors carrying a power of two), "only one convict broke loose from jail," the jail being the "square root." This is an example of where localizing strategies were adopted with students constructed as mathematically disabled. It could be argued that an uneven distribution of mathematics discourse and practice was afforded these students, in comparison with other, "more able" students. Mathematical exposition was produced as a series of disconnections. These students were denied access to the regulating principles of mathematics. Here, esoteric domain text was procedurally elaborated, and alienated student voices were localized within the public domain of mathematics discourse and practice, thereby subordinating these voices. Of interest also, was the way in which the speaker located this practice as an example of the use of "visualization" in "progressive" mathematics teaching practice, thereby subordinating the voice of the audience as well in relation to this practice.

## POSITIONING IN PRACTICE: THE CONSTRUCTION OF DISADVANTAGE

I now elaborate on an empirical study, which I conducted within a secondary school in the Western Cape of South Africa. My discussion invokes a concept of classroom communication that moves beyond a cognition-based analysis of teacher and student utterances in the classroom. In its broadest and deepest sense, the ambit of social discourses is invoked (somatically and discursively) in context and co-emerges to produce subjectivity in the mathematics classroom.

### Aims and Methods of the Study

A small-scale study of a qualitative nature was conducted within a historic and traditional, independent, all-boys secondary school in the Western Cape. The focus of the study was the exploration of subject positions potentially available to the black male students of, what was referred to in the school as, the "Black Scholarship Program" in their study of school mathematics. This included an examination of the particular nature of the schooling ethos and culture, and its role in creating and maintaining boundaries, producing and reproducing forms of power and control, which assisted in holding these black students in positions of subordination. It was proposed that the hierarchical and differentiating rituals and codes within the school context provided the means by which the Black Scholarship students were constructed as disadvantaged.

Particular emphasis was placed on the discourse of mathematics within the school's Academic Support Program designed to "bridge the gap" in these students' academic "knowledge" and "experience." The study also focused on the differentiated nature of the mathematics discourse available to these students within the mainstream program. There was an examination of the power relations between these two discourses and other social domain discourses, which assisted in shaping the way in which these students were positioned in terms of deficit and disadvantage.

Students of the Black Scholarship Program were interviewed in their initial year at the secondary school (Standard 6/Grade 8), as were the two teachers of the Academic Support Program. The discussions were taped and transcribed and formed the basis of a detailed analysis. Field notes were taken of discussions with academic staff within the mathematics department, and school documentation reflecting school policies and discussions within the school were used, where relevant. The study was used to interrogate previous research work in the area of social inequality and mathematics education. It also raised questions about taken-for-granted assumptions, both within the school as well as the wider community, regarding race, social class, language, and cultural difference.

The study sought to investigate and bring into focus how *difference* is created and maintained, produced and reproduced within the context of the school, establishing boundaries to epistemological access, and how this difference is recontextualized into *disadvantage* with respect to this particular socially constructed group of students, in relation to other students and mathematics. It must be noted, however, that the intention of the study was not to establish "the truth" about the black male students at the school and their mathematics, but rather to attempt to explore, through an analysis of their discourse and that of their teachers', whether or not there was resonance between the ways in which they were socially constructed within the school, and the distributions of meaning and mathematics discourse that were made available to these students. In the interests of reflexivity, I acknowledge that my role as researcher and the power relations invested in that role as well as my own positioning within the research context would have necessarily influenced the form and nature of the research. In the interview situations, where it became evident to me that my position within the community and interview role was influencing the discourse, I acknowledged this in the study. The study thus focused on three major descriptive/analytical areas:

1. *School context and disadvantage:* a description of the "stratified" school context and how the institutional rituals and particular cultural ethos of the school assist in producing pedagogic boundaries and establishing "disadvantage."
2. *Pedagogic discourse and disadvantage:* the discourse of mainstream mathematics and its strength of voice in relation to a hierarchy of other instructional discourses, including that of the Academic Support Program, were examined.
3. *Disadvantage realized in pedagogic discourse:* the subject positions marked out by student and teacher discourse and the relationship between these positions/voices and the distributions of mathematics discourse/practice to these voices.

In establishing a methodology, I drew mainly on Bernstein's work in the first two areas and on Dowling's work in the third. As the empirical study began in 1994 and was concluded in early 1998, I drew on their work mainly prior to 1995 in developing a theoretical framework. The study marks a period of unprecedented socio-political and economic change in South Africa, culminating in a new political dispensation for the country, reflected most poignantly in the educational arena.

The study provided empirical examples of teacher and student utterances with respect to school mathematics within the school. It was through these utterances that positions of alienation were established in relation to the students of the Black Scholarship Program and mathematics. These alienated positions work concomitantly with differentiated distributions of mathematics discourse and practice to these positions. The study indicated how certain students associated with "success" were provided with the regulating principles of mathematics within the school, while the Black Scholarship students were denied such access. Social difference and constructed "disadvantage" became instantiated in school mathematics discourse and practice, begetting pedagogic disadvantage. For these scholarship students, mathematics learning was an experience of contradiction and disempowerment, and expressed itself as the "pedagogizing of difference" (Swanson, 1998). The construction of disadvantage does, however, draw on a broader problematic within the social domain and is established within *context.* Therefore, I briefly describe how the cultural ethos of the school, along with ideological discourses from the broader social domain, assist in the (re)production of difference, and how this difference is recontextualized into disadvantage.

## Description of School Context

The research school is an independent, all-boys and historic, church school, steeped in tradition. This tradition is supported by a number of differentiating and consensual rituals (Bernstein, 1975) such as the prefect system, house system, and streaming (tracking) system. The school is considered a prestigious institution with a reputation for high academic and sporting achievement. The school has been open to all race groups since 1978. Nevertheless, student fees are well beyond the affordability of the vast majority of South Africans. A pronounced hierarchy of regulative discourses has been established within the school, supported and (re)produced by codes of social class and culture. The prevailing ethos of the school might be described in terms of a "culture of difference." White patriarchy is hegemonic.

A small group of black students from local township schools were invited to attend the school based on the results of an academic entrance examination in the grade 7 year. Only four students per year "won scholarships" to attend this school and they entered the school in the grade 8 year. This scholarship program was funded by a multi-national corporation as part of its Social Responsibility or Community Investment Program. It was intended to provide educational advantage to "promising" students from "disadvantaged communities". As the students' academic level was considered "weak" by comparison with the "academic standard" set by the school, and as a consequence of their spoken-of "language difference" within this English medium school, they were enrolled in an Academic Support Program. This program was designed to provide educational support and assist these students to "bridge the gap" in their "knowledge" and "educational/life experience," and meet the "standards" set by the school.

These scholarship students were placed in lower academic streams in mathematics, as it was agreed by the schooling administration that: "They would be unable to cope with the language demands of mathematics at a higher level with such a poor standard of English". In the study, my interest was to explain how the low streams were associated with "inability" and educational "lack" within this stratified school, most especially in school mathematics, which carried a strong voice. Discourse on "slow learners," "bottom set" learners (as opposed to "top set" learners) and learners with "difficulties" in mathematics abounded. Strong classification (and framing) (see Bernstein, 1993, 2000) of the mainstream mathematics program, exemplified in the pronounced streaming (tracking) system, assisted in these constructions. While the students were placed in a lower stream as a result of "lack of language skills" and an "educationally disadvantaged background," the low streaming differentially positioned them as "low ability" students. Here, "cultural difference" translated into "educational deficit" in relation to the school, which in turn resulted in the scholarship holders being associated with *inherent* "inability" and assigned a position of "low academic status." A construction of these students in terms of "educational failure" was consequently facilitated. These constructions were established within the stratified school context. Although the positions of alienation were marked out, within language, in relation to other demarcated subject positions as they related to mathematics discourse at the school, it was the *broader school context* and cultural ethos which provided the means by which these positions were established. Constructions of "race," "educational disadvantage," "experiential deficit," "cultural and language difference," were supported by the socio-political discourses within the broader social domain that contribute to the legacies of Apartheid in South Africa.

The Academic Support Program proved to be "weakly classified" (and "framed") (see Bernstein, 1993, pp. 122–123) against the "strong voice" of mainstream mathematics discourse at the school. Consequently, the bridging program did not provide access to the regulating principles of mathematics, as it was designed to do, since it did not possess the "recognition" and "realization rules" (Bernstein, 1993, p. 128) of mainstream mathematics. The rules of evaluation in mathematics were vigilantly policed within the mathematics department, and the upper streams of the mainstream program were provided with greater access to the rules of evaluation than were the lower streams. Consequently, the scholarship students' association with the Academic Support Program served to reinforce their difference on cultural, linguistic and racial grounds, and was ineffectual in providing the expected access to achievement in mathematics at the school.

## Experiential Knowledge versus Mathematical Knowledge

In the interviews, the teachers of the Academic Support Program often spoke of the students in association with "the educational woes in South Africa" and (at that time) "a deficient language policy in education". The students were educationally located within terms of the historical baggage of Apartheid, the history of racialism and poverty. They and their mathematics were spoken about in terms of a "language problem", an "experiential deficit" and "perceptual problems". Mathematics, as an instructional discourse, was spoken about in particular ways which separated it according to dichotomous categories such as "reason" versus "experiential," "taught mathematics" versus "innate mathematical ability." Each of these ways served to position the students in terms of deficiency and disadvantage with respect to mathematics. These constructions located codes of race, class, culture, language, experience and environment. In my asking (as interviewer) what a scholarship entrance examination paper appropriately should look like, one of the Academic Support Program teachers responded:

> You should try and avoid things which are not Maths experience—experiential things which are not maths. You've got questions with bus timetables—travelling distances across countries which some of them have no experience of, I think these sorts of questions are unfair.... Because it's not been in their experience, so you just have to be careful of that sort of question. You want to assess how they have come to their standard in mathematics, on how good they are, and inevitably it depends on how well they have been taught.

Here, "experiential knowledge" is to be avoided, while "mathematical knowledge" is privileged over it. In the same breath, the students are spoken about in terms of an experiential deficit, which further locates them within a mundane and impoverished "real" world outside of the discourse of mathematics at the school. The same teacher says:

> Regarding the scholarship (students) and maths... the main things are, language and... maths is a lot of three-dimensional/experiential stuff. That's something which people would tend to suggest: experiential toys. What do the other (students) here have as birthday presents? Experiential-related stuff. The geography and other people would say that they have a problem with 3D visualization. That would be common. That would be expected. They struggle with that.

On the one hand, mathematics can be located outside of experience, but on the other hand, experience (as constructed in the teacher's terms) impinges on mathematical ability in the form of 3D visualization. "Experience" becomes a "factor" in mathematics achievement when a certain construction of these students is established. This construction situates them in terms of some sort of deficit: poverty, experiential lack or the mundane. When asked, at that time, which Scholarship Program mathematics entrance examination of the past few years might be described as a "good" one, one Academic Support teacher answered:

> If I think of the maths paper that I looked at, what I liked about it was that a lot of it was how well they'd be able to think in the exam situation, using so-called innate maths ability and reasoning. So this particular maths paper didn't have very much on direct syllabus stuff. A lot of it was things that they might know from everyday use like you have R10 and you go off and buy a few boxes of bread or something; so that's talking about their general experience outside the classroom. And then there were quite a lot of things which seemed to me to be in terms of being able to reason. And all like IQ type test situation.... What you are looking at is the child and his ability. So I would say

> you stress more that, OK? You do want to know how good he is and whether he will cope here, but I think I would tend to look more for potential and ability than the level that's been reached, thanks to the teacher.

This teacher talks about "innate maths ability" and "reasoning," separating this from "syllabus" mathematics, but in the same breath incorporates it with "everyday use"—experiential knowledge. It is interesting to note that *this* form of experiential knowledge is considered legitimate mathematical knowledge, as it embeds the scholarship students in the mundane, i.e., the buying of bread. The previous form of experiential knowledge was *not* seen as acceptable mathematical knowledge, as it involved experiences considered usual to "other" students but not scholarship students, i.e., traveling across countries. Consequently, one form of experiential knowledge is privileged over the other, as the scholarship students are seen as possessing a deficit in an area of this knowledge. Further, what is implied in the comment is that "innate mathematical ability" can be assessed through "general experience outside of the classroom," which is the form of experience the scholarship students "have had" and which the Academic Support Program teachers and mathematics department has the authority to judge and is privy to "knowing." The message is one of superiority, and it carries with it the dominant cultural codes of the school, codes of social class.

This association of the scholarship students with experiential knowledge which is mundane (and situates them in terms of "poverty," as they are seen to be buying bread, not computers or motorcars etc., and certainly not in the financial position to be "traveling across countries"), and the linking of mathematical knowledge with this form of experience, is a "localizing strategy" embedding the students' "experience" within the public domain of mathematics discourse. It positions the scholarship students as subordinate in relation to "other" students who are presumed to have access to other forms of experiential knowledge that is not embedded in the mundane. Again, codes of class are evident. These "other" students would be presumed to have closer cultural proximity to the dominant culture of the school and hence access to the associated cultural knowledge embedded in the instructional discourses of the school. They are a majority of students.

Further, "ability to reason" is seen as synonymous with "IQ type questions" which are also, by implication, considered a legitimate component of mathematical knowledge. Implicit in this teacher's comment is the understanding that not only can the ability to reason be assessed, but also it can be assessed through the discourse of mathematics—through questions that are designed to test intelligence. Here, the discourse of mathematics is dominant and possesses absolute authority in determining not only mathematical ability but also reasoning ability and intelligence. These "mathematical" questions are posed as being divorced from cultural, linguistic, and experiential issues, and are presumed not to carry with them any forms of bias which may prevent access to evaluating the criteria for "reasoning" and "intelligence." They are spoken of as unproblematic *indicators* of what is "needed to be assessed"—ability and potential ... innate conditions of being, not social constructs. Again, in the teacher's comment is a message carried by the dominant culture of the school, and it resonates with the strong voice of mainstream mathematics. The fact that ability and intelligence are stressed as being necessary criteria for acceptance into the school, highlights the emphasis of the school culture on exclusivity and "striving towards academic excellence," which is dependent on "ability" and "intelligence."

It is important to note that in fore-grounding the words of an Academic Support teacher as I have done here, my intentions are not to criticize this teacher's pedagogic objectives, which, I am convinced, were intended to serve the best interests of the scholarship students as she saw it, within the prevailing context. My intention, in support of the theoretical framework I am using, is to *describe* how mathematics discourse is spoken about in different and often contradictory ways, which separate it out in accordance with differentiated subject positions, thereby pedagogizing difference. Of equal importance, it is necessary to note that the voices of the Academic Support teachers were *also* subordinated by mainstream mathematics teachers' voices. (For an elaborate discussion, see Swanson, 1998.) Critically, the weak voices of the Academic Support Program teachers, by separating out mathematical knowledge in accordance with the scholarship students' spoken-of experiential "lack," serves, in fact, to resonate with and reproduce the authorial voices of mainstream mathematics teachers. Consequently, these weak voices serve to reinforce cultural and social difference within this schooling context and establish disadvantage, rather than contest or alleviate it.

## Teachers and Students: A Disjunction in Communicative Meanings

There are many extracts in the study that elaborate on different aspects of the construction of disadvantage and provide contextual descriptions of the relations of power and control between discourses within the research school (Swanson, 1998, pp. 86–158). In the space of this text, I present only a few examples to support a partial analysis.

While the scholarship students refer to their "difficulties" with language and understanding in mathematics, mainstream mathematics teachers refer to the students mostly in terms of their socio-economic backgrounds, their previous "inferior" schools and their "lack of effort" and "slothfulness." The students' language becomes contradictory, where they move positions. Occasionally, they contradict their own arguments in ways that resonate with the mainstream mathematics teachers' utterances regarding their learning of mathematics. One example follows:

After two scholarship students, Derrien and Marcus, speak about "language" as being a crucial problem in hindering their understanding of mathematics in the classroom, they continue:

> Derrien: Sometimes I can't understand the work in class on, like, ...
>
> Marcus: ...the tests...
>
> Derrien: ... sometimes we fail. Sometimes I just can't understand the work ... and, like, when the test comes, I just, like, I don't, like, ... um ... learn enough.

It is interesting to see how Derrien changes his position mid-sentence. The suggestion, on the one hand, is that Derrien's "failure" is as a result of "lack of understanding" of the mathematics, while on the other hand, and in the same breath, it is as a result of "lack of sufficient study," resonating with the mainstream mathematics teachers' discourse on "lack of effort." The concepts of "understanding" and "learning"/"studying" are merged or conflated and so are seen in the same light. This has the purpose of masking the reasons for "failure" (and, problematically, the criteria for "failure" are most often hinged to written evaluative instruments, such as tests or examinations), transferring the responsibility for its occurrence from the teacher to the student. Consequently, whether the students speak in resonance with the voices of mainstream mathematics teachers or in contradiction to them, the students construct themselves in terms of disadvantage.

Later, Derrien contradicts his assertion that his "failure" in tests is as a consequence of "insufficient effort," by drawing attention to a gap between the ways in which the mathematics is taught in class and the ways in which it is examined. Derrien highlights a discursive disjunction between the two contexts of the "test" and the "lower stream" classroom. This prevents him from making principled connections between the two contexts in ways that would allow access to mathematics discourse:

> DS (interviewer): So what is it about the test that...?
>
> Marcus: They ask different things ... you expect something like, from ... you don't expect ... they ask tricky questions.
>
> DS: You don't expect tricky questions?
>
> Marcus: ...I always thought, like, they would be questions you deal with in class.
>
> Derrien: Usually in tests they first, ... like, give you a statement and you have to work it out, ... work out what the sum is first and then you can work it out, ... what it was ... so

you waste time on finding out what it means, like ... because ... the language is, like, complicated.... and I still can't understand.

DS: So, the language is complicated and you are trying to find out what...

Derrien: The sum means.

DS: The sum or the statement would actually mean before you....

Derrien: Work it out ... because in class you just work it ... you just work it out in numbers only, you're used to working in numbers only.... In class, you work in numbers, not like questions ... not like words.... They just give you sums and you got to work it out and for tests, they just give you, ... word sums, but you must know what they mean and want.

DS: Don't the teachers do word problems or deal with problems with those statements and what they mean in class?

Derrien: Yes, ... but, ... in tests they are usually true-life questions and, ... they get very complicated.

Derrien and Marcus speak of how the content and expression of the discourse in the lower sets has been sufficiently selective, in relation to the discourse within the context of evaluation (i.e., under test conditions), as to obscure the discourse and hide the meanings. Clearly, the scholarship students' comments are pointing at the unequal distribution of message. They refer to the discourse of mathematics as being constituted differently within the two contexts, and consequently point at not having been granted access to the regulating principles of the discourse necessary to make the connections between the two. The principles of the context of evaluation lie *outside* of the "lower stream" classroom. What becomes evident, as reported in the study, is the difference of message distributed within mainstream mathematics discourse, across the mathematics sets.

In the lower stream classroom, localizing strategies are used which initiate the students into the public domain of mathematics discourse. By "working in numbers only," the mathematics content is being proceduralized and access to the principles of the esoteric domain of the discourse is not granted them, so that these meanings remain hidden and obscured. This, again, refers to an uneven distribution in mathematics discourse, which alienates these students from the discourse, and assists in positioning them subordinately in relation to "upper stream" acquirers of mathematics within the school. The transmitter voice of mainstream mathematics teachers, consequently distributes a different kind of voice across the sets in such a way that the acquirer voice of the scholarship students achieves a construction of "unsuccessful" along with others in the "lower streams."

In the analysis, "a gap" was produced between teachers' and students' meanings regarding what was possible for the scholarship students in their study of mathematics. Teacher constructions of cultural and experiential difference and deprivation, race, social class, and language difference became delimiting criteria in the students' access to mathematics knowledge. On occasions, the voices of the scholarship students resonated with those of the teachers in these constructions, and at other times they did not. Under both circumstances, the scholarship students' voice reflected their position of subordination—either through constructing themselves as unable mathematically, or as lacking access to the regulating principles of mathematics. The dissonance between the voices of teachers and students is most pronounced in relation to the issue of *responsibility* for the cultural and educational "mismatch." While the teachers, for the most part, place the onus on the scholarship students, the school bearing little responsibility for the students' lack of mathematical success, the dissonant voices of the scholarship students place the blame on the school, thereby distancing themselves from the school's interpretation of their lack of success. For the teachers, the problem is interpreted as deficit, and the scholarship students' culture, experiences and language difference are spoken of in terms of this deficit; while for the students, the fault, in part, lies with the differentiated nature of the streaming system and their lack of access to the regulating principles of upper stream mathematics. The scholarship students speak of alienation: linguistic, cultural and experiential as well as pedagogic, and thereby position themselves as subordinate in resonance with their teachers' construction of them.

In this way the construction of disadvantage has different connotations depending on viewpoint: disadvantage with respect to the scholarship students *becomes* their disadvantage in relation to the school, or said differently, the scholarship students carry a construction of disadvantage, which becomes the means by which they are disadvantaged mathematically within the school. This point could, perhaps, best be encapsulated in understanding the construction of disadvantage as working empirically with the "pedagogizing of difference" in the research school. Constructions of disadvantage are recontextualized into the discourse of school mathematics, reinforced by "poor academic achievement" in mathematics, and realized pedagogically. This was exemplified in the following words of a scholarship student, highlighting his alienation and disempowerment, and making visible the dissonance between teacher meanings of successful students and what was possible for these students at the school:

> Like in my last school (referring to a township school), you try to compete with the person and you know he's here and like, you could do better than him. But now you don't know if you'll get to cope there (referring to the upper streams) and you never know what to expect ever, but last year you were the same with everyone and you first compete with each other and, like, that's what I preferred to the sets and stuff here.

At this stage, I will extend my discussion beyond an empirical one and I will move outwards beyond the school context in turning my attention to a broader discussion. My reason is to draw attention to the broader domain of social discourses that impinge on mathematics classroom discourse and practice. This necessitates viewing mathematics classroom communication in the very broadest and deepest sense. Most often, the ways in which the broader socio-political context impacts on the mathematics classroom and school are ill-considered in classroom-based research, thus eliminating consideration of the agencies at play within specific and local pedagogic contexts. Critical examination of the broader complexity and intertextuality of local and global contexts and the concomitant forms of agency produced is crucial before considering any alternatives in approach. Lack of consideration of this broader complex, is, perhaps, one of the most singularly important reasons for policy mismatches and failures in educational contexts.

## DISCOURSES OF DIFFERENCE

As highlighted in the empirical study, it is power and control in-and-between discourses in the broader social domain that produce agency in the mathematics classroom. To refocus within the social domain on the discourses which set individuals and groups apart according to social axes of power, I will broaden the discussion here to include trans-national perspectives: dialogues and discourses within the framework of a global social domain. I will begin with a discussion on "difference" and its particular forms and agency in the production of Western/European master narratives. This is in reference to the dominance of Western/Euro-centric and hegemonic "global" perspectives, that universalize and popularize certain discourses under the banner of "progressivism" in mathematics classrooms. This elicits re-examination of underlying assumptions and contests the socio-political interests they serve. This section also deals with issues of pathology in terms of the "psychologizing of difference." It looks at discourses of "past" and "place" in evoking constructions of difference, and the discourses that assist in this difference-making. I will show the relationship between these discourses and how they provide resources by which subjectivity is established within the mathematics classroom.

## Difference in the Making: The Rhetoric of "Embrace"

Marcia Crosby (1991) discusses "the embracing of 'difference'" in terms of what "seems to be a recent phenomenon in the arts and social sciences" (p. 267), one that has found a place in mathematics education within multicultural discourse. Crosby says that, "as a component of postmodernism," while difference may be formulated in the voices of struggle against hegemony of European "master narratives," it may also be, "in the face of popular culture and an ever-shrinking globe," "a saleable commodity" (p. 267). This commodifying of difference requires examination in terms of the ethical and rhetorical issues of inclusivity, accessibility, and "relevance." How is diversity, as formulated through "popular culture," refigured in mathematics classrooms in different contexts? And, what pedagogic agency is established here?

In reference to the construction of the "Imaginary Indian," Crosby (of First Nations descent) questions whether this is not just Western curiosity or "the ultimate colonization of 'the Indian'" (p. 267). Her point holds equally for "the African," "the Asian," "visible minority," "woman of color," "third world" person, or any construction as "other" established within Western/Euro-centric discourses ("positivist" or "post-modern"). Crosby's argument makes visible the power relations inherent in the apparent legitimizing and embracing of difference. One can extrapolate this to the problematic of the construction of difference as it applies at various levels of discourse, and to its pedagogic implications within diverse or "multicultural" mathematics classrooms.

The patriarchy of Western dominant thought produces representations of "the other" which homogenizes and mythologizes (Said, 1979), rendering it in terms of constructions of nature. This naturalizing and scientizing is not exempt from post-modern discourses, which retain elements of patriarchy and condescension. Crosby describes how "the colonization of images in order to create a new Canadian mythology is parasitic, requiring that the first-order meanings within native communities be drained" (1991, p. 279). For Crosby, this is not an inclusive act, but one which is predicated on exclusion or "otherness." Dangerously, the "embracing of difference" rhetoric is informed by the "salvage paradigm" (Clifford, 1987), which feeds off notions of subordinate cultures in stress, as dying, as needing a savior. It reproduces a cultural construct that homogenizes, collects and reorganizes "cultural" artifacts and presents them in a context which freezes them in the past and situates them *outside* of the context of their production.

This has resonance with certain trends in multiculturalism in mathematics education. In mathematics education research, there is "a growing body of work (which) seeks to celebrate an alreadiness of mathematical content within the practices of different cultural groups" (Dowling, 1998, p. 12), often referred to as *ethnomathematics*. There is much controversy surrounding ethnomathematics in the mathematics education field. Although it has produced a positive research trend in the sense of attempting to provide recognition for different cultural epistemologies and alternative frameworks of reference, (usually focused on "small-scale cultures" however), it has also, for the most part, been reductionist in describing different cultural practices according to the terms and principles of Euro-centered mathematics and Western school mathematics curricula. As in Paulus Gerdes's (1985, 1986, 1988a, 1988b) Mozambican basket and button weaving, hut building and Angolan sand drawings, Western mathematical principles are imputed into the practices of non-European cultures. Gerdes wants to offer these discoveries to mathematics learners, to encourage cultural and psychological confidence, to "defrost" or unlock hidden mathematical knowledge "embedded" in cultural practices and artifacts, and to emancipate—this, in itself, is utopian besides patronizing (see Dowling, 1998, pp. 12–13). Here, Euro-centered mathematics possesses the authority to define the relevance and nature of group culture and how it is to be experienced. It claims the moral authority to do so, under its auspices.

There is strong resonance with Crosby's argument. Crosby's concern for the colonization of images into the realm of Western discourses serves equally for the colonizing of cultural practices into the domain of Euro-centric mathematics. Just as the first-order meanings of aboriginal artefacts are drained in the Canadian context, so the African cultures of Gerdes's "mathematical" artifacts are unable to speak in their own terms but are filtered through the discourse of Euro-centric mathematics, positioning them as mathematical—but separate. They stand outside the "real mathematics" as a repackaged "polycultural" commodity for the consumption of Western tastes. The "defrosting" locates itself within the "salvage" paradigm, Western mathematics and mathematicians acting as the saviors of these "other" cultures or "their" mathematics. Put differently, the rhetoric of embrace appears to speak to the democratic and inclusivist principles of certain post-modern discourses while maintaining, rather than contesting, the mythologizing principles of power found within Western/global discourses. "Other cultures" are then recontextualized and produced as "other" within the framework of these dominant discourses.

"Difference" is dangerous when the hegemonic and deeply embedded social hierarchies reproduce uneven realizations of the produced "difference." Despite its admirable ideals, "embracing" this difference in a context where the power relations are *uneven*, simply accentuates it and reproduces it in the form of *realized disadvantage*.

## Psychologising Difference

Context relies on a range of conceptual features, including social, cultural, political and historical spaces, settings, sites or locations. The historical feature of context, or *kairos*, is highly relevant to social change. Carolyn Miller says:

> As the principle of timing or opportunity in rhetoric, kairos calls attention to the nature of discourse as event rather than object; it shows us how discourse is related to a historical moment; it alerts us to the constantly changing quality of appropriateness. (1992, p. 310)

The motivation to celebrate the mathematical contributions of "other cultural" groups is a response to the consciousness of their cultural devaluation by dominant Western discourses. This cultural devaluation was premised on a historical perception of intellectual or cultural lack, which inhered in the kairos of imperialism and modernism.

Celebration of difference and social pathology are, however, fingers on the same hand. Couched within the rhetoric of "embrace" is the condescension and patriarchy of the salvage paradigm and the commodification of cultural difference. This is, perhaps, a less visible pedagogy than one which produces "lack" and pathological difference, but operates according to similar principles of power and social differentiation.

Modern psychology, until recently, has often been premised on rationalism and individualism. (Lave, 1988, Lave & Wenger, 1991 are some notable exceptions.) Nick Taylor describes this trend in relation to the rise of capitalism in the seventeenth century as being designed to discipline and control the work force and increase productivity (1992). Psychology, in consonance, became the moral guardian of social issues. The individual became viewed as unitary, independent of social circumstance, and rational and consistent across contexts. This became entrenched in Western schooling practices. Valerie Walkerdine argues that the child became the object of psychology, and psychology became the technology of the social (1984, 1988). Some organizing practices of educational psychology pathologize and produce, therefore, the very attributes they wish to measure.

Piagetian child-centered pedagogy is testimony to this self-reproducing mode of discourse. It is deeply entrenched in schooling curricula throughout the world, and maintains its rational-centered, individual-centered form of control pedagogically, through its organizing practices (see Lerman, 1996, Steffe & Thompson, 2000, Lerman, 2000, for debate). Piaget's "stages of development" framework produces pedagogic propensities as biological, "which ensures that the child is produced as an object of the scientific and pedagogical gaze by means of the very mechanisms which were intended to produce its liberation" (Walkerdine in Taylor, 1992, p. 70). The tenets of objectivism find continuity with the aims and principles of Euro-centric mathematics which holds onto this analysis of the child resolutely. The social spaces which are created for individuals (most especially children) and situate them in terms of truth claims about the nature of the reality in which they actively participate, assist in constructing them in terms of innate capacities which often index notions of gender, ethnicity, culture, class, able-bodiedness, among other positions. Consequently, Piagetian psychology is fundamental to many school mathematics curricula and its practices. It has particular agency in separating children according to criteria of age and intellectual ability, the evaluative characteristics of which inhere, most particularly, in school mathematics.

Of concern in much of mathematics education research on equity issues, is how achievement is linked with student demographics to produce educational "factors" which establish conditions of "success" or "failure" for these students. This link produces an educational significance to socially constructed groupings via a "cause-and-effect" process that situates students' experiences in the mathematics classroom in terms of their social backgrounds. There has been a plethora of research into mathematics education with this approach. Bernstein refers to this research approach as "deficit theory," which "attempts to account for educational failure by locating its origins solely in surface features of the child's family and local community. The unit of such theory is a child and the distinguishing features are indices of pathology" (1975, p. 27).

Attributing "mental abilities" to gender or ethnic background, or "achievement" to "socio-economic status" or "societal factors," are just some of the more obvious research examples. In many of these research projects, the main objective has been to observe "difference" in relation to taken-for-granted social constructions, without problematizing these constructions in any way. The model is premised on an assumption of existing difference and the research establishes that difference educationally. The social categories soon become the source of measurable difference and the research focus moves away from "societal factors" to the constructions themselves as carrying some deficiency. This produces an inherent contradiction. Here, the social is not relative. In this research paradigm, no impediment is raised to understanding the social as neutral, measurable and objective. It is a paradigm of thinking whose premise rests on given social structures and no cognizance is taken of the possibility of the relative nature of reality and truth. These studies also shift the burden. Secada says of this research tradition: "Poverty, race, ethnicity, or gender no longer are ways by which our society is stratified; they become causal variables that carry with them their own explanations" (1990, p. 354).

More recently there has been a greater consciousness of the tacit ways in which these studies simply re-label and legitimize the social conditions by which certain students fail. Nevertheless, there is still much research undertaken in this paradigm today, albeit, perhaps, more subtly, justifying gendered, "cultural" or ethnic "ways of learning" or "knowing," establishing "language difference" as an educational difficulty/"problem" or examining how this purported difference "interferes" with learning. Problematically, many of these research "findings" then justify to teachers and administrators their engagement in differentiated practices and communication in the mathematics classroom, which soon becomes associated with constructions of ethnicity, race, class, gender and culture. These differentiated messages assist in separating students according to so-called "educational" criteria, and, consequently serve to provide access to the "regulating principles" of mathematics discourse for some, and prevent access for others. These practices operate in consonance with the socially constructed "attributes" of students and produce and reproduce positions of success or failure.

An exemplar of research conducted in the "deficit" paradigm, (in an African context) is the work of John W. Berry (1985). Berry sets the scene of a "third world" classroom in Botswana where the students (especially Mothibi and Lefa) have "problems" with understanding mathematics at an analytical level and must rely on rote learning to achieve a level of success in "the final examination." He begins to locate the problem in terms of "culture," but it is a culture that stands *outside* of mathematics, impinging on it, so as to "interfere with" or "cause problems" with children's learning. He says: "Scenes like these are repeated in many countries of the third world where teachers are striving, under very difficult conditions, to build an educational system which reflects their own culture and which will serve their society's needs" (Berry, 1985, p. 18).

Berry does not provide any understanding of what the "society's needs" are, or how it relates to the "very difficult conditions." Rather, these simply serve to call on assumptions of poverty and "lack," which situate Lefa and Mothibi firmly in the alienating grasp of a "third world" context. Further, he does not show how either of these situations relates to "the learning problems" in the classroom. Instead, he locates the problem in terms of language difference. He says:

> The meeting point of these problems in the case of both Mothibi and of Lefa is language—specifically the set of learning problems which can be related to the extreme difference in structure between the students' mother tongue (Setswana) and the language of the teacher. (Berry, 1985)

Here, the problems lie in "difference," particularly, language difference, which becomes the object of research. The children and their language are constructed in terms of deficit. The "problem" is spoken of as existing with the child's language, not mathematics, not the teaching practices, nor the broader social context that produces "language difference" as a "problem." What is not considered is a potential rupture in the way in which the student is positioned in the mathematics classroom by the teacher and the mathematics curriculum, and the way in which these students position themselves in relation to others and mathematics.

Berry locates the problem in terms of cognitive psychology. Although he criticizes Piaget's child development theory in that "Piaget and his school have maintained that this progression is invariant across varying cultural and linguistic groups" (Berry, 1985, p. 18), he nevertheless holds onto this mode of development as a necessary tool for mathematical learning, perhaps only with some modification. Further, he speaks of relating "the child's development of mathematical concepts to what are considered a 'normal' progression through the stages of cognitive development" (p. 18), thereby tying mathematics concepts to this cognitive progression, and consequently shaping the nature of mathematics as a "normal" model. He later speaks of "the student's natural modes of cognition," naturalizing this model of cognition in relation to mathematics which is now, too, a natural phenomenon. Language and culture are also organized and naturalized in terms of cognition, but they achieve separation and become "factors" in the way in which they produce learning problems. Berry says: "One can argue, of course, that the causal factors in type B problems are cultural rather than linguistic," although he cannot locate what the exact connections are, other than to say that: "The exact nature of the causal chain linking culture, language and cognition is obviously a deep question" (1985, p. 20).

Even more problematic is that Berry sees the "solution" to "the problem" in terms of *remediation*, thereby further constructing the students and their language in terms of deficit and disadvantage. These "causal factors" produce not only the difference, but also the "problem," which requires remediation through, perhaps, adaptations to the mathematics curriculum and practice. Mathematics can now nurture and correct cultural and linguistic "problems." Berry says:

> Indeed, for mathematical educators the crucial fact is that all these factors, irrespective of the cause and effect relationships between them, have an influence on the way in which learning takes place (or fails to take place). Recognizing this, the educational task is to unravel these influences and to design appropriate remedies for their negative effects." (1985, p. 20)

Ironically, what is worthy of consideration is that the agent of production of Mothibi and Lefa's "problem" with learning mathematics, may lie more in their positioning in relation to mathematics discourse in the context of a "third world" African classroom, than in any "truth" about their language or culture or any "deficit" which this may produce. The crucial problematic lies, instead, in Berry's casual remark describing the "third world" context of learning in Botswana:

> Mothibi's teacher has been instructed to use a curriculum which is essentially a version of the "new math" albeit dressed in a Setswana costume. Originally designed in Britain, it has been carefully translated into Setswana. The examples and illustrations have been carefully changed so as to reflect the environment and outwardly, at least, the culture of the Tswana people, so Mothibi doesn't have to construct one-to-one correspondences between cups and saucers or cricket bats and balls; instead he is to use bags of grain and storage huts, or weaver bird nests and weaver birds. (Berry, 1985, p. 18)

Although Berry views the concern of European mathematics "dressed in a Setswana costume" as somewhat problematic, he does not engage with it in any broader sense, and certainly not in any depth. No structural account is given of the material conditions by which these teachers and students live and learn, nor of the socio-ideological context which assists in producing disadvantage in the Botswanan classroom. Instead, he *shifts* the burden back onto the students, the teachers, their language and their culture as a cognitive problem-producing complex.

Herein lies the rub. Mothibi and Lefa are positioned in terms of a cultural and language difference. This is recontextualized into school mathematics practice through the power of its objectifying gaze. This positions Mothibi and Lefa outside the regulating principles of mathematics discourse, although in a way which situates them in their specific context. Localizing practices are invoked so that contextually produced differentiated mathematics practices are distributed to these students in consonance with their subordinate positions. These assist in holding them to positions of alienation and disadvantage. Their language *difference* becomes read as an educational *deficit*, constructing them as "unsuccessful" in mathematics and producing realized *disadvantage*. This is manifest in Berry's description of Mothibi's "problem:" "Mothibi does very poorly on tests. He can carry out specific tasks when instructed but seems unable to make the transfers which would enable him to apply these skills in new situations" (Berry, 1985, p. 18).

The construction of disadvantage is complete. Rather than providing empowerment through challenging the cultural codes of Euro-centred mathematics and its practices within schools or the ideological messages from the social domain, it embeds learners within the situated account of their socially constructed "experiences." This denies them entry into discursively elaborated mathematics discourse, which provides access to the interrelatedness of mathematical concepts, topics and ideas; what Berry refers to as "transfers" to "new situations," and what Mothibi purportedly "cannot do."

The various discourses of difference co-emerge to produce realized educational disadvantage. The discourses of "embrace," or the "psychologizing of difference" are among social domain discourses that assist in establishing positions of disadvantage/advantage for differentiated learners, and are embedded in such institutional discourses as history, geography and anthropology. While these are separate disciplines whose boundaries are carefully guarded, I refer to them in the sense of their intertextuality, their interconnecting codes and discursive relations. I now elaborate more fully on discourses of *past* and *place* in establishing and pedagogising difference, as realised in its fullest colonialist or imperialist sense, in the form of disadvantage in the mathematics classroom.

## Discourses of Past and Place

The power of past in the historical enterprise of imperialism can never be underemphasized in its contributions to the construction of difference and its ideological investment in the imperialist/colonialist project. In the Hegelian sense, it has held the master to the slave, in many social places and pathways of ideology, but most poignantly in the field of education. It has made so deep a mark on our collective consciousness, our social relations and our lived experiences, that we often no longer recognize it for what it is. We no longer can separate it from what it is not, nor what it might have been. This is not because we have forgotten, as much as that we cannot provide it with its own framework of knowing, never knowing how to know it otherwise than how we do. We establish the relevance of the past according to a Western imperialist framework. We recontextualize that past according to principles of the present (*presentism*), but this is a Western present, where "others" live in the past. It is one that has been well exploited in educational/schooling contexts to produce (re)constructions of past and place in order to support a conception of "relevance" in students' lives.

Past and place work together to situate people throughout the world in relation to a Euro-centered/Western constructed hierarchy. It is critical in understanding how these discourses situate students, teachers and communities within these terms, reproducing the disadvantages or advantages assigned them. John Willinsky (1998) critiques the Western world-view, which espouses that, for large parts of the globe, there exists a differentiated participation in the progress of time, and whose hegemonic principles view the participation in progress as "a privileged mode of being." It is "a differentiating time-space continuum that further ensures the lasting division of the world in the Western imagination" (p. 118). The consequence of this Western historical consciousness is a "world divided among people who live inside and outside history." (p. 118). As Willinsky notes: "If history, modernity, and freedom are defined in nationalist terms, then to be born beyond the coterie of European nations is to live outside the pale, outside the Spirit of the World and its Universal History" (p. 119).

The discourses of past and place work concomitantly to produce a divided world. Their co-emergence is an act of power, as is their institutionalized separations, exemplified by the disciplines of History and Geography. Willinsky notes: "Geography, as a discourse of difference, was about learning to attribute that difference to a people within their landscape" (1998, pp. 13–14), and "culture," investing in this difference-making, "amounts to a collection of practices and artefacts that, in defining difference, effectively distinguishes one group of people from another" (Herbert in Willinsky, 1998, p. 14). This has never been more evident than in the *geographizing* of Africa. There has been much rhetoric (of disavowal and embrace), which has set Africa apart, outside of the "Universal Context" but locked into notions of culture as defined within Western imperialist constructs; just as with Crosby's "Imaginary Indian," so with the "Aesthetic African."

This has been the fate allocated to Africa over its colonial past. It maintains a strong presence in fundamental ways in the educational arena and instantiates itself in school mathematics practices. This is noted in Berry's study describing the "third world" context of Botswanan classrooms in terms of educational deficit and disadvantage. This is not to deny the divisive influences of colonialism, but rather to contest the proclivity to locate these issues in terms of an *educationally significant deficit* within students, their language or a conception of their culture. We are reminded of Berry's remark: "One can argue, of course, that the causal factors in type B problems are cultural rather than linguistic" (1985, p. 20). Berry goes further to arguing that linguistic difference, often determined by geographical proximity, plays a role in the mathematical learning and teaching of students (p. 19).

A deeper issue exists in that the discourses of past and place locate a conception of "culture" in establishing a discourse of "embrace" in mathematics education. This finds its way into mathematics classroom communication and practices. Barton et al. draw our attention to this:

> A further danger which has been observed in New Zealand is the promotion of a false idea of culturally appropriate "Maori learning." In search for cultural identity, it sometimes happens that old ways are mistakenly adopted. Apprenticeship learning, rote learning and algorithmic teaching have all been suggested as culturally appropriate for Maori. However, they may sim-

> ply be based on memories of how people did things in early colonial days, and thus be a reproduction of traditional colonial teaching methods, rather than truly traditional Maori modes. In any event, the removal of any tradition from its total cultural context into a contemporary mathematics classroom is unlikely to be effective. (1998, p. 8)

Graham Hingangaroa Smith speaks of this, in the New Zealand context, in terms of the many sites of neo-colonialism (1999). One such site is "the commodification of Maori language, knowledge and culture as a result of the (free market) economic reforms in New Zealand," while another is "the struggle for the Maori mind through hegemony" (p. 2). What then is being "embraced" in teaching practices in Maori mathematics classrooms?

This has possible resonance with *chanting* and *chorusing* in the more rural African classroom (Adler, 1999). How might we know what the epistemic location of this practice is? How might it have been recontextualized from "past" practices? Is it a form of rote learning derived from past colonial schooling (or Bantu education), or is it a form of oral repetition establishing a "collective memory" which taps into "African ways of knowing" (Dei, 1996)? By embracing this classroom practice as a form of cultural epistemology, is empowerment made possible, or are the students and teachers of these practices rendered obsolete, people placed in the past? Is this difference "a source of strength," a way of "acting together to transform our social and material existence" (Dei, 1996, p. 17), or, instead, is it part of the "uncritical adulation of difference" (Amin in Dei, 1996, p. 16) and just another new fable, a mythological or utopian epistemological land to which we are being lured by the Pied Piper of Western-style multiculturalism?

Further, if these practices *are* of educational benefit by providing access to mathematics discourse for these students, what might we learn from this for the advantage of *all* students? Could this form of knowledge production be made available to all students as equal learners and not only certain groups of students constructed in terms of cultural specificities? What are the contexts that produce advantage for these practices?

On the other hand, there is a critical danger here, in drawing on the past as a means of establishing current cultural validity. It is double-edged. In attempting to establish cultural epistemological credibility, we may well be falling into a trap that Freire defined in terms of the pedagogy of the oppressed (1999). In Freirian terms, the most effective means of colonization is established and its *raison d'etre* realized, when the oppressed begin to colonize themselves. By recontextualizing the past within the present, this may lead to the construction of imaginary difference, which serves to further oppress rather than liberate.

Perhaps our only way out of the coercive, globalizing gaze of Euro-centered/Western colonialist discourse is through a critical examination of our discourses of resistance and change at the highest levels; to establish a platform for reflexive engagement at the center of the panoptic Western imagination; to contest the very structures of its mind. There may well be kairotic spaces available where the discursive possibilities would allow for a critical account of the agency of these dominant discourses and of their epistemic locations. The *raison d'etre* of this book is one such place, which attempts to provide an alternative account of the socio-historical, spiritual and material conditions which produce particular realizations of mathematics classroom communication.

Thus, in returning to the mathematics classroom, what is possibly required is an interrogation of the assumptions that underpin our conceptions of classroom communication. This requires a broader understanding of the politics of context in the configurations, hierarchies and agents at play within its physical, temporal and sociological boundaries. In the forward thrust of progressivism in mathematics education whose momentum draws us into its all-consuming universal Truth, what needs to be addressed? Here, I provide some questions addressing these assumptions: Who envisions the ideal universal mathematics classroom? What are the prerequisites of such a classroom? What do the students look like in this classroom? How is the classroom imaged? What constructions of good practice are established within this classroom and what then is the ideal student?

On the one hand, we need to be aware of the socio-political premise on which school mathematics is based. It is not "neutral." It is based on epistemological foundations whose ideological imperatives produce different realizations in different contexts and divide students in ways that have social and educational ramifications. Here, we carefully need to consider Whorf's remark: "Western culture has made, through language, a provisional analysis of reality and, without correctives, holds resolutely to that analysis as final" (1956, p. 244). On the other hand, we need to be careful about our claims on mathematics, not only as a schooling discourse, but also as a socio-cultural and political one. In an attempt to contest the Western/Eurocentric dominance embedded in school mathematics discourse and practice, the proclivity may exist to dress school mathematics in the clothing of "other" cultures, to create imaginary difference, or to commodify it for the consumption of the socio-economic ends which inhere in certain aspects of progressive education. This is a trend that needs careful examination in consideration of the power principles invested in the contexts of its production. For, it is a strange truth that makes us unable to know what we have lost. That only in the losing, do we begin to understand what it could have been and what it might have meant to all humanity. But our melancholy cannot allow us the luxury of sinking into despair, nor can we afford to recruit an illusory history in order to recreate it for ourselves in the present. This may be the death knell of our human and self-understanding, the acme in the discourse of difference.

## CONCLUSION

As exemplified in the empirical study, the mathematics classroom is a pedagogic context in which, through the practice and discourse of school mathematics, certain constructions of "success" or "failure" are established. These constructions are particularly pronounced given the strength of voice of school mathematics and its position of privilege in the hierarchy of the social division of labor of discourses within schools. These constructions position students, teachers, and parents within the context of the school according to a hierarchical array. These positions are inscribed within language and may often index discursive resources within the social domain such as those that refer to social class, gender, ethnicity, cultural difference, linguistic competence, intellectual capability, and others, so as to establish these positions. The broader socio-political context provides the resources by which these differentiated positions are established. In mathematics classrooms throughout the world, these are contextually realized in the pedagogizing of difference. If we are to affect any change that may assist in less divisive/differentiated practices in school mathematics, it is necessary to view the activities of the classroom as inextricably interconnected with discourses from the broader social domain. This means, we must take account of the power structures within the classroom, as well as without, and take cognizance of the divisive nature of constructed "difference."

Finally, we need to view school mathematics discourse and practice in terms of the social imperatives these imply for students and teachers in different socio-political and geographical contexts. Certain questions need to be fore-grounded in critically examining the product of our practices. Some may be: How does presenting mathematics as a culturally specific and fallible discourse address social inequity, or provide a structural account of the material conditions by which students, teachers and their communities live? Does presenting mathematics as a socio-cultural and historic construct provide a greater opportunity for all students as equal learners to gain access to the regulating principles of mathematics discourse and challenge or improve the social conditions by which they live, as is often claimed? Or, is the dominant position which mathematics enjoys within schools and the broader social domain in the form of the rhetoric of technology and economic progress, (upheld by the social division of labor of discourses), the site of an act of pedagogic Apartheid? For whom is the heuristic and "progressive" teaching methods designed and how is "success" measured; who gains access to the regulating principles of mathematics, and who gets to "just learn the rules"? For whom is mathematics made "theoretical," and who gets the "technical" or "applied" version? What are the criteria determining who gets the "academic" or "vocational" mathematics? Which students (and teachers) are provided with generalized mathematics, and for which of these is mathematics a situated experience couched within the rhetoric of "relevance"? Consequently, is learning mathematics one of access and empowerment, or is it one of contradiction, arbitrariness and dislocation for some students and not others?

It could be argued that there are vested interests in many policy reforms and educational trends that simply provide for the repackaging of "mathematics" in a way which perpetuates the *status quo*—a commodifying of mathematics to serve the utilitarian interests of particular economic models embedded within Euro-centric/Western discourses, and which hides or obfuscates the agency of its cultural product. We are reminded of Bernstein's remark: "in whose interest is the apartness of things, and in whose interest is the new togetherness and new integration"? (1993, p. 122).

## ACKNOWLEDGMENTS

I would like to recognize Paula Ensor for her invaluable guidance and advice with my empirical study.

## DEDICATION

To Terry, Grace, and my parents for their love, and to all my students whose lives have so enriched my own.

## REFERENCES

Adler, J. (1999). What counts? Resourcing mathematical practice in the South African school classroom(Plenary Lecture 4) Delivered at the Canadian Mathematics Education Study Group Annual Conference, cf. Brock University, St. Catherines, Ontario. *Proceedings of the Canadian Mathematics Education Study Group Annual Conference* (pp. 45–56).

Barton, B., Fairhall, U., & Trinick, T. (1998). Tikanga Reo Tatai: Issues in the development of a Maori mathematics register. *For the Learning of Mathematics, 18*(1), 3–9.

Bernstein, B. (1975). *Class, codes and control (Vol. III).* London, UK: Routledge and Keegan Paul.

Bernstein, B. (1993). Pedagogic codes and their modalities [Special Issue]. *Hitotsubashi Journal of Social Studies, 25,* 115–134.

Bernstein, B. (2000). *Pedagogy, symbolic control and identity: Theory, research, critique.* New York: Rowman and Littlefield.

Berry, J.W. (1985). Learning mathematics in a second language: Some cross-cultural issues. *For the Learning of Mathematics, 5*(2), 18–22.

Clifford, J. (1987). Of other peoples: Beyond the salvage paradigm. In H. Foster (Ed.), *Discussions in contemporary culture.* Seattle, WA: Bay Press.

Crosby, M. (1991). Construction of the Imaginary Indian. In S. Douglas (Ed.), *Vancouver anthology: The institutional politics of art.* Vancouver, B.C.: Talon.

Dei, G.S. (1996). *Anti-Racism education: Theory and practice,* Halifax, Canada: Fernwood.

Dowling, P.C. (1992, May). Pedagogic voices, pedagogic messages: a sociological algebra, Presented at *Research into Social Perspectives in Mathematics Education,* South Bank Polytechnic, mimeo available from the author.

Dowling, P.C. (1993). *A language for the sociological description of pedagogic texts with particular reference to the secondary school mathematics scheme SMP 11–16.* Ph.D. Dissertation, Institute of Education, University of London.

Dowling, P.C. (1998). *The sociology of mathematics education: Mathematical myths/pedagogic texts.* London, UK: Falmer.

Dowling, P.C. (2001). Mathematics education in late modernity: Beyond myths and fragmentation. In B. Atweh, H. Forgasz & B. Nebres (Eds.), *Sociocultural research on mathematics education: An international perspective.* (pp. 19–35). New Jersey: Erlbaum.

Ensor, M. P. (1991). *Upsetting the balance: Black girls and mathematics,* Unpublished Master's Dissertation, Institute of Education, University of London.

Ensor, M. P. (1996, July). "Learning to teach" in the new South Africa. In K. Morrosin (Ed.), *Proceedings of the second national congress of the association for mathematics education of South Africa.* (pp. 112–126). Cape Town, South Africa: Peninsula Technikon.

Ernest, P. (1991). *The philosophy of mathematics education.* London: Falmer.

Freire, P. (1999). *Pedagogy of the oppressed.* New York: Continuum.

Gerdes, P. (1985). Conditions and strategies for emancipatory mathematics education in undeveloped countries. *For the Learning of Mathematics, 5*(1), 15–20.

Gerdes, P. (1986). How to recognize hidden geometrical thinking: A contribution to the development of anthropological mathematics. *For the Learning of Mathematics, 6*(2), 10–12.

Gerdes, P. (1988a). On culture, geometrical thinking and mathematics education. *Educational Studies in Mathematics, 19*(1), 137–162.

Gerdes, P. (1988b). On possible uses of traditional Angolan sand drawings in the mathematics classroom. *Educational Studies in Mathematics, 19*(2), 3–22.

Lave, J. (1988). Cognition in practice: Mind, mathematics and culture in everyday life. Cambridge, UK: Cambridge University Press.

Lave, J. & Wenger, E. (1991). *Legitimate peripheral participation.* Cambridge, UK: Cambridge University Press.

Lerman, S. (1996). Intersubjectivity in mathematics learning: A challenge to the radical constructivist paradigm?. *Journal for Research in Mathematics Education, 27*(2), 133–150.

Lerman, S. (2000). A case of interpretations of *social:* A response to Steffe and Thompson. *Journal for Research in Mathematics Education, 31*(2), 210–227.

Miller, C. (1992). Kairos in the rhetoric of science. In S. P. Whitte, et al. (Eds.), *A rhetoric of doing: Essays on written discourse in honour of James L. Kinneary.* (pp. 310–327). Carbondale, Ill: Southern Illinois University Press.

Said, E.W. (1979). *Orientalism.* New York: Vintage.

Secada, W.G. (1990). Needed: An agenda for equity in mathematics education. *Journal for Research in Mathematics Education, 21*(5), 354–355.

Skovsmose, O. & Valero, P. (2001). Breaking political neutrality: The critical engagement of mathematics education with democracy. In B. Atweh, H. Forgasz & B. Nebres (Eds.), *Sociocultural research on mathematics education: An international perspective.* (pp. 37–55). New Jersey: Erlbaum.

Smith, G.H. (1999). *New formations in the colonisation of Maori knowledge—New forms of resistance: The view from down under.* Paper presented for Noted Scholars Forum, Faculty of Education, University of British Columbia, Vancouver, B.C.

Steffe, L.P. & Thompson, P.W. ( 2000). Interaction or intersubjectivity? A Reply to Lerman. *Journal for Research in Mathematics Education, 31*(2), 191–209.

Swanson, D.M. (1998). *Bridging the boundaries?: A study of mainstream mathematics, academic support and "disadvantaged learners" in an independent, secondary school in the Western Cape, (South Africa).* Unpublished master's dissertation, University of Cape Town, South Africa.

Swanson, D.M. (2000). Teaching mathematics in two independent school contexts: The construction of "good practice." *Educational Insights, 6*(1), http://www.csci.educ.ubc.ca/publication/insights/online/v06n01/swanson.html).

Taylor, N. (1992). Difference and discourse: Valerie Walkerdine and the sociology of mathematical knowledge. *Perspectives in Education, 13*(1), 69–79.

Walkerdine, V. (1984). Developmental psychology and the child-centered pedagogy: The insertion of Piaget into early education. In J. Henriques, et al. (Eds.), *Changing the subject: Psychology, social regulation and subjectivity.* London, UK: Routledge.

Walkerdine, V. (1988). *The mastery of reason.* London, UK: Routledge.

Walkerdine, V. (1989). *Counting girls out.* London, UK: Virago,

Whorf, B. (1956). *Language, thought and reality.* John B. Carroll (Ed.), Cambridge, UK: MIT Press.

Willinsky, J. (1998). *Learning to divide the world: Education at empire's end.* Minneapolis, MN: University of Minnesota Press.

CHAPTER 10

# CRITICAL COMMUNICATION IN AND THROUGH MATHEMATICS CLASSROOMS

**Tony Cotton**
***Nottingham Trent University, UK***

## ABSTRACT

This chapter examines the idea of a competency of critical communication and its development through mathematics education. A pedagogy for critical communication is examined and described. In tandem the issue of research methodology and of the communication of research is problematized. How should we act as researchers if we ourselves aim towards "critical communication"? The chapter offers a view of "pedagogy for critical communication" synthesised from the research in the areas of gender, ethnicity and social class as well as from data gathered in English classrooms. This is not a model to be universally applied; rather it is a work of "utopian realism" against which we can measure current practice and develop future practices.

*Challenging Perspectives on Mathematics Classroom Communication*, pages 295–320

## BACKGROUND TO THE STUDY

*Once the affirmative nature of a (critical) pedagogy is established, it becomes possible for students who have been traditionally voiceless in schools to learn the skills, knowledge and modes of enquiry that will allow them to critically examine the role society has played in their own self formation. More specifically they will have the tools to examine how this society has functioned to shape and thwart their aspirations and goals, or prevented them from even imagining a life outside the one they presently lead.*

—Giroux, 1989, p. 106

Communication, and the teaching and learning of mathematics are linked in many complex ways. Long gone is the notion that a mathematics teacher unproblematically "communicates" his or her own mathematical knowledge and understandings in order to enable his or her learners to follow in the teacher's footsteps. In this chapter, I offer my view of the nature of a mathematics teaching which would see critical communication as the basis for personal and social development through mathematics education. Here I consider critical communication as a key competency to develop within our learners—a competency which allows them to take control over their lives, their life choices and the development of the communities in which they live. It is also a communicative understanding, an understanding of how we have arrived at the place in which we find ourselves, how we are constructed by the contexts we find ourselves in and how we construct these spaces around us. As a mathematics educator I am interested in the role that the learning and teaching of mathematics plays in all this.

Keiko Yasukawa (2002, p. 31) citing Tine Wedege (2000, p. 195) defines a competence through a four part model. A competence is always linked to a subject, it is something that "we have" and that we can use. Such a competence consists of a readiness for action and thought based on specific knowledge and is a result of personal learning, either through everyday practices or through formal education. Finally such competence is only seen when it is linked to a specific situation in which it is applied. So, for example, someone with the competence of "critical communication" in mathematics would use their mathematical understandings to critique reports that base their arguments on the use of mathematical data in the media. Similarly they could draw on mathematical argument to counter claims from those that hold power over them. They would be able to negotiate the educational processes in which they found themselves to increase their mathematical understandings, and they could analyze the administrative structures that have constructed a common sense view of what it is to learn mathematics. Perhaps most importantly they would not think, "I don't understand this because it is mathematics," or "It is my fault I didn't learn mathematics at school," rather they may respond "this mathematics does not describe my world" or "in order to learn more mathematics I can."

At the opening of the chapter, Henri Giroux describes a situation in which classroom practices have created voiceless learners, and in many cases, voiceless teachers. We hear echoes of critical theory and social justice, and are offered a measure against which we can judge the success of our teaching in mathematics classrooms. Giroux suggests that we will be successful if our pupils can, and do, consciously shape their social world outside the school during their time in formal education, and after they have left us behind.

Throughout the chapter, I follow a Bernsteinian approach to justify the foregrounding of pedagogy while backgrounding those other important forces that shape a learner's mathematical experience such as assessment and curriculum. My aim is to look at the difference that individual teachers can make in their classrooms: what is it that teachers can do on a day to day basis to develop their learners as critical communicators? As researchers we notice differences in individual teacher's classrooms. We see some teachers operating in mathematics classrooms steeped in critical communication. What is it that they do that we can draw on for our own practice, both in terms of teaching and in terms of research? The chapter then tightens this general focus on pedagogy to examine the work of critical pedagogues in the areas of gender, ethnicity, and class. These are areas in which the academy has worked hard to develop pedagogical approaches which remove the silence that these competing oppressions have brought with them. Through such a tightening of gaze I suggest ways through which mathematics teachers can develop critical communication in all learners, ideas formed by synthesizing the research from inside and outside of mathematics education, but developed to meet the demands of mathematics education.

Bernstein's work offers me justification for treating pedagogy as an important site for both educational research and political action. It also supports me in treating pedagogy as a single part of the experience of schooling, not an essence that can be viewed separately, but along with curriculum and evaluation one part of a triad that defines schooling for our learners. He alerts me to the necessity not to simplify issues of pedagogy. He reminds me that I must "distinguish between pedagogic practice as cultural relay and pedagogic practice in terms of what that practice relays—in other words pedagogic practice as a social form and as specific content." (Bernstein, 1990, p. 63). His comment seems particularly relevant at a time when in classrooms across the world centralized control of pedagogic practice both as form and content are in the forefront of many teachers' minds.

Working on pedagogy is of practical importance to me in laying open to the pupils with whom I work the ideology present in their classrooms. Such ideology can create the silence many pupils feel within their mathematics classrooms. Often they find it extremely hard to look outside their own experience, to explain the reasons why teachers act as they do, or to offer alternative courses of action (see Cotton, 1998). One of the most important aspects of the research described here was to find ways of exposing to pupils that which had previously remained hidden, rendering the invisible, or unquestioned, visible and problematic. In this way there was a possibility that they could see through the choices made on their behalf by teachers, administrators, and politicians, choices that lead to a differentiated educational experience and differentiated outcomes. Such awareness is a key facet of critical communication and such differentiation often occurs along lines of gender, race and class. To draw on Paulo Freire,

> pedagogy which must be forged with, not for, the oppressed (be they individuals or whole peoples) in the incessant struggle to regain their humanity. This pedagogy makes oppression, and its causes, objects of reflection by the oppressed, and from that reflection will come their necessary engagement in the struggle for their liberation. And in this struggle this pedagogy will be made and remade (Freire, 1972, p. 25).

This view of forging pedagogy with learners, rather than for learners, has been further developed by Giroux. If the rallying cry of the feminist movement has been "the personal is the political" perhaps educational researchers and mathematics teachers working for justice in and through schools should adopt the slogan "the pedagogical is the political," and follow the advice offered by Giroux and his colleagues,

> as critical teachers working in schools, we can make the pedagogical more political by clarifying how the complex dynamics of ideology and power both organise and mediate the various experiences and dimensions of school life ... fundamental (to this) would be the opportunity for students to interrogate how knowledge is constituted as both a historical and a social construction. (Giroux, 1989, p. 33)

## PEDAGOGY AS SOCIAL CONTROL

In this chapter, I separate pedagogy from assessment and curriculum. These three areas form a triad that Bernstein used to define the ways in which education reproduces current social orders. The three facets are of course inextricably linked. However, I argue, it is possible to explore ideas of pedagogy with an awareness of the shadows of assessment and curriculum. Even if the styles of assessment and the content of curriculum are dictated, we can change what we "do" in the classroom. Bernstein separated these "essences" of education as follows, "curriculum defines what counts as valid knowledge, pedagogy defines what counts as valid transmission of knowledge, and evaluation defines what counts as a valid realisation of that knowledge on the part of the taught." (Bernstein, 1973, p. 85). He used these "message systems" alongside his concepts of classification and framing to analyze the ways in which schools reproduce the social order. His message was that the overseers of the validity of both knowledge and the transmission of knowledge excluded the experience of learners from working class backgrounds. His analysis has been critiqued as excluding issues of race and gender and more recently developed to include these issues within it (see Sadovnik, 1995). However I find the ideas useful to develop my analysis of the pedagogic practice present in mathematics classrooms and use this analysis to explore future possibilities.

Pedagogy is a key relation in social reproduction as learners "learn" how to be a pupil in the classroom while simultaneously teachers are regulated by rules that lay down what it is to be a teacher. There are three features that regulate the relationship between teacher and pupil, or transmitter and acquirer, as Bernstein describes them. These are, first, features of hierarchy, rules that define the hierarchy contained in the roles of teacher and pupil. They describe what is allowable in pupil behavior and similarly describe the boundaries that define our role as teacher. The second set of rules are those of sequencing. These rules regulate progression in terms of both content and pedagogy. Sequencing rules may appear in the form of a rigid curriculum which teachers feel obliged to follow closely, or a set of text books that regulate the order in which the curriculum is exposed to learners. The *National Numeracy Strategy* in England offers a paradigm case of sequencing, both in terms of content and pedagogy, as it gives detailed advice on both curriculum and modes of delivery of the curriculum. This strategy is described for teachers in schools within a set of teacher resources which prescribe the content to be taught on a day to day basis. All teachers have been expected to attend a series of training sessions, the content of which is common. This training "teaches" them how to "deliver" the National Numeracy Strategy. All of this is then monitored through a rigorous system of inspection which expects to see both the content and form of the strategy present in every mathematics lesson. The third feature of Bernstein's hierarchy are those criteria implicit in schools that determine and monitor acceptable behaviors. These criteria are initially described by teachers and later internalized by pupils who regulate themselves within the system of schooling.

Before we look at pedagogical approaches which may support the development of critical communication within our classrooms, we need to be aware of how we may ourselves be acting to silence learners and teachers of mathematics, following Bernstein in that a "traditional" pedagogy from the learner's point of view, as well as from the teacher's point of view, consists of those methods that are "allowable" within our classrooms. The limits that are placed on us as teachers may differ from country to country and from school to school, but within each setting there are well defined boundaries outside of which we cannot step without controversy. We know when we have breached these boundaries as our learners may well question us, telling us, "our real teacher doesn't do it like that;" our managers may have a quiet word with us, advising us to stick more closely to institutional policies and practices, but most of all we can sense as we plan that we are "taking a risk." In fact stepping outside these boundaries is perhaps the start of developing critical communication. We can respond to our managers and to our pupils. We can explain why we have chosen to teach mathematics in this "different" way and offer to evaluate it with them.

So a "traditional" classroom in this sense, is one which closely adheres to a predetermined curriculum content, or unquestioningly follows approved teaching methods. This "tradition" can work against critical communication as the space for intervention either by teacher or pupil is removed. Here both learners and teachers are following "orders," either from government administrators or from published resources.

The rules of hierarchy, sequencing and criteria can be seen in either "visible" or "invisible" pedagogies. Visible pedagogies are those in which the teacher is explicit both about how the "rules" operate and how they are used to describe the performance of the child in the classroom. Within 'invisible' pedagogies the rules are implicit, but decoding and following the rules are still defining features of failure or success in school. Invisible pedagogies can silence those who feel excluded by educational processes. I would argue that the most important shift towards critical communication is through making visible the ways in which hierarchy, sequencing, and criteria operate in mathematics classrooms. This is perhaps a move away from viewing pedagogy as a choice between "traditional" and "progressive" towards seeing the choice as "transparency" or "opaqueness."

## PEDAGOGY AND MATHEMATICS TEACHING

During three years of fieldwork in one multi ethnic city primary school with five hundred pupils and the four secondary schools to which these pupils progressed, I carried out both participant and non-participant observation in addition to focus group interviews with pupils, to explore the nature of the pedagogy offered in their classrooms. One of the ways I worked with these pupils was through a pupil focus group of eight children. This group was selected by the pupils as being representative of the diversity of ideas and backgrounds within the primary school. Over three hours of group interviews we discussed the teaching styles that dominated their primary classroom. The group defined three different ways of learning/teaching mathematics. I use the pupils' own headings for these three styles:

1. *Maths Groups:* In the "maths groups," pupils worked in groups sorted by a teacher notion of "ability," which focused on a particular area of content of the mathematics curriculum. The pupils would describe themselves as being in the "top group," or the "bottom group." They were less clear however, as to how decisions were made as to which group was appropriate for which pupil.
2. *Tables races and Mentals:* This mode of teaching required pupils to work in all attainment teaching groups. "Tables races" and "mentals," were tests to encourage memorization of number facts. The pupils characterized the purpose of this teaching and learning strategy as "helping us get quicker."
3. *"HBJ"*: HBJ is the acronym for the commercial mathematics program used in the school, Harcourt Brace Jovanovich. Working within an all attainment group, the pupils worked on mathematics planned by the teachers through drawing on detailed teachers resources and pupils support material. This was seen by the teachers as an innovative approach to mathematics, as it took a thematic, inquiry based approach to mathematics that encouraged open investigation and group work. This is in contrast to many other commercial mathematics schemes that followed an individualized, hierarchical approach to mathematics.

In order to analyze these data, I used Paul Ernest's work while holding on to the important notion of visible and invisible pedagogy. This allowed me to shift from the the generic Bernsteinian view to a view particular to mathematics education (see Ernest, 1991, pp. 138–139). Ernest suggests that the challenge for "critical mathematics pedagogues" is to move from "traditional" pedagogical approaches to inquiry methods within our classrooms. Such a move would signal a clear shift in the ways in which we would view communication within these classrooms. For Ernest, "traditional" approaches are characterized by an underpinning conservative ideology and a view of mathematics as an infallible body of knowledge. A shift away from the "traditional" would involve therefore, a shift in the political project presently underpinning our systems of schooling. Similarly, a shift in teachers' views of mathematics is also needed. In my analysis I avoid using the labels "traditional" and "progressive" preferring to describe pedagogical approaches as "visible" or "invisible." As I suggested earlier, it is the problematization of pedagogical approaches with and for learners that opens up the possibilities of critical communication within the classroom.

The labels "traditional" and "progressive," as all other simplistic "commonsense" labels, are driven by ideology, and are constructs of particular and peculiar political and social contexts (see Walkerdine, 1998).

Table 10.1 offers an analysis of the focus group data against the Ernest criteria—the shift towards what Ernest describes as a "critical pedagogue" can be seen as a movement down the table. I have also classified the "categories of teaching mathematics" as visible or invisible. Visible approaches are ones for which both pupils and teachers could articulate a shared purpose. For example, the pupils described "tables races" as important because they increase your speed and accuracy in arithmetic. They talked about maths groups as the place where the teacher could concentrate on their particular content needs, and in which they worked with "people of the same ability." However they describe the sessions when the teacher used the HBJ program materials in a less secure way. They described it as preparing them for their national examinations, or in terms of covering content without reference to the style of organization or learning approaches. In this sense the purpose of the pedagogical approach was "invisible." (For more detail see Cotton, 1998).

This analysis suggests that an aim for building a pedagogy for critical communication would be to move the pedagogy described by pupils as HBJ from an invisible pedagogy to a visible pedagogy—in other words to validate this practice for the pupils.

**Table 10.1. Invisible and Visible Pedagogies in the Primary Classroom**

| | *Visible* | *Invisible* |
|---|---|---|
| Drill no frills | Tables races and mentals | |
| Skill instructor—motivate through relevance | Maths groups | |
| Explain, motivate—pass on structure. | Maths groups | |
| Facilitate personal exploration. | | HBJ |
| Discussion, conflict, questioning of content, and pedagogy | | HBJ |

The cohort of final year primary pupils kept learning journals during the first two years in which I worked with them. These 55 journals offered me evidence of the ways in which these learners described visible pedagogies. Pupils would give their latest score in their tables and mentals tests, often describing this in terms of success or a failure. In a similar way pupils would inform me of how many "gold spots" they had been awarded during the week, the "gold spot" being a reward for "good work." An entry from Rupa is typical:

(5/10/95) I have done some HBJ pages. Some were hard, some were easy. We had a tables race, I got the gold star first. I was slow on my times tables now I am fast. We had mental maths, they were hard so I got 10 out of 20.

Here Rupa is clear about the purposes and expectations of tables races and mentals but less sure about HBJ. She does not quantify her success/ failure in the same way. The authors of the HBJ program are clear in their teachers notes as to how success may be measured. They suggest that:

> HBJ Mathematics provides challenges at which all pupils will be able to work, although the outcomes will be different for each child. Each (activity) offers an opportunity for all pupils to extend themselves and to show what they are able to do. This will ensure that no child needs to be regarded as unsuccessful at mathematics. At the same time, there are no limits to the level of response so that all children can operate at their optimum level. (Kerslake et al., 1992, p. 4)

Rupa's response suggests that she has noticed a contradiction between the pedagogic practice embodied in "tables races and mentals" and in HBJ. The invisible nature of the HBJ pedagogic practice for her does not offer her any way of measuring her success. She is therefore reduced to saying that she has completed tasks.

This contrast between visible success in tables races and a less secure view of the purpose of HBJ can also be seen in the following two entries:

> (Duwayne 15/11/95): "I am pleased with my work. I am not sure if (my teacher) will be?" and (Sangita 9/11/95) "I have done my HBJ. I have done my planning sheet. I want to do my (?). I am pleased with it—(the teacher) said."

These entries suggest that the pupils interpret success in HBJ by using the teacher as a mediator. Once the focus group had discussed the pedagogical approaches to mathematics within their classroom, we moved on to explore the role of the teacher within these approaches. The focus group defined these roles as "explainer," "evaluator," and "tutor." An example of teacher as explainer would be "fractions was a bit hard at first but when (my teacher) explained it to the class I knew how to do it (a bit)." (Journal entry 21/9/95—Sadia). An example of teacher as evaluator can be seen in "My teacher said I was a superstar." (Journal entry 21/9/95—Sairah) Finally an example of teacher as tutor could be "The coach said that (my teachers) were looking out for pupils to play basketball at Granby halls," (Journal entry 12/10/95—Mehnaz). A tutor here is a coach, someone who teaches a child by modeling.

Table 10.2 shows how members of the focus group described their teacher's roles. The pupils in the focus group were drawn from two tutor groups and so had been taught by two different teachers. Their descriptions allow teachers to move between roles depending on the context of the mathematics lesson. I analyzed the learning journals and interview transcripts from the pupils in the focus group at the end of the year, to explore how they described the different roles they attached to these two teachers. I was surprised at how rarely pupils mentioned their teachers at all. A total of 14 descriptions of the way in which their teacher operated out of 58 pages per pupil of journal entries and 15 hours of interview transcript.

**Table 10.2. Teacher Roles as Assigned by Focus Group**

| | *Sadia* | *Kenny* | *Imran* | *Sairah* | *Mehnaz* | *Jenny* | *Rupa* |
|---|---|---|---|---|---|---|---|
| Explainer | 2 | 0 | 1 | 1 | 0 | 0 | 0 |
| Evaluator | 1 | 0 | 0 | 1 | 0 | 0 | 0 |
| Tutor | 0 | 4 | 0 | 0 | 4 | 0 | 0 |

This apparent invisibility of the teacher during primary education contrasts with the experience of the pupils in secondary school, as does the invisible pedagogy of HBJ. Focus group interviews after the transfer to secondary education centered on clear descriptions of their teachers and their pedagogical practices. This suggested to me that for these pupils a shift from "invisible" to "visible" pedagogies occurred at the transition from primary to secondary school. Perhaps it is this shift that supported the pupils in their positioning of secondary pedagogy as of major importance and their devaluing of primary pedagogy as immature. On occasions the secondary teachers made this positioning overt: "Try and be grown up enough to do it on your own without constantly looking at your neighbor's work." (Teacher comment to pupil—noted in my journal 30/5/97.)

During the pupils' first year in four different secondary schools I observed thirty three lessons taught by nine different teachers. My classroom observations and my discussions with the focus group of school students suggested that here the pupils in my study predominantly experienced a pedagogy which was both "visible" and "traditional" in Ernest's sense, that is, underpinned by conservative ideology and an infallible view of mathematics. This experience contrasted to the approach they had experienced in primary school. Of course within this definition of "traditional" lie several styles. Table 10.3 shows the number of lessons, which followed a particular version of a "traditional" pedagogy.

**Table 10.3. "Traditional" Teaching Styles**

| *Traditional Styles* | *Number of Lessons Following Style (out of 33)* |
|---|---|
| Brief teacher introduction followed by exercises | 24.0 |
| Individualized work with teacher focused on administration | 2.5 |
| Class test | 3.5 |
| Teacher exposition to whole class | 2.0 |
| Mixture of exposition, group work, individual work | 1.0 |

Such a classification illustrates how the pupil focus group could share a common sense about how mathematics should be taught and learned in secondary school. They described these ways of working in the classroom unproblematically, it was assumed that this was the "correct" way to teach and learn mathematics, a clear case of "pedagogic practice as cultural relay." Although the secondary classrooms described above conformed to a particular view of a "traditional" pedagogy, other data from the primary classrooms showed evidence of teachers of mathematics viewing learning as a shared endeavor, shared between pupils, and shared between teachers and pupils. I have evidence of teachers exhibiting openness and honesty within their mathematics classrooms and expecting and/or encouraging the same from their pupils. There are examples of democratic, collaborative, and inclusive decision making within mathematics classrooms, and teachers showing trust of learners and giving care and responsibility to learners. Also, I have evidence of teachers expecting pupils to learn through challenge and variety; here mathematics teaching and learning included acquiring and using critical thinking skills.

Such data suggest to me that moves towards critical communication through mathematics education are possible when teachers open up their practices to critique and render the invisible, visible. Nevertheless there are clear constraints on the development of critical practices. Data showed that occasionally teachers abdicated responsibility for their teaching; by blaming external forces, by not challenging inappropriate behavior, by offering mundane tasks to keep learners busy or just "leaving it to the kids." Teachers occasionally saw their roles in administrative terms, either as administrators of schemes (both commercial and school based), or of school testing and setting practices. This resulted in teachers who were simply not teaching. Responsibility for communication was given over to the scheme, to the institution of the school, or even to the pupils. In such an atmosphere, the hidden pedagogy of which Bernstein warns us was left to prosper. Those pupils most able to interpret and discover the hidden pedagogy flourished—others were left to wither and fight it out for scraps. This abdication of responsibility, which may be due to the pressure of external constraints on teachers, can lead to a deskilling, in terms of pedagogy, for the teacher.

Against the requirements of critical communication, teachers also acted as the fount of all knowledge (both mathematically and administratively). The teaching role here was to offer answers, offer techniques that will be successful, and act as controller of the classroom. Controller in this case usually meant that learners were quiet and occupied. This was evidenced for example, in teacher's questioning style—the decision made on whether to use "hands up" or not. It could also be seen when the only teacher interventions were to explain what the "book meant" when pupils asked for help, or when teachers controlled pace through setting arbitrary targets. These practices became routinized: 10 minutes introduction, getting on on your own with textbook or worksheet, and 10 minutes at the end marking the work. The pupils responded to this routinized behavior and described it as "usual" when questioned on how one might learn mathematics.

**Table 10.4. Teacher Actions and Critical Communication**

| *Actions Compatible with Critical Communication* | *Actions Acting Against Critical Communication* |
|---|---|
| Seeing learning as a shared endeavor. | Abdicating responsibility for teaching approaches to schemes or imposed curricula. |
| Exhibiting openness and honesty. | Not challenging inappropriate behavior. |
| Planning for democratic, collaborative, and inclusive decision making within the classroom. | Offering mundane tasks to keep learners busy. |
| Viewing pupil control of learning as positive. | Viewing the teaching role in administrative terms. Either administrators of schemes (both commercial and school based), or of school testing and setting practices |
| Showing trust of learners. | Acting as source of all knowledge within the classroom. |
| Giving care and responsibility to learners. | |
| Expecting pupils to learn through challenge and variety. | |

Table 10.4 contrasts those actions I observed in classrooms which are compatible with the development of critical communication, and those which act to discourage critical communication. The description of the constraints observable within classrooms should not be read as a blaming of teachers. This set of constraints can be seen in a positive light; if it is our fault we can fix it. And we can fix it by following the clear examples from

teachers operating with the values of critical communication: values of openness, honesty, vulnerability, inclusive decision making, developing power with their learners, developing a pedagogy of care with responsibility for safety in learning, and a pedagogy which expects learners to succeed in complex tasks demanding critical engagement with mathematics. The classrooms I observed and analyzed in terms of actions supporting or acting against critical communication often evidenced conflicting themes. Through analyzing our classrooms, and exploring the constraints which make us act in ways which cut across our beliefs with our learners we both open ourselves, and impose practices, to critique as well as act within the realms of possibility.

## THEORIES OF PEDAGOGY FOR CRITICAL COMMUNICATION

If the preceding account describes one experience of pupils in UK schools, what other work can we draw on as we move towards critical communication in and through our teaching? Much has been done in our schools to develop a mathematics pedagogy which addresses the needs of girls. Although there has been less work in the area of race and mathematics education, there is a wealth of literature exploring more general issues of race and education and this can offer us fruitful paths to follow. Similarly Marilyn Frankenstein's work in the USA (see Frankenstein, 1989), developing curriculum and pedagogical critique is useful. Here I attempt to draw together common threads from this literature, and use this synthesis to explore strategies which may allow us to build a pedagogy for critical communication through mathematics education. It is necessarily a small slice out of a large cake. This particular selection is used as the authors describe directly the ways in which their research has translated into pedagogical practice. They have also developed this practice with a clear focus on working with previously silenced groups. My task is to find the common themes which run through this work.

### A Socially Just Pedagogy for Girls

Let us first look at how Pat Rogers suggests we put the theoretical stances of feminist pedagogues into practice (Rogers, 1995). She suggests that Carol Gilligan's (1982) work can offer us a link between "maths avoidance" and maths pedagogy. Pat Rogers uses the idea of "connectedness" to argue for a pedagogy which allows women and girls to see themselves as producers of mathematics as opposed to consumers of a predetermined and absolute body of knowledge. The shift from consumer to producer represents the shift from voiceless to communicator. She then offers us descriptions of the methods she employs to encourage "direct access and engagement, free creative expression, and ownership" of mathematics. These can be summarized as follows:

- *Lecturing for a purpose:* sparing use of lectures to introduce new material or to draw together ideas rather than lecturing being the norm.
- *Think-write-pair-discuss:* a method of response to new ideas involving independent response on paper, which is then shared in pairs before being taken to the whole group.
- *Whole group dialogue:* such a dialogue follows assigned reading or think-write-pair-discuss. The dialogue proceeds through questioning to involve and challenge members of the group.
- *Board work:* students as teachers taking control at the board to explain solutions or offer questions.
- *Brainstorming:* first in pairs and then as whole group to offer support.
- *Problem posing:* problems that are both posed and solved by students.
- *Reading exercises:* developing mathematical reading skills as an essential step towards independence in mathematics (Rogers, 1995, pp. 179–181)

Here I would like to offer an illustration of critical communication in practice, a way of describing to learners the reasons we make particular choices. We may say, for example, "I am lecturing to you now, as I wish to draw the threads of that discussion together;" or "I want you to form a 'working pair' as a way of thinking through a risky idea with a supportive colleague before trying it out on the whole class." Such an uncovering of pedagogy is a part of the move from "power over" to "power with." In this way our learners may challenge us—let us know when our techniques are not succeeding. Pat Rogers also offers a way to obtain immediate feedback. At the end of each session students write down on a piece of paper one or two points about which they remain confused. These are anonymous and can be used to plan future sessions. I have used all of the above techniques with a group of teacher education students who were extremely nervous about their mathematical "ability." I used the above list and we analyzed every session using the list as our audit tool. Their evaluations offered me a view as to how the students' view of teaching and learning mathematics developed during the sessions.

In the same collection as the chapter by Pat Rogers described above, Sue Willis alerted me to problems in implementing such a pedagogy for girls in our classrooms (Willis, 1995). Problems become obvious as we bring transparency into our pedagogy. For example, one teacher in Sue Willis's study introduced into her mathematics classroom materials that demanded more reading than previously. Other teachers in the school reacted by passing on complaints from the students that there was too much reading. The teachers were asked to find out which groups of students were challenging this new approach. The response was as follows,

> ... the teachers returned rather abashed. They had failed to notice that the original complaints had derived from a vocal minority of high-achieving boys, and a few high-achieving girls ... the pupils who the teachers regarded as less-able insisted that they did the reading ... moreover, it was predominantly the high achieving male students who were most reluctant to participate in the group work that was an essential part of the course. (Willis, 1995, p. 195)

There are links here both to Basil Bernstein's descriptions of invisible pedagogy and a Rawlsian view of social justice. The purpose for which the teacher introduced the reading tasks remained hidden from the pupils, it cut across their common sense view of what mathematics teaching should be like, and was rejected by them as "progressive." Those who were heard in this situation were the most "powerful," the high attaining boys. Those who were silenced were the "invisible" majority. Rawls would suggest that we should take note of the views of the "least powerful" in any individual situation, otherwise those already privileged in and through our pedagogical practices continue to dominate. On many occasions it is the "least" powerful, who fail to communicate, indeed the lack of control means that however loudly they speak their voices are not heard.

## A Socially Just Pedagogy for "Learners of Color"

Earlier in the chapter I took the Freirean view that pedagogy should be developed with the members of the many communities, which constitute our learners. This illustrates the issue that we cannot take a simplistic view—we are not searching for a single style which suits learners of color, or girls or white working class boys. Beverly Gordon reminds us, "at the present time it is probably more helpful to focus our attention on separating content from pedagogy than talk about a unique learning style for Black children." (Gordon, 1993, p. 234). However this is not to suggest that teachers of color do not operate in ways, which can be seen as distinctive from the practices of many of their white colleagues, and not to suggest that particular pedagogical approaches advantage particular groups of learners. It is the synthesis of approaches developed for learners of color with those developed by feminist pedagogues, which proves fruitful for this project.

Michele Foster describes relationships marked by social equality, egalitarianism, and a mutuality stemming from a group rather than an individual ethos as forming a part of an African-American ethos within teaching and learning (see Foster, 1993). Christine Callender (1997) has carried out research on "Black teaching style" in the UK. For her, even though such a teaching style may seem at odds with the aims of her white liberal colleagues, it is effective for the Black pupils with whom she works. Her research can be seen as a means of "validating" such a pedagogy in the Bernsteinian sense. I would argue that a just education system would have a teaching force representative of those communities that make up its learners. While this remains a vision, the pedagogical styles which such teachers would bring with them remain on the margins. We are in a closed loop: students from minority ethnic groups continue to experience discrimination in and through education. This can lead to an unwillingness to train as a teacher, or to a lack of qualifications that would enable such students to become teachers.

The pedagogical approaches above share the theme of connectedness with a feminist pedagogy. This connectedness is demonstrated overtly or covertly in Black teachers' interactions with Black children, either occurring explicitly as a part of classroom discourse, or implicitly as a part of the teachers' hidden curriculum (Callender, 1996). Similarly, elements of what has been described as "African-centered pedagogy" (Gordon, 1993) echo the work of Pat Rogers described in the previous section. Such an "African-centered pedagogy" would be expected to legitimize African stores of knowledge, exploiting productive community and cultural practices. It would build on indigenous language practices and describe a world that idealized a self-sufficient future for one's people without denying such rights to others. It would also support cultural continuity while promoting critical consciousness.

## A Socially Just Pedagogy for Working Class Pupils

I offer no apology for describing groups of pupils in our schools as "working-class." It is clearly a simplistic label, as are all labels. I would argue it is no more simplistic than the labels of "gender" or "race." It is however, a label which many of our learners carry with them through education. It is also a label which is too easily transformed, almost unproblematically into labels such as "less-able" or "disaffected." I reject the argument that working-class pupils are less-able or disaffected but I would argue that they are over-represented within the groups of pupils labeled in this way. I concur with Basil Bernstein when he says:

> I find myself in the position of stating that working class children, especially lower working class children relative to middle class children, are crucially disadvantaged, given the way class relations affect both the family and the school, but that their forms of consciousness, their way of being in the world, must be active in the school, for without such expression there can be little change in the children or in society. The steps required to validate this inclusion will also, in the end, involve a restructuring of education itself. (Bernstein, 1975, p. 28)

That is the point of critical communication, but what form should such a pedagogical restructuring take? One of the few mathematics educators to take a class view of developing mathematics pedagogy is Marilyn Frankenstein. In her book *Relearning Mathematics: A Different Third R—Radical Mathematics,* a text book written for those labeled maths failures by school, she outlines her aims (Frankenstein, 1989). These can be summarized to offer a view of a pedagogy for working class pupils which works towards critical communication. Such a pedagogy would involve teachers drawing on learners' present mathematical knowledge, both as starting points, and to acknowledge the mathematical skills always present in our learners. This supports the development of critical communication through building learners' confidence in themselves as learners of mathematics, and helping them to take control of their own learning process. The pedagogical process also offers alternative views of what makes mathematics and what constitutes "ability" in mathematics. The aims of such a course would be to show mathematics as enjoyable, but also to use it as a political tool to increase the learners' background knowledge about key social, political and cultural issues. There is an epistemological challenge in this approach as it counters the view of teachers as expert and builds problem posing skills in individuals. In addition there is a social transformation challenge as an overt aim is to motivate learners to fight for progressive societal change. These aims and objectives offer a stark contrast to the experiences of the children in the classrooms I described earlier and allow me to move towards an alternative model.

## TOWARDS A PEDAGOGY FOR CRITICAL COMMUNICATION

My task here is to draw on the previous sections, pulling out the common threads from pedagogical approaches designed to support those previously silenced in mathematics classrooms, and to bring them together to form a model which would work towards critical communication as a key competency to be developed in all our learners. I also draw on the experience of the pupils I encountered through my fieldwork to offer a model which is practical (in the sense that I would find it possible to work with), and visible (in the sense that my learners could see it as acceptable).

My fieldwork and the synthesis of the ideas described above suggest that such a pedagogy would be anti-competitive. It would work against the domination of mathematics classrooms by particularly advantaged groups of learners, or communities, and would be non-exploitative of both learners and teachers. Such a pedagogy could not and would not lead to cultural selection or cultural imperialism. Within such a pedagogy individuals would feel empowered as learners and as individuals, and would feel secure in their environment. Through their mathematical experiences they would become active members of their communities within the classroom, and equal partners in the building of a culture of learning.

The pedagogical purpose is to move towards critical communication in an open and overt way. Here pedagogy is both the teaching style and the political/educational aim for working in this way. Thus, brainstorming in itself is not a pedagogy for critical communication, but using brainstorming as a means of validating and valuing cultural knowledge is a facet of such a pedagogy. Table 10.5 draws together these ideas in terms of the themes, which are addressed through particular pedagogical approaches, the processes through which we may address these themes, and the contribution which is made to developing the competency of critical communication.

Earlier in the chapter I used data from empirical work to describe the educational experiences of a group of pupils within schools in the UK. I only briefly touched on the methodological approach I used and want now to spend time developing these ideas further.

## DEMOCRATIC RESEARCH METHODS AND EMPOWERMENT

I set myself twin goals for this chapter. One goal was to describe pedagogical approaches which may support us in developing critical communication through our mathematics teaching when we are teaching mathematics. The other was to describe the research process itself as a practice which can develop critical communication within ourselves as teachers, researchers, and readers of research. For me, the research process should also be dialogical, research with rather than research on. My research methods should fit the model I have created for myself and offered for you above.

**Table 10.5. Features of a Pedagogy for Critical Communication**

| *Theme* | *Process* | *Contribution to Critical Communication* |
|---|---|---|
| Acknowledge and build on learners heritage, valuing and emphasising cultural practices and knowledges. | Brainstorming, pair work, student choices, different entry points within activities. | Culture creating, challenges domination. Emphasises connections. |
| Move from informal to formal languages of mathematics to develop oral, formal and informal traditions. | Whole group dialogue, two-way pedagogies including learner presentations, discussions, journals and learning logs, debriefings, interviews and conferences. | Culture creating, anti-competitive. |
| Promote collaborative and mutually supportive ways of working. | Working pairs, collaborative activity, development of connected relationships within which learners feel valued. | Anti-competitive, challenges domination, builds security in the learning environment. |
| Encourage learners to become active in both interpreting and changing their worlds, develop critical consciousness through maths activity. | Reading exercises, projects and performances that use the methods of a field study and represent a whole piece of work within that field, mathematical studies of political and social issues. | Empowerment through mathematics. |
| Moving towards learner autonomy, emphasis on power with not power over. | Learner board work, problem posing, self evaluation, scaffolding a process of successive conversations, learning experiences that take learners from different starting points to proficient performance. | Challenges domination, culture creating, anti-competitive. |

The following is an extract from my personal journal which is a starting point for this discussion. Holly is my stepdaughter, age 11 at the time of this incident. Helen is Holly's mother and my partner.

> Tuesday 30 January
>
> Holly came home in tears today because she has been moved into what she describes as 'the bottom group with the kids who can't add two and two'. She is particularly distressed as she knows she can 'do' maths except for the bits that demand quick number recall. She is also confused because she knows that Helen and I don't measure her ability by which group she is in and disagree with the whole idea of setted groups. I guess she can't explain to herself why her self-esteem is so damaged in this case. She has also been disempowered by the way this move took place. (Her teacher) asked her if

> she would mind if she moved down a group. This meant that she had to agree to something she didn't want to do—making her an accomplice in her own distress.

My first reaction was to cry with her, my second to resign immediately from the governors, stop my research and move her to another school. Eventually Helen and I agreed that the best thing we could do was to enable Holly to cope with this act of unfairness—try to support her through it, aim to empower her to survive school as unfortunately many schools are like this and many children will share this experience. This also confirmed to me the point of carrying on with my research and emphasised the need to work towards empowering children—my focus from now on will be particularly on the children.

As often happens Holly very perceptively described the problems of such ad hoc groupings (as all sets are). She said, "I don't feel as though I belong anywhere, in the top group the work is too hard and they go too fast, in the bottom set the work is too easy." I pointed out that the vast majority of children probably felt the same way—that is what is fundamentally and morally wrong with setting; it excludes the many, it makes children feel as though they do not belong, and through loyalty to their teachers, and a misplaced sense of trust in the system, they blame themselves.

I use the extract above as it offers a vignette which clearly and powerfully describes the settings in which I was working. It also communicates a sense of the standpoint from which I began the work. As I reread it, explanations for several choices I made, choices that had a direct influence on the research, become clear. Holly is placed in a competitive situation in which she feels the loser. She is disempowered and feels unable to voice her feelings of rejection from a community of learners in which she had previously felt secure. More than this, the teacher asks her if she will tacitly agree with his assessment—she becomes implicated in her own failure and has no way of communicating this feeling of failure, or feeling of rejection. The extract also throws up many of the difficulties that face me in the research process. My focus is the voice of the pupils, which relates back to Rawlsian views of social justice suggesting we can only move towards a more just society by listening and accounting for those occupying positions which carry little power. In the extract, for this moment, Holly was placed in such a position, as am I as a parent. The voicelessness described by Henri Giroux at the opening of the chapter is illustrated. Both Holly and I can tell of the situation, but we cannot voice it in the sense that we can play no role in altering the situation. We are placed in a lose–lose position, and have no way of negotiating a win–win position. Critical communication is absent.

The extract further signals a personal commitment to exploring a research process that would in itself work towards empowering the children who were involved in the process. Empowerment here satisfies the demands of critical communication in that learners are becoming active in both interpreting and changing their worlds. Later I came to hold the view that every action I took to move my research forward should in itself be a worthwhile activity for the children involved. The ends and the means should become one. For me, this is the real meaning of praxis. Thus the extract illustrates two facets of research for critical communication. One is a research process, which has as an aim the empowerment of those with whom we work, and a direct focus on this in the instant. A second is the idea of research as praxis—the research process in itself should develop the skills of critical communication within the researcher and within those with whom the researcher works.

My growing awareness of my personal standpoint led to the choice of reflexive interviewing techniques over individual, scripted interviews. It also supported the choice I had already made that my role for the first year of the research should be that of participant observer, this decision would place me "inside" a particular discourse. Using classrooms in which I was well known, either as teacher, co-worker, or parent placed me in different discourses from which to reflect on the everyday practices and experiences of the pupils. The use of pupil journals and group interviews supported me in this. One of my aims in choosing pupil journals and interviews was that they allowed me data which could support me in introducing other "voices" into the text.

## EMPOWERMENT AND RESEARCH

In a chapter exploring communication as a way of being heard both within mathematics classrooms and in wider society, the idea of empowerment is key. The term "empowerment" is a contentious one. For Barry Troyna, empowering research must illustrate how it has transformed social and political relations (Troyna, 1994). He differentiates between "empowerment" and "giving a voice." The notion of empowerment through involving students as researchers is also problematized by Bill Atweh and Leone Burton (1995). For them, empowerment for pupils is affected by the ways in which the pupils themselves access the process of research, coming through collaborative approaches to research involving a sense of ownership. For Atweh and Burton, the impact of the information gathered through the research process can also be empowering.

J. Rappaport describes how empowerment can be seen as promising an alternative view to that of "prevention" in terms of community psychology

(1981). Here communities often regarded as having "problems" actually have very well developed systems for survival, and indeed development, in a system which is loaded against them. Working with the solutions that are already in place rather than developing new solutions from our positions as experts may both offer us a better chance of success, however we measure it, and also lead to an empowered community. He suggests that,

> empowerment rather than prevention is far more promising both as a plan of action and as a symbolic ideology.... by empowerment I mean that our aim should be to enhance the possibilities for people to control their own lives.... Empowerment presses a different set of metaphors upon us. It is a way of thinking that lends itself to a clearer sense of the divergent nature of social problems. (Rappaport, 1981, p. 15)

He also offered me a measure against which I could judge the methods I had chosen if I wished to work towards empowerment of the pupils with whom I was working.

> There are at least two requirements of an empowerment ideology. On one hand it demands that we look to many diverse social settings where people are already handling their own problems in living, in order to learn more about how they do it ... it also demands that we find ways to take what we learn from these diverse settings and isolations and make it more public, so as to help foster social policies and programs that make it more rather than less likely that others now not handling their own problems in living or shut out from current solutions, gain control over their lives. (p. 15)

So, for me, empowerment means exploring the solutions which are already in place, but often hidden, as we, as teachers and researchers, impose our own solutions on top. As a researcher interested in critical communication, I needed methods which allowed me both to see these solutions and discuss them with the pupils without imposing my own interpretation. I should then use this discussion as a yardstick against which to measure my own interpretations of the situation. More than that though, any final report should aim to make the learners' views public, in order to support the process of pupils gaining control over their own lives. This indeed supported the idea of reflexivity which is so important to critical theorists.

This view of empowerment supports notions of democracy that have already been discussed in mathematics education, particularly Ole Skovsmose's idea of *mathemacy* (1994). This offers a view of the unique and vitally important role that teachers of mathematics have.

> Mathemacy as a radical construct, has to be rooted in the spirit of critique and the project of possibility that enables people to participate in the understanding and transformation of their society and, therefore, mathemacy

becomes a precondition for social and cultural emancipation. Could mathemacy be actively involved in naming and transforming these ideological and social conditions that undermine the possibility for forms of community and public life organised around the imperatives of a radical democracy. (Skovsmose, 1994, p. 27)

Mathemacy is offered by Skovsmose as a complementary notion to that of literacy and is, in a similar way composed of several competencies. He suggests these competencies include: mathematical skills and competencies to construct new mathematics, the competency to apply mathematics, and finally an understanding of how to reflect on the process being employed in order to see how the mathematics is constructing the knowledge we are developing, and how this knowledge may have been different if we had used our mathematics differently.

## RESEARCH AS PRAXIS

If my research was to work within the constraints of praxis I had imposed on it of being worthwhile in its own right, first as an educational activity, but more than that as an empowering, democratic educational activity, it seemed to me that I needed to be working with pupils in classrooms, exploring new ways of working, and finding ways of working with them on an analysis of what I was doing and why.

Participant observation, my own research journal, and pupil journals developed a preliminary record. An initial impression of what was happening. The focus group and reflexive interview techniques allowed me to question this initial view and offered me what Phil Carsprecken (1996) terms *dialogical data generation.* Here, "dialogical data generation" involved developing my initial theoretical ideas by trying them out on the pupils, or as a colleagues termed it, "bouncing the theory off the practice." Later non-participant observation allowed me to work in more detail, to collect "thicker" data, before returning again to the focus group for further "dialogical data generation."

Of course the role of participant observer does not come without pitfalls. I will use an extract to illustrate these pitfalls as well as to illustrate the use of pupil journals as producers of data in more detail. This is from a girl whom I will call Liz.

(Extract from Liz's journal 19/1/96)

(My teacher) has just said that we might not do gym later because Tony might be coming in. I got 1 g.s. (gold star) for finishing my weather work first and don't normally so M was pleased with me. M's birthday is on Monday. I'm

> going to get her a present. I hope Tony doesn't come because I would really like to do P.E. and if Tony does come in I hope we still do P.E. because I find maths so bouring that I would rather go and watch world wide wrestling federation and that is bouring. Oh god, Tony has come (help me, please), just this minute he is sitting with Jagdeesh, Richard and Simon.

I have left spelling and punctuation unaltered. As I read the entry I wondered to what extent I was being used by the class teacher as an excuse not to have to do PE. The teacher and the pupil knew I was coming in to school on that day. I looked back in my journal to check. I found that the whole term was planned during the previous holiday. The dates of my visits were arranged together with the activities which we would work on. I am also interested in the way Liz writes. The journals were written with the knowledge that I would read them—so Liz wrote this knowing I would read it. So there we have it. I am more boring than the World Wide Wrestling Federation. However this also suggests that there is a form of communication taking place—the journals offer a forum through which Liz can voice her feelings. She will also expect a response, in this case a conversation with me about the issues she raised in the extract. I would argue that she has found a way in which she can both describe how she is feeling and influence future outcomes.

## LOOKING BACKWARDS—LOOKING FOR(E)WARDS

> Utopian realism ... is the characteristic outlook of a critical theory without guarantees. "Realism" because such a critical theory, such a radical politics, has to grasp actual social processes to suggest ideas and strategies which have some purchase; "utopianism" because in a social universe more and more pervaded by social reflexivity, in which possible futures are constantly not just balanced against the present but actively help constitute it, models of what could be the case can directly affect what becomes the case. (Giddens, 1994, p. 250)

I stated earlier in the chapter that I aim for my work to be practical both during the process (praxis) and as a result of the process. Another measure of success is the extent to which I have created alternatives for myself and for my learners in terms of the ways in which we communicate within the mathematics classroom.

My initial question for this chapter was how can we work towards critical communication through our work in mathematics classrooms? I worked on this question taking note of Giddens' view of a critical theory without guarantees (see Giddens, 1994), accepting the notion that my work would simultaneously describe and construct realities in mathematics classrooms.

My aims demanded that I describe facets of a pedagogy of critical communication through which I could observe classrooms and against which I could measure progress. This way of working draws heavily on critical pedagogy in terms of its view of agency. However the work does not suggest that this journey has an endpoint—a fixed conclusion.

The knowledge that I seek does not aim to order or interpret the "reality" of mathematics education in schools, nor does it try to understand the lived experiences of pupils. Rather I am searching for ways of producing knowledge that allows me and those with whom I work to fight for a more "just" world as described by the notion of critical communication. This form of knowledge involves a different attitude towards meanings as well as an alternative view of communication. My aim is not to describe and understand but to criticize and transform. Results take on a different meaning. I do not aim to answer questions or to solve problems. I aim to pose interesting ones, and then ...

## REFERENCES


Atweh, B. & Burton, L. (1995). Students as researchers: Rationale and critique. *British Educational Research Journal, 21*(5), 561–575.

Bernstein, B. (1973). *Class, codes and control. (Vol.2).* London, UK: Routledge.

Bernstein, B. (1975). *Class, codes and control (Vol. 3). Towards a theory of educational transmissions.* London, UK: Routledge & Kegan Paul.

Bernstein, B. (1990). The structuring of pedagogic discourse (Vol. 4). Class, codes and control, London, UK: Routledge.

Callender, C. (1996). *Cultural styles in multi-ethnic classrooms—the case of African Caribbean teachers and pupils.* Paper presented to the BERA Conference, Critical communication Symposium.

Callender, C. (1997). *Education for empowerment: The practices and philosophies of Black teachers.* Stoke on Trent, UK: Trentham.

Carsprecken, P.F. (1996). *Critical ethnography in educational research: A theoretical and practical guide.* London, UK: Routledge.

Cotton, T. (1998). *Towards a mathematics education for social justice.* Unpublished PhD. thesis. Nottingham trend University.

Cotton, T. (1999). Flipping the coin: Models for social justice in the mathematics classroom. *The British Society for Research in Learning Mathematics Yearbook* (Vol. 1, pp. 67–86) York, UK: John Bibby.

Ernest, P. (1991). *The philosophy of mathematics education.* London, UK: Falmer.

Foster, M. (1993). Educating for competence in community and culture: Exploring the views of exemplary African-American teachers. In M.J. Shujaa (Ed.), *Too much schooling, too little education: A paradox of black life in white societies.* (pp. 221–244). Trenton, NJ: Africa World Press.

Frankenstein, M. (1989). *Relearning mathematics: A different third R—Radical maths.* London, UK: Free Association.

Freire, P. (1972). *Pedagogy of the oppressed.* London, UK: Penguin.
Giddens, A. (1994). *Beyond left and right—The future of radical politics.* Cambridge, UK: Polity Press.
Gilligan,, C. (1982). *In a different voice.* Harvard. Harvard University Press.
Giroux, H.A. (1989). *Schooling for democracy: Critical pedagogy in the modern age.* London, UK: Routledge.
Gordon, B. (1993). African-American cultural knowledge and liberatory education: Dilemmas, problems, and potentials in postmodern American society. In M.J. Shujaa (Ed.), *Too much schooling, too little education: A paradox of black life in white societies.* (pp. 57–80). Trenton, NY: Africa World Press.
Kerslake, D., Burton, L., Harvey, R., Street, L. & Walsh, A. (1992). *HBJ Y3 Mathematics teachers resource book,* Collins. London, UK: Educational.
Rappaport, J. (1981). In praise of paradox: A social policy of empowerment over prevention. *American Journal of Community Psychology,* (*9*), 1–25.
Rogers, P. (1995). Putting theory into practice in equity in mathematics education. In P. Rogers & G. Kaiser (Eds.), *Equity in mathematics education: Influences of feminism and culture.* (pp. 175–185). London, UK: Falmer.
Sadovnik, A.R. (Ed.) (1995). *Knowledge and pedagogy: The sociology of Basil Bernstein.* New Jersey: Ablex Publishing.
Skovsmose, O. (1994). *Towards a philosophy of critical mathematics education,* Dordrecht, The Netherlands: Kluwer Press.
Troyna, B. (1994). Blind faith? Empowerment and educational research, *International Studies in the Sociology of Education, 4,* 3–24.
Walkerdine, V. (1998). *Counting girls out: Girls and mathematics.* London, UK: Falmer.
Wedege, T. (2000). Technology, competencies and mathematics. In D. Coben et al. (Eds.), *Perspectives on adults learning mathematics.* (pp. 191–207). Dordrecht,The Netherlands: Kluwer.
Willis, S. (1995). Gender reform through school mathematics in equity in mathematics education. In P. Rogers & G. Kaiser (Eds.), *Equity in mathematics education: Influences of feminism and culture.* (pp. 186–199). London, UK: Falmer.
Yasukawa, K. (2002). Mathematics and technological literacy. In P. Valero & O. Skovsmose (Eds.), *Proceedings of the third international mathematics education and society conference* Vol. 1. (pp. 30–42). Denmark: Centre for Research in Learning Mathematics, The Danish University of Education.

# part III

## REFLECTIVE COMMENTARIES

This last part plays the role of an epilogue. Instead of having a single chapter devoted to the conclusions of this work, we decided to offer three distinctive reflective commentaries, written by experienced and acknowledged researchers in the field of mathematics education: Anna Sfard, Ole Skovsmose and Helle Alrø, and Steve Lerman. These people, although working within quite different theoretical and methodological frameworks, have approached communication as an essential element of mathematics education and classroom instruction. They offer us reflections on the topic of the edited book as a whole, and on individual chapters, as well as overall reflections concerning issues of theory, methodology, and practice.

In this way, we aim to present a story about research on communication in mathematical classrooms that does not (and cannot) have one end, but has many possible ends. This story continues, and can possibly be re-written by each one of us, perhaps in different ways, depending on individual purposes, values, starting points, and intentions.

COMMENTARY 1

# CHALLENGING DISCOURSE

**Anna Sfard**
***The University of Haifa, Israel***

Anna Chronaki and Iben Maj Christiansen readily admit to having ventured into an ambitious project. According to their own testimony, the two editors asked the prospective contributors to provide texts which would "directly challenge existing perspectives, theories or practices," and would bring "novel perspectives" and "novel methodological approaches" instead. This pressing need for renewal is only natural in view of the dissatisfaction of the creators of the book with what they see all around them, not only in schools and in research, but also in wider society. Research in mathematics education, the editors complain in the introductory chapter, although socio-cognitively oriented and thus seemingly heading in a desirable direction, is dominated by a handful of theoretical approaches that restrict its potential for useful insights and limit its power to generate badly needed social change. While expressing their hope that prospective authors will adopt a "socio-political approach to classroom communication," the editors make clear that the change they are seeking is supposed to go beyond vocabularies and discursive routines. In fact, they are asking for nothing less than a paradigm shift.

The idea of novelty has much appeal. Replacing the safety of well trodden paths with the risks of trailblazing is not without its rewards. Except for the intoxicating sense of freedom, a change in perspective is likely to bring a change in seeing. According to Erickson's often quoted claim, such change or, more precisely, the ability to see the familiar as strange is the

*Challenging Perspectives on Mathematics Classroom Communication*, pages 323–338

precondition of truly insightful research in any domain, but above all in the study of humans and their actions (Erickson, 1986, p. 121). The editors' idea of loosening the confines of old conceptualizations seems, therefore, to be a move in the right direction.

On the other hand, the call for change, be it as justified as it might, is not without its dangers. Research, like any other collective activity, requires a language in which the participants may effectively communicate with one another. Indeed, what's the use in sharing the eye-opening results of one's investigations if those who speak and those who listen may be using the same words in different, possibly incompatible, ways? In this situation, how can one build on the work done by others, and what is the chance that the story collectively told will make any sense at all? The question of the effectiveness of communication within the community of mathematics education at large, and within the *Challenging Perspectives* team in particular, is the point of departure for the rest of this commentary.

In what follows, I try to make sure that the challenge to what the editors call "the mainstream discourses of mathematics education" does not come at the expense of the coherence of communication. Because the object of my analysis is the *discourse of research,* my present task is similar to those I perform routinely as an investigator of mathematical teaching and learning. Indeed, since I agree with those of the *Challenging Perspective* authors who seem to view mathematics as a special type of communication (see mainly chapters 4 and 9), analyzing discourse is what I am supposed to do in both cases, whether acting as explorer of mathematical thinking or reflecting on researchers' thinking about mathematical thinking. In light of this, the method I am using in investigating mathematical discourse can be employed also in studying its meta-discourse. A brief introduction to the *communicational* perspective (Sfard, 2001) is in order before I apply this method to the discourse of *Challenging Perspectives.*

According to the basic assumption underpinning communicational inquiry, thinking is a special case of activity of communication and, as a result, scholarly learning is the activity of developing specialized discourses. Such different discourses as mathematics, history, or physics can be distinguished from one another according to a number of well-defined characteristics. Two of these features will be considered in my analysis. First, I take a look at the nine chapters trying to identify their specialized *vocabulary,* that is, their keywords and these keywords' uses. Second, I turn to the *endorsed narratives* typical of the nine chapters. Endorsed narratives are propositions which are regarded by a given speaker or a given research community as particularly helpful in making sense of human experience. Known also as "factual statements," or as "truths about the world," endorsed narratives are the ultimate reward for researcher's pursuit. In the search for methodological innovation I scrutinize discursive routines used by the *Challenging Perspectives* authors in producing and substantiating the narratives they offer in this book.

Let me add a disclaimer regarding the motives behind this project. It must be understood that rather than trying to grade the editors' success in reaching the target they set for themselves, I reflected on the way their call guides the implementers. My aim is not unlike the one with which the two editors presented the contributing authors. Although I do not try to "directly challenge" the perspectives offered in the nine chapters, I do submit these offerings to careful scrutiny. The intended result should be read not as criticism or praise, but as this commentator's reflection on what has already been done and what remains to be done if the socio-political is to become what the editors would like it to be: a new powerful tool for sense making and for social betterment.

## VOCABULARY

While gauging the degree of lexical consolidation of the nine chapters, I was guided by two questions. First, I tried to identify specialized keywords characteristic of all the texts. At this point, I also asked how different this vocabulary was from that to which we have become accustomed. Second, I tried to find out whether all the writers used the same keywords in the same way, and whether the uses of different keywords were consistent with each other.

Thanks to the fact that the text of *Challenging Perspectives* was sent to me in electronic form, I could use a computer to check how many times different words appeared across the nine chapters. To begin with, I chose a set of frequently used terms that appeared to be good candidates for the specialized vocabulary of teaching and learning at large, and of teaching and learning mathematics in particular. Table C1.1 presents those among the most visible learning-related words that appeared the largest number of times.

**Table C1.1. Socio-Cognitive Keywords and their Frequencies**

| *Rank* | *Word* | *Score*[a] | *Rank* | *Word* | *Score*[a] |
|---|---|---|---|---|---|
| 1 | communication | 550 | 6 | language | 287 |
| 2 | social | 382 | 7 | understanding | 278 |
| 3 | practice | 334 | 8 | conceptualization | 254 |
| 4 | knowledge | 313 | 9 | discourse | 243 |
| 5 | construct | 297 | 10 | culture | 230 |

[a] *Score* is the number of appearances in the volume

With the words *communication, social, practice, discourse* and *culture* featuring prominently among the nine winners, the list has a distinct "participationist" slant, that is, it demonstrates that the authors in this book tend to conceptualize learning as an initiation into a certain kind of collective endeavor (practice, culture or discourse) rather than as enhancement of individual possession (as is the case with such traditional expressions as "acquisition of knowledge," or "construction of mental schemes"). Since a more detailed count has shown that the prominent keywords are distributed quite evenly among the nine chapters, I conclude that the participationist tendencies are common to all the authors. This finding is reinforced by the conspicuous scarcity of such distinctively "acquisitionist" terms as *scheme* (19 appearances), *conception* (19) and *acquire* or *acquisition* (33), which prevailed in many comparable texts only a decade or two ago.

The participationist flavor of all nine chapters, be it as prominent as it might, is not enough to satisfy the editors' requirement for novelty. The target discourse was described as *socio-political*, but the keywords identified so far are typical also of the more traditional, *socio-cognitive* discourse that the new one was meant to supersede. In order to check the degree of their politization, I returned to the nine chapters and counted such words as *power, ideology, disadvantage, colonialization* and *difference* (with this last term used in the sense of social or cultural disparity), all of which are politically tainted. I soon discovered that while ordered along the political dimension, the chapters split into two contrasting groups, with the last three being much more saturated with politics-related terms than all the rest (see Table C1.2).

**Table C1.2. Politics-Related Keywords and their Frequencies**

| | *Score* | | *In Percent* | |
|---|---|---|---|---|
| | *Chapters 2–7* | *Chapters 8–10* | *Chapters 2–7* | *Chapters 8–10* |
| power, empowerment | 5 | 95 | 5 | 95 |
| colonial, colonialize | — | 68 | 0 | 100 |
| advantage/disadvantage | 13 | 47 | 22 | 78 |
| difference | 27 | 83 | 27 | 83 |
| ideology, ideological | 2 | 18 | 10 | 90 |

Needless to say, the simple word count is not enough to ascertain the nature of communication. And yet, the present statistics do give a general idea about the object of the analyzed discourse. Whatever is claimed here on the basis of the numbers will be subjected to further examination when I deal with the issue of production and substantiation of narratives. For now, let me formulate a tentative answer to the first question asked above: Did *Challenging Perspectives* result in a well-defined discourse, at least as far as can be judged by its vocabulary? It seems that in spite of a certain diversity in the degree of politization, there is enough lexical homogeneity to say that the nine texts passed this first, very crude test for consolidation.

The next thing to do is to check whether different interlocutors use each of the identified keywords in the same manner and whether the way they do it is in tune with the more explicit claims of the authors about the nature of their endeavor. In fact, this question can be broadened to the current discourse of mathematics education in its entirety, perhaps even to all the educational discourses. Since it is not often that we scrutinize our professional communication for effectiveness and consistency, we cannot be certain that these features are developed to a satisfactory degree not just in this volume, but across all the educational publications.

The easiest way to answer the question of consistency would be to consult the authors' definitions of the terms under consideration. Since, however, definitions are mostly unavailable or, if given, they are not necessarily operational, I had to rely on indirect clues in my attempts to decide how different writers interpret the different words. Let me begin with the ubiquitous term *social*. In view of its strong, seemingly incontestable presence across the current literature, there seems to be no reason to worry about the clarity and consistency of its use. And yet, since the habit could make us complacent, a measure of precaution is definitely in order. "Although 'the social' is regarded important within many schools of research in mathematics education, the connection between social interaction, communication and learning has been approached in rather different ways," observe the editors in their introductory chapter. While sharing this opinion, I wish to claim that rather than being a result of incompatible worldviews, the noted diversity in the researchers' views on the role of "the social" in human development may be a simple consequence of unacknowledged disparities in the different authors' uses of the term *social*. The ambiguity, made possible by the absence of explicit definitions, is exacerbated by the fact that in spite of its being an adjective, the "social" often appears standalone, as a noun (see above). Such use frees the speakers from ontological commitments at the expense of unequivocality.

As a result of the unacknowledged conceptual ambiguity, the numerous claims about "the primacy of the social over individual" that can be found in the current literature—and the present volume is no exception—can be interpreted in more than one way, with each of the possible interpretations leading to different conclusions. A closer look shows that those who contrast "the social" with "the individual" may have two different distinctions in mind. In one of these dichotomies the term *social* means that whatever is described with this adjective has been done or attained through interaction with others. In this case, the word *social* could probably be replaced by *collective.* This substitution may be in place in statements such as: "those who supported the notion that cognition is a social process [thought that] interpersonal activity precedes the development of intrapersonal cognition" (Chapter 2). More generally, it is the opposition between *collective* and *individual* that is usually intended whenever the learning activities and their nature are in the focus. This interpretation of the claim about the primacy of social over individual may be called the *weak* participationist vision of cognition (some other writers denote this position with the term "cognition plus"—see e.g., Lave, 1991).

The other dichotomy that hides behind the opposition *social versus individual* regards not so much the "technicalities" of the processes of learning as the nature of what is being learned. This time, the issue at stake is that of the ontological-epistemological status of knowledge, with the word *individual* functioning as almost synonymous with *natural* or *biologically determined.* According to my reading, it is this version of the dichotomy that lies at the heart of the famous controversy, and of the eventual irrevocable split, between Piagetians and Vygotskians. While Piagetians view knowledge as having its sources in the world itself and as something that, at least in principle, could be learned directly from '"the nature," sociocultural thinkers sustain that what we learn is inherently a product of human communication and, although definitely constrained by reality, would not exist for us if not for our being a part of the human community (recall Vygotsky's famous claim, quoted in the introductory chapter, according to which learning involves a transition from the inter-personal, that is the social, plane to the intra-personal plane). If the terms *individual* and *social* are interpreted as referring to *what is being learned* (knowledge, concepts) rather than to the process of learning as such, the declaration on "the primacy of the social over individual" becomes a statement on phylogeny rather than on the ontogeny of human knowing. In this case, it is a suggestion on how to conclude the time-honored "nature versus nurture" debate: What is being learned is inherently social and, as such, resides between people rather than in the world itself. Similarly to Wittgenstein who insisted on the untenability of the idea of "private language," those who interpret the claim about primacy of social over individual as a statement on the nature of what is being known sustain that there is no such thing as "private knowledge," resulting from a socially unmediated interaction between an individual and the world. I will call this stance the *strong* participationist view of human cognition.

We seem to be so much used to this double use of the *social versus individual* distinction that sometimes, imperceptibly to ourselves, we move from one dichotomy to another within the same text. Such unnoticed shifts are not without consequences. How we interpret the term *social* does make a difference. For example, those who adopt the weak participationist position speak about "social dimensions" of learning, knowledge or communication, thus implying that each of these activities or products has non-social dimensions as well. But for those who use *social* in the sense of "man-made" and as the antonym of "biologically given," learning may be non-collective, but it cannot be non-social, not even when a person is learning alone. Thus, the adherents of the strong participationist position must be somehow suspicious of texts in which the word *social* appears with high frequency. If it is true that knowledge is inherently social, speaking about "social dimensions of learning" actually undermines this message, as it implies the possibility of non-social aspects of learning.

The frequent occurrence of the term *social* in *Challenging Perspectives* may thus be a sign that the controversy between weak and strong participationist doctrines is not yet resolved. More than that, this controversy seems mostly unacknowledged. Many users of the word *social* remain unaware of its diverse interpretations in the ongoing educational discourse. Since the different interpretations are bound to entail different answers to fundamental questions about cognition, a tacit difference in the use of basic words may end up in debates that could be avoided if the interlocutors cared to be more explicit about these uses. I will go so far as to claim that even in the famous argument between traditional Piagetians and Vygotskians, much of the ado was due to unrecognized terminological ambiguities.

The fear of invisible epistemological and ontological inconsistencies compels me to go even further in my critical scrutiny of the discourse of *Challenging Perspectives*. According to my reading, it is the strong participationist discourse on learning that was promoted by Vygotsky, embraced by situationists, and now advanced even further by those who are heading toward socio-political discourse. This is therefore the discourse that can be expected in the mouths of those who join the socio-political project. And yet, our discursive ways may be slower to change than our declared allegiances. As happens only too often at times of major discursive upheavals, our deeply rooted linguistic habits may persist even if they are incompatible with principles that underlie the new discourse. By careless adherence to certain time-honored word uses we may be sustaining dichotomies that are part-and-parcel of the learning-as-acquisition metaphor but are inadmissible within the strong participationist discourse.

Thus, for example, problems can be lurking behind such seemingly innocent pairs of words as *knowledge* and *learning*, *concept* and *representation*, *meaning* and *symbol*, *thought* and *communication* (or *language*, or *speech*), all of which have been featuring prominently in the educational discourse ever since its inception. It is enough to take another look at Table C1.1 to see that the discourse of *Challenging Perspectives* is no exception. No wonder. One can hardly imagine educational discourse without the ubiquitous expressions involving one or more of these words. On the other hand, perhaps we should. Our habitual ways of speaking portray the two elements of each pair as, in a sense, separate, self-sustained entities. The distinct essentialist flavor of such rendering disagrees with the strong participationist claim about the social nature and origins of human knowing and learning. Consider, for example, such familiar, common expressions as "linguistic representation of knowledge," "communicating ideas," "passing information in words," "representing concepts," "acquiring knowledge," all of which can be found in *Challenging Perspectives.* These utterances imply a dichotomy of a human-made tool (representation, language) or activity (communicating, passing, representing, acquiring) on the one hand, and of an entity (knowledge, idea, information, concept) that is the object of this activity or tool use, on the other hand. This entity is implied to be primary to the discourse itself in that it is supposed to preserve its identity across different discursive "renderings," "expressions," or "representations." But the strongly participationist discourse does not admit of such objectification and disembodiment. Human activities of getting to know, even though inspired and constrained by an extra-discursive world, are discursive through and through and do not involve super-discursive entities, as implied in the objectifying statements on "knowledge" or "concept." The discourse is all they produce. Or, to put it differently, knowledge does not "reside" in factual propositions such as $2 + 2 = 4$ or $(x^2)' = 2x$—it is tantamount with these propositions and their use.

In view of all this, the distinction between *communication* and *thinking* or between *communicating* and *knowing* seems similarly inadmissible in the strong participationist discourse. If knowledge is discursive formation, then learning is a change of discourse. When understood broadly as an activity that can, but does not have to, be either verbal or collective, communication is not a separate process that merely "helps" or "improves" learning—it is the very thing that is being learned! In this context, the editors' remark that the current studies consider "communication as an integral part of pedagogy and didactics in mathematics classrooms" should not be misunderstood as saying that communication might have ever been absent from schools. This comment only emphasizes the researchers' new awareness of the fact that deliberate fostering of interpersonal communication may promote the student's mathematical communication at large, and this includes her communication with herself, that is, her mathematical thinking.

In summary of the above analyses, I can say that the socio-political discourse offered collectively by the authors of *Challenging Perspectives* does have its distinct vocabulary and that its word use is imbued with a unifying, strongly participationist flavor. And yet, at this early stage in discourse development our old discursive habits still tend to interfere with the implicit rules of the new language games. To repeat Benjamin Whorf's insightful statement, quoted by Dalene Swanson in Chapter 9, "Western culture has made, through language, a provisional analysis of reality and, without correctives, holds resolutely to that analysis as final." The attempt to free ourselves from the confines of this old "provisional analysis" can be found in this volume. Consider, for example, Marcelo Borba's thoughtful rejection of the dichotomy between humans and technology. Indeed, since strong participationist approach presents thinking and knowledge as but other names for different forms of human communication, it is only justifiable to view "humans-with-media as the basic unit for production of knowledge." (Chapter 2). If knowledge production is a form of communication, any attempt at understanding technology-assisted learning by treating the respective contributions of humans and machines as separable would be like trying to understand the workings of a gear by analyzing the movements of each individual cogwheel in isolation from all the others. Borba's contribution is a promising beginning of the type of conceptual analysis that is necessary to increase the consistency and power of socio-political discourse.

## ENDORSED NARRATIVES

In this additional attempt to identify distinct characteristics of the evolving socio-political discourse of mathematics education, I now focus on endorsed narratives produced by all authors throughout the research presented in this volume. I first concentrate on the thematic features of these narratives, and follow with a few remarks on the ways in which they have been substantiated.

Let me begin with presenting a selection of narratives from the different chapters. The samples are in the form of isolated propositions taken directly from the original texts. My criteria for choosing these particular sentences from among the many possibilities were as follows: I was looking for generalizing statements that, first, presented the gist of the authors' position, and second, were illustrated, substantiated or used as a point of departure for a pedagogical proposal presented throughout the respective chapter.

Chapter 2: [T]he use of technology, together with pedagogies that are in resonance with this new medium, reorganize thinking.

Chapter 3: [T]he prioritization of vision in geometry classrooms is culturally constructed, constituting certain practices as more "appropriate" in systematically forming the object of which they speak... [T]he integration of different modes of communication may be able to challenge existing ways of communication in mathematics classrooms.

Chapter 4: Teachers have an obligation to teach students the mathematics register but need also to make explicit how appropriate linguistic choices are made.

Chapter 5: [A]rgumentation can be a means to develop and systematize knowledge

Chapter 6: The need for the students to interact with more experienced members of the mathematical community is ... crucial

Chapter 7: When a student sees learning as taking on board the perceived social practices within the mathematics classroom, rather than cognitive considerations of the inherent mathematics, this can lead to conflict and a lack of success at mathematical tasks.

Chapter 8: We are colonized by numbers.

Chapter 9: The various discourses of difference co-emerge to produce realized educational disadvantage.... [I]t is power and control in-and-between discourses in the broader social domain that produce agency in mathematics classroom.

Chapter 10: Pedagogy is a key relation in social reproduction as learners "learn" how to be a pupil in the classroom whilst simultaneously teachers are regulated by rules that lay down what it is to be a teacher.

Paraphrasing Tony Cotton, one can say that educational research is an important site for both pedagogical and political action (to return to Tony Cotton's original statement, just substitute the expression *educational research* with *pedagogy*). A close look at the nine quotations above reveals that the authors of *Challenging perspectives* are willing to engage in both types of actions, the pedagogical and the political. Indeed, the endorsed narratives listed above can be divided into two distinct sets according to their themes.

The narratives taken from Chapters 2–7 can be described as *pedagogically oriented* because they deal with the explicit agenda of mathematics classrooms or, using Tony Cotton's formulation again, on the *visible* pedagogy. The focus is on mechanisms underlying the learning and teaching of mathematics, and the question is that of their possible improvement. These narratives are in the form of caveats and pedagogical advice, and they are all in the spirit of our times: they speak about integration of pedagogy and technology-mediated ways of communicating, they stress the importance of social and emotional context, and they do not refer to the time-honored norms of mathematical discourse as sacred.

The second set of narratives, those taken from Chapters 8, 9, and 10, can be called *politically oriented.* Narratives of this type deal with the hidden social agenda of the classrooms. The focus is on how the norms and values of the wider society infiltrate the classroom discourses, and how these discourses collaborate in positioning the students within the existing network of social relations. It is this set of narratives that collides directly with those listed by the editors as representative of mainstream mathematics education, the narratives that treat classrooms as independent cultures, that portray communication as rational process, and that uncritically accept the "mathematical content" dictated by the existing curricula.

In spite of their being divided into two groups, the nine narratives are unified by a distinct common feature: They all have an unmistakably emancipatory tenor. The term *emancipatory* is used here to describe narratives that aim at alleviating the oppressive impact of educational discourses. It is in this sense that both groups of authors engage in the discourse of emancipation. The first group demands loosening the confines of traditional mathematical discourse in the classroom. The outcry against the notorious rigidity of this discourse stems from the belief that a greater flexibility will improve learning and allow for a more rewarding intellectual experience. The other group of authors reaches beyond the walls of kindergartens, schools and universities in the protest against the use of mathematical discourse as a "universal ruler" with which to measure, gauge, and order human beings. This kind of use, say the authors of the last three chapters, is a hallmark of Western culture. It turns mathematics into a safeguard of the social order that, in its inner workings, rests heavily on a variety of splits and divides. This order would be in danger without the possibility of telling the "able" from "disabled." Proficiency in the school mathematics seems to be the perfect yardstick for such labeling.

Narratives of this latter type strive to attain emancipatory effect simply by uncovering certain hidden aspects of the educational discourses' inner workings. They portray the process of initiation into mathematics as a form of institutionalized activity of social positioning, categorizing and identity-building. The critical classroom discourse will make the students aware of how, through this very process, "society function[s] to shape and thwart their aspirations and goals, or prevent... them from even imagining a life outside the one they presently lead" (Giroux, 1989, p. 106, quoted in Chapter 10). By making the hidden political agenda of mathematics classroom readily visible, this kind of emancipatory narratives paves the way toward *critical pedagogy* (or to *critical classroom communication,* as proposed in Chapter 10) which aims at rendering agency to the hitherto passively collaborating subjects of the educational process. The emancipatory effect of turning the familiar into the strange, the liberating impact of being able to see the invisible is, indeed, the aim of all the authors, whether they have an operative political agenda or are just trying to see how conflicting social interests infiltrate the ostensibly "neutral" discourse of the mathematics classrooms.

Unless they are substantiated according to the norms adopted by the given community, generalizing narratives cannot count as endorsed by anybody except, perhaps, their authors. Theoretical and empirical substantiations that can be found in the different chapters of the present volume are meant to convince the reader to follow in the author's footsteps and endorse the narratives that have been listed above. In my attempt to identify research methods typical of *Challenging Perspectives*, I am particularly interested in empirical substantiations. These substantiations are yet another set of endorsed narratives, except that this time the story is of particular students and particular teachers rather than of students and teachers at large. It is from these propositions that the generalizing narratives are to be derived. It is the researcher's obligation to perform the derivation in such a way as to ensure that the final endorsement is granted. Below, I offer a brief list of substantiating empirical narratives chosen at random from different chapters.

> Chapter 2: I tried to emphasize the role of technology in communication in the classroom... [The students'] patterns of communication were reorganized in the sense that the students communicated with peers, and with the machine or the software design of the graphing calculator.
>
> Chapter 3: From our analysis it became evident that the integration of representational media helped students to become more conscious of the processes of organizing coherent texts to communicate mathematical understandings.
>
> Chapter 6: We found that [students'] difficulties rested on a metatheoretical level.
>
> Chapter 8: I tell the story of the 'repeat' students who are using number and redefining it to work within their experience and to help them both resist and reproduce the educational system. ...[T]he repeat students have colonized number to re-create social relations.
>
> Chapter 9: Teacher constructions of cultural and experiential difference and deprivation, race, social class and language difference became delimiting criteria in the students' access to mathematics knowledge.

Each of these narratives summarizes the authors' interpretation of what they saw in the particular classroom in which their study took place. As such, each of the narratives is, in itself, a derivative of many other more specific narratives about classroom events. For all their thematic diversity, the five substantiations share two distinct properties: They are all grounded in one particular case; and all of them, but one, have a form of a story told in the first person. Let me consider these two features in more depth.

All the substantiating narratives at hand are summaries of stories about specific conversations between specific interlocutors. The traditional researcher would certainly wonder how such a restricted number of cases can become the basis for any generalization. This doubt is certainly justified as long as one aims at finding "general laws" and believes in the possibility of deriving such laws directly from empirical data. And yet, none of the thirteen authors of this volume seems to be heading toward this type of generalization. Rather than proposing narratives beginning with the general quantifier *for all*, these researchers conclude their studies with statements containing the existential quantifier *there is*: "There is something we have not seen before even in the most familiar of classrooms," or "There is an interpretation of familiar situations that makes much sense and about which we have not thought before." Theirs is interpretive research that tries to draw a high-resolution picture of well-known situations and see these situations' inner workings rather than forcing the reality into a straightjacket of a universal ordering structure. This research allows us to see things that have been overlooked in spite of the thousands of years of our presence in the classroom.

But let me make a disclaimer. We must be careful of the slogan "making the invisible visible" as these words can be interpreted as implying that the discursive constructs created in order to make sense of our experience reside in the reality itself. In the present discussion, the term *invisible* applies not to the world as such but to its plausible, and so far unacknowledged, interpretations. I have already discussed the dangers of old discursive habits that die hard. Alternative interpretations, once constructed, shake our complacency. Multiplicity of interpretations is thus a worthy goal.

To make room for a new interpretation one has to make the familiar strange. In the recognition of this need some of the authors of this volume ventured far beyond their own culture. It is in such culturally distant places that, as stranger, one is "able to hear words and watch their usage and impact" and "to see social relations, classroom communications and actions differently than in classrooms where [one is] insider" (see Chapter 8). Indeed, as Bakhtin put it, "In the realm of culture, outsideness is a most powerful factor in understanding. It is only in the eyes of another culture that foreign culture reveals itself fully and profoundly" (Bakhtin, 1986, p. 7).

In the light of the interpretive nature of the research, the already noted personalized character of the endorsed narratives contained in this volume is only natural. Interpretation is a story, and being a story, it has an author. The researchers in this volume do not make any secret of their authorship of the many narratives that can be found in these pages. Like any writer, they seek the reader's endorsement, but in contrast to many others they would not try to attain this goal by removing the human storyteller and by implying that their narratives are dictated by "how things really are." Rather, these researchers invite the reader to share their vantage point and to appreciate the richness and usefulness of the insight thus afforded. Wenda Bauchspies (in Chapter 8) may be speaking for all the authors when she makes the following declaration:

> I ask my readers not to read my stories as "their" stories but to read them as "my" stories or perhaps "our" stories. They are stories in a socio-cultural context. I do not wish to remove them from that context by claiming authority, objectivity, validity or Truth. The authority, objectivity, validity and Truth does not exist from me the writer, but from the reader when the words resonate with a story from the reader's world and provide the reader with a clearer understanding of his/her own socio-cultural context.

To sum up, the typical socio-political narratives, as found in *Challenging Perspectives,* are emancipatory, interpretive and personal. This is yet another set of common features that adds to the consolidation of this discourse. Although interpretive research is not exactly new, it does not yet have a well established presence in mathematics education, and it certainly may be seen as a challenge to some of the "oldtimers."

## FINAL REMARK: ON THE IMPORTANCE OF DISCURSIVE INNOVATION

As this commentary nears its conclusion, I can say with full conviction that there is such thing as socio-political discourse of mathematics education and that this discourse is, in many respects, quite unlike the discourses to which we have been used. The authors of this volume achieved a lot in their attempt to create it. To be sure, what can be found on these pages is a beginning of a journey that, as Tony Cotton reminds us, has no end. But even if we have a long way to go before we are ready to tell ourselves that we have been traveling long enough, the new discourse is already here.

One question, however, still remains unanswered: Why should we desire discursive innovation in the first place? In the beginning of this commentary I claimed that the editors did not seek conceptual and methodological novelties for their own sake, nor did they launch their project merely for the sake of a handful of new insights. The innovative explorations were undertaken in the hope that they would lead to consequential deeds. The new discourse allows the telling of new stories not only about the past, but also about the future, and this latter type of narratives can have a direct impact on what will happen to us from now on. Because, as Tony Cotton reminds us in the words of Anthony Giddens, "what could be the case can directly affect what becomes the case" (Giddens, 1994, p. 250).

Most certainly, this impressive collection of thoughtful essays may influence classroom practices. The pedagogical ideas presented in this volume can revolutionize school mathematics in ways that, even if not quite pleasing to traditional mathematicians, would turn the intellectual life of children into something richer and more rewarding. To quote Tony Cotton one more time, even if none of the innovative pedagogical schemes implemented and investigated by different authors is "a model to be universally applied," these innovations can certainly be seen as "a work of 'utopian realism' against which we can measure current practice and develop future practices" (Chapter 10).

But the prospective contribution of the socio-political discourse goes well beyond that. The new discourse of mathematics education does not divide the child into mathematics student and "all the rest," with "all the rest" believed to reside outside mathematics classroom and outside the researcher's account. In this discourse, there is room for all the ingredients of the child's life and identity, as well as for wider society's norms and values. The participants in the new mathematics education discourse are well aware that the relevance of what society as a whole has to say does not evaporate at the doorstep of mathematics classrooms. Armed with new discursive lenses, one asks not only how people learn mathematics, but also why they do this and what this learning does to them.

But the most important lesson I am taking with me from this collection of thoughtful essays is that discourses in general, and educational discourses in particular, may become dangerous if left unchanged for long periods of time. Their ostensible innocence, their reputation for being "just words," endows discourses with more power to hurt than can be found in many overt weapons or in declared adversaries. Discourses may turn us into oppressors, even if we are acting with the best of intentions. This, it seems, was the fate of the "weak voices" in Dalene Swanson's story: Unknown to themselves, these well-meaning teachers became collaborators of an oppressive system. For this to happen, nothing more was necessary than their participation in educational discourses into which they were born and to which they knew no alternative. We are living in the world ordered by discourses inherited from those who were here before us and who, in their turn, inherited these discourses from those before them. As long as these inherited discourses are the only ones we have, we remain their captives. With the help of new discourses, such as the one developed in this volume, and by means of new interpretations that these discourses bring with them, we alleviate oppression and, not of least importance, we free ourselves from the role of oppressors.

## REFERENCES

Bakhtin, M.M. (1986). *Speech genres and other late essays.* (Trans.) V. W. McGee, C. Emerson & M. Holquist (Eds.) Austin, TX: University of Texas Press.

Erickson, F. (1986). Qualitative methods in research on teaching. In M. Wittrock (Ed.), *Handbook of research on teaching* (3rd Ed.). (pp. 119–159). New York: Macmillan.

Freire, P. (1972). *Pedagogy of the oppressed.* London, UK: Penguin.

Giddens, A. (1994). *Beyond left and right—The future of radical politics.* Cambridge, UK: Polity Press.

Giroux, H.A. (1989). *Schooling for democracy: Critical pedagogy in the modern age,* London, UK: Routledge.

Lave, J. (1991). Situating learning in communities of practice. In L.B. Resnick, J.M. Levine, & S.D. Teasley (Eds.), *Perspectives on socially shared cognition.* (pp. 63–82). Washington, DC: American Psychological Association.

Sfard, A. (2001). There is more to discourse than meets the ears: Learning from mathematical communication things that we have not known before. *Educational Studies in Mathematics, 46*(1/3), 13–57.

COMMENTARY 2

# CHALLENGING PERSPECTIVES

**Helle Alrø**
***Aalborg University, Denmark***

**Ole Skovsmose**
***Aalborg University, Denmark***

It has been a wonderful challenge to read about the different perspectives on mathematics classroom communication presented in this book. Challenging, because of very different approaches. Challenging, because of the absence of fixed answers to the question of the relation between classroom communication and the learning of mathematics. Challenging, because of recognition of issues that are discussed from perspectives different from our own. And not least, challenging to our perspectives on communication and mathematics education.

Being challenged means accepting the invitation to reflect on the presented perspectives in relation to your own understanding and pre-understanding. Receiving the invitation to reflect presupposes involvement, so intention and reflection are closely related terms that operate when you let yourself be challenged. Challenging is an essential part of a dialogue that aims at developing new insights and critical actions. In short, challenging is essential for learning critically. In our book *Dialogue and Learning in Mathematics Education—Intention, Reflection, Critique,* we have developed a theoretical concept of learning mathematics that emerges from the connection between the notions of dialogue, intention, reflection and critique. In this short chapter we comment on some key concepts, presented by the

*Challenging Perspectives on Mathematics Classroom Communication*, pages 339–347

authors of the present volume; concepts that question and widen our concepts. (We are not trying to be "fair" in the sense that we comment on each of the contributions. As will also become clear in what follows, this does not signify any evaluation. This is just a consequence of limitations of space.)

The contributions in this volume combine theory and practice in the study of learning mathematics. Thus, we are invited into many classrooms to meet teaching practices and examples of communication that have actually taken place. And by studying these examples it becomes obvious that practice is complex and not always in line with only one theory. The development of theoretical concepts and ideas is enriched and challenged by practice, and on the other hand practice can be enriched and challenged by theory.

In Chapter 6, *Understanding mathematical induction in a cooperative setting: merits and limitations of classroom communication among peers*, Inger Wistedt and Gudrun Brattström discuss university students' understanding of mathematics when cooperating in groups. The claim is made that students' learning of mathematics has to be related to the context of learning. The notion of context is understood in constructivist terms not as the situational setting as such but as the individual student's interpretation of this setting. This interpretation relates to different aspects of context, and the intentions of the students are seen as decisive for students' contextualization, i.e., for their focus on cognitive, situational or cultural aspects and for understanding what is the purpose of their activities. The instructional task is not to change students' conceptions but to help them pay attention to the relevance of alternative ways of thinking and acting. In this process, communication with peers and tutors is claimed to play a major role.

In one of the groups, the authors observe an "almost Socratic dialogue" between some students, but they do not elaborate further on the concept of dialogue. However, certain communicative elements are identified as supportive of the students' collaboration. One communicative pattern that seems to support the students' collective reflections is "stepwise hypothetical reasoning," in which the students explore understandings on the basis of assumptions and imaginary aspects. Another communicative element that is emphasized is the rephrasing of utterances that highlight comments and suggestions that have just been put forward in order to reflect on them collectively. Further, the "opening of interpretations" in the form of "challenge to taken-for-granted notions" is stressed as an important student activity in the learning process.

We recognize all of these elements from our *Inquiry-Cooperation Model* (IC-Model), which is an indicator for mathematics classroom communication when the students, with or without the teacher, are engaged in an inquiring dialogue with the purpose of getting to know. This model is a cluster of dialogic acts such as: getting in contact, locating, identifying, advocating, reformulating, challenging and evaluating. Stepwise hypothetical reasoning can be seen in relation to our notion of locating perspectives characterized by hypothetical "what-if" questions; rephrasing of utterances is a direct parallel to our notion of reformulating, and challenging taken-for-granted notions is similar to our notion of challenging perspectives. We find it thought-provoking to be introduced to similar communicative features in this chapter.

Another challenging point is that there are not only similarities but also strong differences. In the communicative field, Wistedt and Brattström stress the importance of "conflicting views" to the learning process. A supportive style, so they claim, may lead to submission and agreement, so that the students would miss the chance to elaborate on the perspectives put forward. We concur that this would be an unfortunate consequence of a supportive style. However, we have not observed such a consequence in any of our studies. On the contrary, we have observed how students who are mutually supportive in their cooperation are urged to struggle with problems and to continue their inquiry. We have also observed a lot of reluctance and unproductive quarrelling, when the students seem to refute each other's contributions and stick to fixed perspectives. Finally, we have observed the fragility of dialogic acts; thus, a "reformulating" could easily turn into a "correction."

That individual students act differently and learn different things in the same situational context is emphasized in a number of studies including our own. The big challenge for us is in the conclusions that the authors draw in this case: "Communication may function as a means to appropriate a 'mathematical attitude,' and through communication we can ensure that the individually acquired knowledge will fit with the mathematics constructed by others" (see Chapter 6). Communication in this sense is not a means of collective reflection, inquiry, and production of mathematical knowledge. It is seen as a means to ensure that what the students get to know is in accordance with already constructed mathematical knowledge. This is a challenge to our understanding of learning through dialogue. Furthermore, it is emphasized that student interaction with "more informed members of the mathematical community" is needed "in order to overcome the limitations in their personal views of the subject" (Chapter 6).

Thus, the big question remains: Can students gain mathematical knowledge on their own or when working together in groups? We do not take the radical view that teacher input and interference is a bad influence on students' cooperation and learning of mathematics. On the contrary, our studies show that students may miss important aspects and mathematical ideas if they are not supported and challenged by a "more experienced member of the mathematical culture." The main difference is that we do not consider the teacher's role as an ambassador for mathematics but rather as a facilitator of students' learning of mathematics.

In the mathematics classroom you may expect some learning of mathematics, but many other things are going on as well. Besides the cognitive challenges of the mathematical content, the students have to cope with social challenges like being friends with and being accepted by others. Add to that the emotional challenges of getting hurt or feeling "not-as-good-as." Every student faces a complex set of interwoven challenges in the classroom. How to manage?

In Chapter 7, *Conflicts and Harmony Amongst Different Aspects of Mathematical Activity*, Dave Hewitt introduces us to the concept of *Zone of Opportunities (ZoO)*, which operates when students pay attention to different aspects of the challenges they meet in and outside the classroom. Thus, the ZoO is a metaphor for the opportunities that are available to the student in the context. Hewitt does not use the notion of contextualization, but like Wistedt and Brattström he also emphasizes that learning is related to cognitive, social, and emotional aspects of the context in which a person is situated. Thus, a student may concentrate on cognitive mathematical issues, but he or she may equally be occupied with other aspects of the context.

A teacher cannot control the students' attention but he or she can arrange and offer material, and if the students pay attention to this instance of "offering" and begin to work with it, it is likely that learning will take place. We are tempted to draw a parallel to our consideration of how students engage in cooperation for the purposes of inquiry. A teacher can invite students into a landscape of investigation, but he or she cannot decide for them to accept it or force them to enter it. In this sense we see "invitation" and "offer" as similar concepts. But offering is more than a teacher invitation. Offers can come from fellow students or from something that is happening outside the window. Offers can be intentional or they can be unintentionally directed to the student. The important thing is that they are available to an individual student, whether working alone or with others, and that the student's *choices* as to where to place the attention becomes a determinant of what is learned.

Unlike the Vygotskian Zone of Proximal Development, the ZoO is not a comparative zone.

> The Zone of Opportunities is a zone—determined by the student's attention—where offerings can be turned into material: Potential opportunities for learning (offerings) can become actual opportunities for learning (material). A student would then need to work with that material for learning to take place. (Chapter 7)

Thus, Hewitt focuses "on the dynamics within an individual (in a social setting) rather than within a group of individuals (which constitutes a social setting)" (Chapter 7). He does not discuss whether cooperating in groups can widen the field of offering and in that way widen the ZoO for the individual student as well as for the group as a collective. Social interaction is seen as only one source of offering among others that students can pay attention to in the ZoO. This counts for teaching as well. A teacher can prepare and arrange some opportunities, and he or she can try to call upon students' attention, but it is only the students themselves who can make choices of to what they pay attention.

We find the notion of zone of opportunities inspiring but we want to challenge it. The word "opportunity" has positive connotations. Something is possible, and this something is not fully examined or known yet. Something that is not quite examined or known may be uncertain and unpredictable. It may be combined with running a risk. A defining element in our concept of dialogue is "running a risk." A dialogue is unpredictable; there are no fixed answers to questions beforehand. This creates the opportunity to produce new insights and ways of acting. But it also means that people engaging in a dialogue must be willing and able to accept moments of insecurity. We think this applies to the ZoO as well. Paying attention to specific offerings may be a risky affair. Learning may be risky.

Attention is a precondition for being occupied with something, and this means that attention has to be a crucial element in learning. We have emphasized the importance of students' intentions-in-learning. Intention and attention is not the same thing, but they can be seen as related. Thus, we can understand attention as a source of intention. Attention in that way is a precondition for being able to put intentions-in-learning. But it can also function the other way around. Intention can be decisive of to what a student wants to pay attention. Does the student intend to learn mathematics—consequently paying attention to the teaching of the subject matter? Does he or she intend to be a nice fellow student—consequently paying attention to the two fellow students at the desk behind? Or does the student have a dual purpose? Attention and intention are both important for the students' involvement and for maintaining the ownership of their learning process.

In Chapter 2, *Humans-with-Media: Transforming Communication in the Classroom*, Marcelo Borba presents the notion of humans-with-media. He suggests that a certain environment may transform communication in the classroom. We agree. Here, however, we shall concentrate first of all on the very notion of humans-with-media. Borba considers *humans-with-media* as "the basic unit for production of knowledge in the classroom" (Chapter 2). And he adds: "from this perspective, knowledge will be viewed as not solely a product of humans, but rather as a product of a humans-with-media unit" (Chapter 2). In *Dialogue and Learning in Mathematics Education—Intention, Reflection, Critique*, we distinguish between dia-logical and mono-logical theories of learning. While mono-logical theories, as for instance exemplified by radical constructivism, consider the individual as the learning subject, dia-logical learning theories consider the learning subject as constituted by the relationship between several persons. Naturally, it is possible from a radical constructivist point of view to discuss communication and interaction. However, from a radical constructivist perspective, communication has to be considered as a facilitator for individual processes of construction. The epistemic processes take place in the individual; such processes resemble introspection. Such individualism is marked by the proposed nature of processes of assimilation and accommodation, as suggested by Piaget. In our conception of learning as well as of learning mathematics, we find that communication, interaction, and dialogue constitute the very processes of learning. This means that we can consider the learning subject not simply as an individual but as a group. This idea is reflected in Borba's conception as he talks about "humans" in plural.

Borba proposes a further addition to this idea. The available tools or media (Borba uses "media" as a broader concept than "tool") play a role in the representation of mathematics and therefore also for the learning of mathematics. When concentrating on a geometric problem, Archimedes greeted a Roman soldier with the remark: "Do not step on my circles!" We can imagine that Archimedes had made diagrams in the sand, which served as an always available sketchpad. We doubt that Archimedes would have been able to acquire the same mathematical insight had he not made use of any such sketches, but merely of mental images. We could claim that the learning subject in this case was not Archimedes as an individual, but Archimedes-with-his-tools-for-drawing. It seems that learning mathematics has been so intimately connected to tools for making diagrams and representations of numbers, formulae, and mathematical ideas in general that we do not even consider that learning mathematics could occur with human beings not having access to paper and pencil. New tools have been constantly added and explored. The information and communication technology (ICT) has made a remarkable addition to the tools available for the learning of mathematics. ICT as a tool is not only a simple supplement to the human mind, but a medium that reorganizes the way we come to deal with mathematical issues. In this sense we arrive at the idea that the learning subject can be considered as humans-with-media. And not only "humans" is plural, "media" is plural as well. This is an important idea to be considered for any dialogical epistemology. And we should not forget that Archimedes could draw on an abundance of mathematical ideas and proposals. He was not alone in exploring the diagrams.

As inspiration for this conception Borba refers to Tikhomirov, who again is inspired by the Vygotskian approach, paying special attention to the context, and also to the tools available for the learner. This brings Borba to conclude that ICT provides a reorganization of the whole process of learning, including a reorganization of the patterns of communication. This point, however, needs some further consideration. A context can set the scene for the communication, and we have referred to some such scenes as "landscapes of investigation." But a change of context does not by itself bring about a change. The way students are invited into landscapes of investigation opens up new patterns of communication. No new tool as such, not even ICT, provides any reorganization of the communicative pattern. It is the humans-with-media that are reorganized but also reorganizing (the media themselves are not the reorganizers, nor are simply the "humans").

In *Dialogue and Learning in Mathematics Education—Intention, Reflection, Critique*, we make a distinction between mono-logical and dia-logical epistemologies. We also make a distinction between non-critical and critical epistemologies. Stated briefly, a non-critical epistemology would consider the aim of a learning theory of mathematics to be able to grasp and explore the way students may come to develop or construct mathematical concepts, ideas, techniques, and competencies. The formulations can be very different, depending on the overall perspective of learning. However, an underlying assumption is that the learning theory should address *mathematical* ideas and notions. If the main concern of mathematics education is to serve as an ambassador for mathematics, any mathematics teaching should facilitate the students' learning of mathematics, and theories of learning mathematics should provide conceptions that could assist this learning.

A critical theory of learning has to broaden this perspective. First, a critical consideration has to address the possible roles of mathematics in society. It could refer to mathematics as being part of technology, as not only producing "wonders" but also technological "horrors." It might be difficult to identify any separation between such horrors and wonders. Any technique or technology that includes mathematics faces an uncertainty about what might be done by means of this technique or technology, and this uncertainty raises an ethical demand in terms of responsibility. Second, the importance of a critical theory of learning mathematics emerges when we consider what social and cultural processes might be included in or supported by a particular teaching-learning program in mathematics. Mathematics education might facilitate social processes of inclusion as well as of exclusion. From the perspective of gender, this possibility is marked by the title of Valery Walkerdine's study *Counting Girls Out*. Issues of inclusion and exclusion have also been addressed in ethno-mathematical studies that question the assumption that mathematical rationality represents a universal human rationality. Instead, mathematics is seen as a culturally rooted and context-bound rationality, which can be expressed and exercised in many different forms. However, a political and cultural expression of power and control becomes exercised when one such form of rationality becomes celebrated as the proper form of mathematics. Such considerations raise the question of what in fact might take place in mathematics education. It might be highly problematic to discuss the learning of mathematics as if simply the learning of *mathematics* is at stake.

This brings us to consider issues with respect to mathematics and colonization as addressed by Wenda Bauchspies in Chapter 8, *Sharing Shoes and Counting Years: Mathematics, Colonalization and Communication.* The important point here is that processes of learning mathematics need not be significantly different from processes of colonization. This is a strong and very direct way of indicating the need for a critical epistemology in mathematics. It is not possible to assume that processes of learning mathematics can be analyzed adequately in terms that concentrate on processes of coming to grasp mathematical notions. There are many processes which could be related to learning mathematics: processes of inclusion and exclusion, signified by the whole system of exams, evaluation, organizing students in categories according to their performance. Learning mathematics can also mean learning what could count as an authority (an issue that has been addressed by the discussion of the "ideology of certainty" in Skovsmose, 1994). Many and seemingly highly different social processes are part of the processes of learning mathematics. And in some such contexts the learning of mathematics and colonisation go together.

Bauchspies emphasizes:

> The colonized nation was colonized by a foreign educational system. Students of the colonized nation were colonized by the teachers and administrators of this foreign educational system. Or education colonizes students. I am purposively using the term colonization over socialization or enculturation because I want to emphasize issues of power and culture." (Chapter 8)

The question is: "What does this mean for mathematics educators and researchers?" And Bauchspies adds:

> Mathematics is the one western discipline that is held to be neutral, abstract, objective and pure. Western culture values describing the world by the use of numbers, for example in the behavior of fluids, atoms and disease. An equation is considered pure, unquestionable and trustworthy when telling us about the best gasoline, compound or medicine for a particular purpose. We believe in numbers. They provide answers. We are colonized by numbers. (Chapter 8)

This not only applies to the African context to which Bauchspies refers, but is part of overall processes of globalization. If mathematics provides a seemingly neutral and objective tool for description and decision-making, then the power exercised by making this tool universal has implications for every person who comes to handle this tool, and certainly also for those who are effected by this tool. What might appear as colonization to students in Africa, might serve as a similar repression to students in Denmark.

Therefore, it becomes important for us to consider the perspective suggested by Tony Cotton in Chapter 10, *Critical Communication in and through Mathematics Classrooms.* He talks about the learning of critical communication in mathematics classrooms. By doing so, he broadens the aim of mathematics education far beyond what is normally considered important by non-critical epistemologies. Critical epistemologies also have to consider broadly which general competencies a particular subject matter like mathematics might provide. The issue is not simply to learn mathematics. Educational concerns can also address competencies like literacy (in the form suggested by Paolo Freire) or mathemacy (Skovsmose, 1994). There could be a concern for the learning for democracy. This is what Cotton shows us when he pays special attention to the development of critical communication. Cotton's analysis brings forward one more idea, namely that there is a connection between dia-logical and critical epistemologies. For us it is important to emphasize that a critical epistemology may develop from a dialogical epistemology, while we do not expect a mono-logical epistemology to facilitate critical considerations.

We cannot ignore that the movement towards broader considerations about mathematics, and towards critical and dia-logical epistemologies is difficult from a theoretical point of view as well. One danger lies in losing sight of content matter issues. Such issues have been more clearly marked by other presentations in this book, but also by some of those on whom we have commented, in particular the articles by Wistedt and Brattström and the one by Borba. Whenever we want to include a critical dimension in the learning of mathematics, the discourse develops into new areas outside the traditional scope of mathematics, and this is important. However, there is no point in losing sight of mathematics. This is a basic challenge to us and to anybody who tries to develop a critical theory of learning mathematics. The chapters on which we have commented, but also the chapters we have not commented on, represent this challenge.

## REFERENCES

Alrø, H. & Skovsmose, O. (2002). *Dialogue and learning in mathematics education: Intention, reflection, critique.* Mathematics Education Library, Dordrecht, The Netherlands: Kluwer Academic Publishers.

Skovsmose, O. (1994). *Towards of philosophy of critical mathematics education.* Dordrecht, The Netherlands: Kluwer Academic Publishers.

Walkerdine, V. (1989). *Counting girls out.* Written together with The Girls and Mathematics Unit, London, UK: Virago.

COMMENTARY 3

# CHALLENGING RESEARCH READING

**Stephen Lerman**
***South Bank University, UK***

The chapters in this collection take as their starting point the notion that language is more than the externalization of one's private thoughts or a tool for giving structure to thought. To varying degrees the authors argue that language (in all forms, verbal, gestural etc.) shapes how people think; indeed it shapes who people are. As the editors indicate in their *Introduction* there is no attempt to be exhaustive of the field in these chapters—an impossible task. Instead we have evidence in this collection of research on communication that is new and challenging to the mathematics education research and teaching community. I will not repeat here the review of the chapters that appears in the introduction but an example can be seen in Nadia Douek's work on linking communication and argumentation in the classroom and their role in the development of scientific concepts.

A focus on language and communication has a reasonably long pedigree in mathematics education. We can look back to Pimm's (1987) book as a key text in bringing such resources into the field and to more recent discourse analysis in its various forms (e.g., Morgan, 1998, Evans, 2000). The chapters in this book add to and push forward the body of work in the community.

Writing a final reflective chapter is a difficult task, especially given that the introductory chapter serves a similar purpose. Although read first (per-

*Challenging Perspectives on Mathematics Classroom Communication*, pages 349–357

haps) and certainly appearing first in any book, it is actually written last and provides its own overview and reflection. I will not, therefore, make detailed comments about individual chapters or the particular perspectives they offer. Instead I will make some remarks about the importance of these kinds of studies. I will then take a step back and make some meta-comments about this particular turn in research in mathematics education.

## STUDIES OF CLASSROOM COMMUNICATION

I have set out my own stall regarding communication elsewhere (Lerman, 2001, p. 107). I have already discussed many aspects of becoming mathematical as being able to think/speak mathematics as it is legitimized in the classroom by the teacher's framing of what is considered legitimate mathematical texts. The cultural, discursive view places practice in place of objective reality; the social practices of mathematics are constitutive of its meaning (Solomon, 1998). Thus the child is not expected to arrive at the objective reality of the structures of mathematics by herself or himself, pulling herself or himself up by the bootstraps of reflective abstraction and being pathologized if she or he cannot manage to arrive at those structures alone. Learning school mathematics is nothing more than initiation into the practices of school mathematics, hence the central role of the initiator, the teacher. The phrase "nothing more than" is not to play down the great difficulties children experience in learning mathematics, it is merely to emphasize that there is nothing beyond. Learning mathematics or learning to think mathematically is learning to speak mathematically. What constitutes an acceptable grammatical construction, in mathematics, is what is approved of within the discourse. Over time, studies of the development and increasing sophistication of students' language in mathematics indicate their becoming mathematical.

We can say, first, following Vygotsky, that communication drives development and, second, that communication and the meanings conveyed in communication precede us; we are enculturated into those meanings in a range of context-specific situations. By focusing on those context-specific situations as the source of meanings we are continuing a line through Vygotsky, Bakhtin, Wittgenstein and Bernstein, to the recent work on situated cognition and communities of practice. Of course there are substantial differences between these ideas (see Lerman, 2000 for an account of the recruitment of a range of social theories into mathematics education research) but what they have in common is the focus on what people do, on culture, on meanings, and on social practice as the media through which to look at the development of identity. Thus when we look at mathematical enculturation we cannot but focus on mathematical communication.

At the same time we must take note of the heterogeneity of cultural voices in the classroom. As we are "being spoken by" discourses we each "speak out" from those discourses in individual and creative ways. Some of the chapters in this book study the ways in which individuality, the unique intersection of each person's social and cultural identities, manifests in the mathematics classroom.

## THE LINGUISTIC TURN IN MATHEMATICS EDUCATION RESEARCH

I have argued elsewhere (Lerman, 2000) that we can discern a shift in mathematics education research towards the social, encapsulated in many ways by what we could term a linguistic turn. This book is yet further evidence: Indeed in their introduction the editors argue that the purpose of the book, as its title indicates, is to "challenge predominant perspectives and to highlight... what are still unanswered questions, the silenced issues and the research challenges." Although I cited many references in support of that claim (Lerman, 2000) I would now want to ask a number of questions that require a more systematic study. How can we "read" the field in order to be able to make claims about its orientations and shifts? What leads to changes in a particular field of research? What effects do changes have on new researchers, in teacher education, and in schools and, again, how can we "read" such changes? How do new orientations find a voice, given that the process of reviewing for journals and for books (as well as for research funding applications, research studentships, examinations of dissertations) draws on researchers who are well established in the field?

In what follows I draw on the work undertaken in the project entitled *The production of theories of teaching and learning mathematics and their recontextualisation in teacher education and education research training*, which commenced in October 2001. (The project is directed by the author and by Dr. Anna Tsatsaroni, the researcher is Guorong Xu, and it is supported by the Economic and Social Research Council in the UK, Ref: R000 22 3610. Initial findings of the project have been published in the following papers: Lerman, Xu & Tsatsaroni, 2003a; Tsatsaroni, Lerman & Xu, 2003; Lerman, Xu & Tsatsaroni, 2003b).

I know that the editors of this collection made great efforts to draw upon as wide a spectrum of authors as they could, across the world and across alternative theories, but I also know how hard it is to open up texts like these to other voices. For the most part one can make contact with some people from countries under-represented in the literature when they are studying in other, usually richer countries but it is not necessarily the case that these potential authors are producing other than "mainstream" ideas.

To talk about changes over time in a field of research in terms of changes in the priorities, understandings, and interpretations given by people in these positions of power is, at the same time, to acknowledge a number of aspects:

1. the structures and social relations constituting the field as well as, perhaps, the changing strength of the boundary separating this sub-field from other research subfields within education research (curriculum studies, educational research methods, science education, etc.);
2. changes in the relations between education research and other fields within the overall arena of research production (sociology, psychology, etc.);
3. the wider picture of power and control relations which affect the (relative) autonomy of the intellectual field of knowledge production, establishing certain forms of social relations between, on the one hand, the official policy agencies and, on the other, agencies and agents in the field (in our case of mathematics education) of knowledge transmission, dissemination, use and reproduction. (Bernstein, 1990; Morgan, Tsatsaroni, & Lerman, 2002)

It is also most important to recognize that most mathematics education research takes place in University departments whose principal role is that of initial and in-service teacher education whilst retaining at least a historical role in the improvement of education. All of these roles (researcher, teacher educator, actor in developments in school teaching) are affected by the overall political context (cf. Dale, 2001) and result in the tensions of competing demands in particular of having to act as a career academic and as a teacher educator.

The picture is further complicated once we consider that educational publishing has a key role here. It is a crucial agency in the process of validation, authorization, and dissemination of research productions. It has multiple dependencies resulting from: its symbolic control function (specialization in discursive resources which shape consciousness); its location within the cultural field but driven/constrained by economic imperatives; and its hierarchical location in the division of labor—as not only diffusers but also shapers of knowledge, in the sense that they influence what counts as developments or changes within sciences (Bernstein, 1990, pp. 133–145). Furthermore, developments towards electronic publishing might further change the relations between publications, publishing, and its publics. Given the electronic networks that are now possible and that can reach a large majority of any particular research community, setting up new journals and other forms of communication become possible for individuals. It is hard to see how some processes of regulation would disappear such as the reviewing by established researchers and the need for well-known names of the editors and editorial board. There is an in-built conservatism in peer evaluations of research productions of academic research prevalent around the world, that is likely to inhibit any changes in those aspects that determine the acceptance or rejection of articles for publishing. However new strands of thought that find it hard to gain space in the field might well find the effort of setting up an electronic journal worthwhile.

All the above affect the degree to which new orientations and interpretations become current in the field. They affect the degree of authorial agency, the author as intellectual researcher and as somebody who balances the priorities/demands of the publications, the official policy or research accounting procedures, while keeping her/his credibility amongst teachers (Nixon, 1999).

I will repeat here the questions above and offer some responses that relate to this book and the move to emphasizing communication for researching mathematics teaching and learning.

*How can we "read" the field in order to be able to make claims about its orientations and shifts?* This is a challenging research question. While reviews of particular aspects of the development of theories, ideas and information have always been a feature of research, to review the whole field for its shifts and changes over time calls for new analytical tools. Drawing on Bernstein's work on intellectual fields and knowledge structures (Bernstein, 1999, 2000) we (Anna Tsatsaroni, Guorong Xu, and myself) have developed just such a tool and applied it to the analysis of a sample of papers from the Proceedings of the International Group for the Psychology of Mathematics Education (PME), from *Educational Studies in Mathematics* (ESM) and from the *Journal for Research in Mathematics Education* (JRME) from 1990 to the present.

The tool consists of a series of questions to interrogate the data, the criteria for making judgements being a challenging methodological issue. Key questions include: what theories do the authors draw on, are they made explicit, and do they come from other intellectual fields; does the paper have a theoretical orientation or an empirical one; is there a pedagogical model and if so how can we classify that model; and what is the focus, teachers, pupils, policy, or pedagogy.[1]

Of course any choice of data sources for analysis defines and delineates the research: other choices might well lead to different results. Our choice of these sources was based on their role as among the leading arenas of publication in mathematics education research in terms of status, and the choice of years was made to enable us to bring the study up to today.

Extending the arenas would have required a much larger project. Among the many very interesting findings we can certainly see the proliferation of theories drawing from other sub-fields of educational research and other intellectual fields beyond education. The growing interest in the role of communication is certainly evident, from much the same range of theories as are represented in this present collection.

*What leads to changes in a particular field of research?* There are some structural features of the field that play a key part in enabling and constraining the growth of research and directions of research. In particular, the relations between what we can call, paraphrasing Bernstein (2000, see also Morgan et al., 2002), the official discourses (policy-makers, politicians) and the unofficial discourses (mathematics education researchers, textbook writers, teachers). The former are influenced by such groups as parents, media, and influential lobby groups, and in some places at some periods by the latter, by unofficial discourses. When there is relative autonomy between the two fields there is much greater opportunity for the proliferation of theoretical frameworks. In our analysis we chose international publications. We have found relatively few papers in ESM and PME that address policy issues, suggesting that when people write for those publications they exhibit relative autonomy from official discourses. Interestingly, although JRME is an international journal too, it is perceived by many of its authors to address, in the main, the USA community and, as a consequence, we have seen much greater engagement with policy issues. Other features of the field also come into play of course, in changes emerging within a field. One of these is the book series, as we have seen in our own field by Kluwer, Falmer, Lawrence Erlbaum, and the series in which this book is being published. The senior editors and advisory panels have the possibility to encourage new directions and new authors. We have not analyzed these sources, so I will not make further comments. To some extent, special issues of ESM and JRME (the monographs) are also a field for the potential development of new directions.

*What effects do changes have on new researchers?* and *How do new orientations find a voice?* Here we can only speculate as, inevitably, we are not able to analyze those articles and papers that have not been accepted for publication; without this dimension our comments have just one side of the story as their basis. There is a perception, we conjecture, that if one wants to make a mark and carve out a space for oneself, one needs to develop or use a theory that is new to the field. At the same time, there is a tension in the opposite direction, arising from the knowledge that dissertations, articles, papers, and chapters will be accepted or rejected, and solicited or not, by those established in the field. It is evident that dominant developments, such as the Standards (NCTM, 2000) in the USA, and what has become known as the reform, has opened a space, especially within JRME, for new researchers to produce and publish research. One can observe similar spaces created by other innovations worldwide (TIMMS is another example, its findings having led to research within countries in attempts to explain performance, as well as strong criticisms of the study itself (Keitel and Kilpatrick, 1999). A similar effect of creating spaces for new research and new researchers is the gradual build-up of a body of theory. We would argue that Bernstein's work is one such space (e.g., Brown, 1999; Cooper & Dunne, 2000; Dowling, 1994, 1998; Ensor, 1999; Lerman & Tsatsaroni 1998; Morgan, Tsatsaroni, & Lerman, 2002 in mathematics education); communication/discourse is another, to which this book contributes substantially. Our analysis shows that the appearance of articles focusing on language and discourse have appeared only in recent years in our sampling of PME, ESM and JRME. Authors of chapters in books have the scope to report more substantially on their research and the coherence of a book such as this one can have a proportionally greater impact.

Finally, *What effects do changes have in teacher education, and in schools and, again, how can we "read" such changes?* At the time of writing this reflection we have not reached the stage of looking at the recontextualization of research texts into teacher education materials, and we are aware that this will require modifications of the tool of analysis that we have developed for research texts. In producing materials for teacher education courses, both pre-service and in-service, selections are made both of which research texts to include and of what elements of those texts to include. Such selections open spaces for the play of values and ideologies. The further stage of selection for school texts is also subject to the same influences.

The chapters in this book address many of the issues and concerns I have raised here and many offer proposals of the possible impact of their research for teachers and students. All of the authors focus on classroom communication and I would say that it is because they take as their basic assumption that mathematical thinking is produced in classroom interactions that their analyses offer fresh insights into the teaching and learning of mathematics and the field of research in mathematics education. Swanson's study of how different forms of classroom communication produce and reproduce inequities and Cotton's use of the same Bernsteinian perspective to argue for a critical position are both strong examples of this theme. Similarly Douek's study of the forms of classroom discourse that facilitate the development of mathematical argumentation, Borba's notion of the power of mediation with media and the ethnographic focus of Bauchspies indicate how alternative perspectives on pedagogic orientations transform what constitutes school mathematical thinking. Triandafillidis and Potari's and also Meaney's focus on the textual productions of students in the mathematics classroom highlight the importance of examining what Bernstein would call the framing of the classroom interactions in the dialectic of mathematical understanding and the contexts of culture and situation. Finally Hewitt's study unifies social, cognitive and affective elements of learning in what he calls zones of opportunity; the latter element is not often given sufficient attention in research on classroom communication.

I feel sure that this book will make an important contribution to the shift in orientation towards communication, discourse, and sociocultural theories in general.

## NOTE

1. See http://www.sbu.ac.uk/cme/Paperforpeercomment.doc for an early version of the tool.

## REFERENCES


Bernstein, B. (1990). *Class, codes and control, (Vol. 4), The structuring of pedagogic discourse.* London, UK: Rutledge.

Bernstein, B. (1999). Vertical and horizontal discourse: An essay. *British Journal of Sociology of Education, 20*(2), 157–173.

Bernstein B. (2000). *Pedagogy, symbolic control and identity: Theory, research, critique.* Maryland: Rowman & Littlefield.

Brown, A. J. (1999). *Parental participation, positioning and pedagogy: A sociological study of the IMPACT primary school mathematics project.* Unpublished PhD Thesis, University of London Library.

Burton, L. (Ed.): (1999). *Learning mathematics: From hierarchies to networks.* London, UK: Falmer.

Cooper, B. & Dunne, M. (2000). *Assessing children's mathematical knowledge: Social class, sex and problem-solving.* Buckingham, UK: Open University Press.

Dale, R. (2001). Shaping the sociology of education over half-a-century. In J. Demaine (Ed.), *Sociology of Education Today.* Basingstoke, UK: Palgrave.

Dowling, P. (1994). Discursive saturation and school mathematics texts: A strand from a language of description. In P. Ernest (Ed.), *Mathematics, Education and Philosophy: An International Perspective.* (pp. 124–142). London, UK: Falmer.

Dowling, P. (1998). *The sociology of mathematics education: Mathematical myths/pedagogic texts.* London, UK: Falmer.

Ensor, P. (1999). *A study of the recontextualising of pedagogic practices from a South African university preservice mathematics teacher education course by seven beginning secondary mathematics teachers.* Unpublished PhD dissertation, University of London.

Evans, J. T. (2000). *Adults' mathematical thinking and emotions: A study of numerate practices.* London, UK: Falmer.

Keitel, C. & Kilpatrick, J. (1999). The rationality and irrationality of international comparative studies. In G. Kaiser, L. Eduardo, & I. Huntley, (Eds.), *International comparisons in mathematics education.* (pp. 241–256). London, UK. Falmer.

Lerman, S. (2000). The social turn in mathematics education research. In J. Boaler (Ed.), *Multiple perspectives on mathematics teaching and learning.* (pp. 19–44). Westport, CT: Ablex.

Lerman, S. (2001). The function of discourse in teaching and learning mathematics: A research perspective. *Educational Studies in Mathematics 46,* 87–113.

Lerman, S. & Tsatsaroni, A. (1998). Why children fail and what mathematics education studies can do about it: The role of sociology. In P. Gates (Ed.), *Proceedings of first international conference on mathematics, education and society (MEAS1).* (pp. 26–33). University of Nottingham, UK: Centre for the Study of Mathematics Education.

Lerman, S., Xu, G. & Tsatsaroni, A. (2003a). Developing theories of mathematics education research: the ESM story. *Educational Studies in Mathematics, 51*(1–2), 23–40.

Lerman, S., Xu, G. & Tsatsaroni, A. (2003b). An analysis of PME research: Theories, methods and the identities of academics. To appear in *Proceedings of Twenty-Seventh Meeting of the International Group for the Psychology of Mathematics Education,* Hawaii.

Morgan, C. (1998). *Writing mathematically: The discourse of investigations.* London, UK: Falmer.

Morgan, C., Tsatsaroni, A. & Lerman, S. (2002). Mathematics teachers' positions and practices in discourses of assessment. *British Journal of Sociology of Education, 23*(3), 443–459.

Nixon, J. (1999). Teachers, writers, professionals. Is there anybody out there?. *British Journal of Sociology of Education, 20*(2), 207–221.

Pimm, David (1995). *Symbols and meanings in school mathematics.* London, UK: Routledge

Solomon, Y. (1998). Teaching mathematics: Ritual, principle and practice. *Journal of Philosophy of Education 32,* 377–387.

Tsatsaroni, A., Lerman, S. & Xu, G. (2003, April). *A sociological description of changes in the intellectual field of mathematics education research: Implications for the identities of academics.* Paper presented at the American Educational Research Association, Chicago, IL.

# ABOUT THE CONTRIBUTORS

**Helle Alrø** is an Associate Professor in the field of interpersonal communication, Department of Communication, Aalborg University, Denmark. Her research on Dialogue and Learning relates especially to mathematics education, but also to other fields of interpersonal communication such as coaching, supervision and conflict transformation. Together with Ole Skovsmose and Paola Valero she is involved in the research project: "Learning from Diversity" that has a multicultural perspective on communication and conflict in mathematics education.

**Wenda Bauchspies** is an Assistant Professor in the STS (Science, Technology and Society) Program and the Women's Studies Program, as well as the coordinator of the STS Graduate Studies. She did her Ph.D. studies at Rensselaer Polytechnic Institute, Troy New York. Her research and publications are in the areas of sociology and anthropology of science and knowledge, women's studies, African studies, and education in transnational contexts. She is an interdisciplinary scholar who focuses on issues of knowledge, power, difference, and gender.

**Gudrun Brattström** is an Associate Professor at the Department of Mathematics, Stockholm University, Sweden. Gudrun's main research interests lie in the areas of number theory and, more recently, in statistical genetics. In the latter field she has particularly focussed on linkage analysis of complex diseases. She has participated in several projects concerning children's and students' learning in mathematics.

*Challenging Perspectives on Mathematics Classroom Communication*, pages 359–363

**Marcelo C. Borba** is the chair of the Graduate Program of Mathematics Education at the State University of São Paulo, Rio Claro in Brazil. He is a Ph.D. graduate from Cornell University, U.S.A., 1993. His research and academic work are in the areas of ethnomathematics, computer technology in mathematics education, and research methodology. He is a member of varied editorial boards of journals and research centers worldwide and has been a reviewer for different PME meetings. Currently he is a member of the ICME-2004 scientific committee.

**Leone Burton** is Professor Emerita, the University of Birmingham, UK and Visiting Professor in Mathematics Education at King's College, London, UK. She is the series editor for this series and also editor of the book *Which Way Social Justice in Mathematics Education?* published in the series in 2003. Her next book, published by Kluwer in 2004, is called *Mathematicians as Enquirers—Learning about Learning Mathematics.* Her professional obsessions have been in enquiry-based learning of mathematics, social justice, and most recently, mathematical epistemology.

**Iben Maj Christiansen** works at the School of Education, Training and Development at the University of KwaZulu-Natal, Pietermaritzburg in South Africa. Before that, she was an Associate Professor in mathematics and science education at Aalborg University in Denmark. There, she also served as director of the Centre for Educational Development of University Science, which offered staff development and coordinated action research projects at seven higher education institutions. Iben is involved in research concerning: tacit knowledge in mathematics education and teacher training; teaching and learning mathematical modeling; mathematics education for democracy; and developmental work in higher education in South Africa.

**Anna Chronaki** is an Assistant Professor at the University of Thessaly, in Greece and teaches courses on mathematics education, technology and research methods. She has worked at the School of Education of the Open University and of the University of Bath in the UK, as well as the Computer Technology Institute, University of Patras, Greece. Together with Brian Hudson and Hannele Niemi, she convenes the ECER network OPENnet (Open Contexts, Cultural Diversity, Democracy). Her main research interests are in sociocultural and anthropological perspectives on teaching and learning mathematics and technology in the social contexts of school communities.

**Tony Cotton** is a Senior Lecturer at the School of Education at the Nottingham Trent University and the course leader for the BA Primary Education. He has worked as a maths teacher, a maths educator and an advisory teacher for multicultural education. He has published materials for pupils and teachers as well as academic work. He is a regular contributor to the *Journal of the Association of Teachers of Mathematics.* His current research projects involve issues concerning the recruitment and retention of teachers from minority ethnic backgrounds and the raising achievement of minority ethnic pupils.

**Nadia Douek** is a maths educator at the Caen IUFM (University Institute for Teacher Training) since 1989 (then at Creteil). She has a Ph.D. in pure mathematics and is currently involved in a postdoctoral research (supervised by G. Vergnaud & P. Boero) concerning the process of argumentation and concept construction in the fields of experience of everyday life and school tasks at the primary school level. Her main research interests lay within the area of linguistic and representation aspects of teaching and learning of mathematics.

**Dave Hewitt** is a Senior Lecturer at the School of Education, University of Birmingham. Prior to arriving at Birmingham, he taught in secondary schools. Dave has an interest in researching issues related to the notion of the economy of personal time and effort in the teaching and learning of mathematics. This general theme has manifested itself with more particular areas of research interest including the teaching and learning of algebra; use of technology; and viewing the mathematics curriculum in terms of arbitrary and necessary. He has also developed various pieces of software.

**Steve Lerman** was a secondary mathematics schoolteacher for 15 years in Israel and the UK before beginning research. He joined London South Bank University in 1987 and became a full Professor in 1998. Steve was President of the International Group for the Psychology of Mathematics Education from 1995 to 1998 and Chair of the British Society for Research in Learning Mathematics from 1994 to 1996. His research interests are in sociocultural research on teaching and learning mathematics.

**Tamsin Meaney** is a researcher at the Educational Assessment Research Unit, University of Otago. She has worked as a teacher and curriculum developer in various settings (e.g., working with ESL students at a Technical and Further Education College in Sydney, with Aboriginal students at schools in remote communities in the Northern Territory of Australia, with curriculum developers and teachers in the Republic of Kiribati and with parents and teachers of a Maori immersion school in New Zealand). Her present research involves analyzing the talk that Year 4 and Year 8 New Zealand students used in discussing mathematics problems.

**Despina Potari** studied mathematics at the University of Athens, Greece. In 1987 she completed her Doctorate in Mathematics Education at the University of Edinburgh. Since 1989 she teaches Mathematics and Methods courses at the Department of Primary Education at the University of Patras, Greece. Her research focuses on geometry and teacher education.

**Anna Sfard** is a Professor at the University of Haifa, Israel and Lappan-Phillips-Fitzgerald Professor at the Michigan State University. With a formal background in mathematics and physics, Anna Sfard specializes today in mathematics education, focusing her research on the intricacies of human learning and creative thinking. The overarching theme of her work is the constitutive role of language and investigates the implications of the assumption that human thinking as a particular case of communicative activity. She is a member of varied editorial boards in journals and books related to Mathematics Education such as the Mathematics Education Library published by Kluwer Academic Publishers.

**Ole Skovsmose** is a Professor at Aalborg University, Department of Education and Learning. He has been a member of BACOMET, an international research group in mathematics education and the co-director of the BACOMET-4 project. He is also a member of the editorial boards of various journals as well as of the Mathematics Education Library published by Kluwer Academic Publishers. He has been co-director of The Centre for Research of Learning Mathematics, a co-operative project between Roskilde University Centre, Aalborg University and The Danish University of Education. He has a special interest in critical mathematics education.

**Dalene Swanson** earned her Ph.D. in Mathematics Education in the Curriculum Studies Department at the University of British Columbia, Canada, where she teaches and researches. She is also a graduate of the University of Cape Town and has taught school mathematics in South Africa and Canada for more than 15 years. Her research interests are in sociological perspectives in mathematics education, and social justice. She focuses on how forms of social difference are recontextualized into disadvantage through the discourse and practice of school mathematics; the "pedagogizing of difference." She is particularly interested in how disadvantage might be differently realized and established across a range of pedagogic contexts.

**Triandafillos Triandafillidis** studied mathematics at the University of Athens, Greece. In 1993 he completed his Doctorate in Mathematics Education at the University of Edinburgh. In 1994 he was a visiting scholar for a year at the School of Education at the University of Michigan. Since 1996 he teaches Mathematics and Methods courses at the Department of Primary Education at the University of Thessaly, Greece. In his research he examines the mathematics classroom through the lenses of semiotics and has a special interest in geometry.

**Inger Wistedt** is Professor in the Department of Education at Stockholm University and Visiting Professor at Växjö University, Sweden. Inger's main research focus lies within the field of mathematics learning, and she has conducted studies of students' mathematical reasoning in cooperative settings ranging from studies of very young children to university students. She has also taken an interest in gender studies and is presently engaged in studies of gifted education in mathematics, another aspect of her general interest in equity issues. Her research often has a cross-disciplinary character and is carried out in cooperation with researchers in mathematics.

# INDEX



*Challenging Perspectives on Mathematics Classroom Communication*, pages 365–371

## O



## P



## R



## S

## Z

9 781593 11151